Cultural Heritage Science

The preservation and interpretation of our cultural heritage is one of the major challenges of today's society. Cultural Heritage Science is a highly interdisciplinary book series covering all aspects of conservation, analysis and interpretation of artworks, objects and materials from our collective cultural heritage. The series focuses on science and conservation in three main fields

- Art technology
- Active conservation and restoration
- Preventive conservation and risk management

The series addresses conservators and conservation scientists at museums, institutes, universities and heritage organizations. It also provides valuable information for curators and decision makers at museums and heritage organizations. All titles in the book series will be peer reviewed. Titles will be published as printed books and as eBooks, opening up the opportunity to include electronic supplementary material (videos, high-resolution figures, special data formats, and access to databases).

Anna Tummers • Robert G. Erdmann

Frans Hals or not Frans Hals

Connoisseurship, Technical Analysis and Digital Tools

With Contributions by Andrei Anisimov,
Silvia Centeno, Joris Dik, Nouchka De Keyser,
Roger Groves, Babette Hartwieg, Erma Hermens,
Katja Kleinert, Annelies van Loon, Dorothy Mahon,
Claudia Laurenze-Landsberg, Vassilis Papadakis,
Arie Wallert

Springer

Anna Tummers
Ghent University
Ghent, Belgium

Robert G. Erdmann
University of Amsterdam
Amsterdam, The Netherlands

ISSN 2366-6226 ISSN 2366-6234 (electronic)
Cultural Heritage Science
ISBN 978-3-031-59488-5 ISBN 978-3-031-59489-2 (eBook)
https://doi.org/10.1007/978-3-031-59489-2

This publication was supported by the Netherlands Organization For Scientific Research (NWO)

This Springer imprint is published by the registered company Springer Nature Switzerland AG
The registered company address is: Gewerbestrasse 11, 6330 Cham, Switzerland

Contents

Chapter 1
Introduction

Anna Tummers

Abstract This book examines the criteria that experts have used to determine whether a painting can be attributed to Frans Hals (1582/83–1666); tests if new technical analyses combined with observations with the naked eye and a close reading of primary sources can replace and/or expand these criteria; and explores how data visualisation tools can facilitate close comparisons and thus aid in the analysis.

Assessing the characteristic qualities of a painting with the aim of determining who made it and when, known as 'connoisseurship', is one of the most difficult tasks art historians have set themselves. Of all the seventeenth-century Dutch painters, Frans Hals (1582/83–1666) is the most controversial in terms of the exact scope of his work. The most prominent twentieth-century experts on Hals, Harvard professor Seymour Slive (1920–2014) and his German colleague professor Claus Grimm (b. 1940), disagreed on no less than one-third of Hals's oeuvre. While Slive accepted 222 paintings as authentic works by Hals in his survey of 1974–1977, Grimm limited the oeuvre in 1989 to 145 paintings (of which 135 were also in Slive's selection) (see below, Chap. 2). Since then, Slive and Grimm have continued to refine their selections, while no other scholar has published a new Frans Hals oeuvre catalogue. Moreover, some of the most heated public debates and legal battles about attributions in this field concern paintings in the style of Frans Hals. Both the first and second survey exhibitions of Hals's paintings, in 1937 and 1989/90 respectively, were overshadowed by sharp debates about the validity of the attributions of the paintings on display. In addition, the earliest court case in the Netherlands in which chemical evidence was used to investigate a possible forgery (1925) involved a portrait of a man in the style of Frans Hals and caused a bitter controversy (see below, Chap. 2). More recently, in 2016, the auction house Sotheby's decided to refund $10 million to a private collector who had purchased a painting as a Frans Hals, stating that technical analyses showed that the painting was not an original but a forgery (BBC News 2016). However, a British art dealer involved in the sale

A. Tummers (✉)
Ghent University, Ghent, Belgium
e-mail: Anna.Tummers@UGent.be

A. Tummers, R. G. Erdmann, *Frans Hals or not Frans Hals*, Cultural Heritage Science, https://doi.org/10.1007/978-3-031-59489-2_1

strongly denied that the work was a forgery (Noce 2019; Siegal 2019). The painting subsequently became the subject of three lawsuits, the largest of which is still pending in France.[1]

Although Frans Hals is considered one of the three most important Dutch seventeenth-century painters, along with Rembrandt and Vermeer, relatively little technical research has been done on his paintings. His popularity, the lack of technical reference material, and differences in opinion among experts as to the exact scope of his oeuvre have made works in his style prone to doubt and misattribution. This was one of the main reasons for initiating two research projects which form the backbone of this publication: *Frans Hals or not Frans Hals: Defining the oeuvre of the painter Frans Hals (1582/83–1666)* (NWO Museumbeurs 2016–2018) and *21st Century Connoisseurship: Smart Tools for the Analysis of Seventeenth-Century Paintings* (NICAS Seed Money Grant 2018–2022).

The goal of the *Frans Hals or not Frans Hals* project was to investigate the criteria used by experts to determine whether a painting can be attributed to Hals, and to test whether new technical analyses, combined with observations with the naked eye and a close reading of primary sources could replace and/or expand these criteria. Three case studies have been selected to explore different attribution issues. One case study evaluates the respective use of five non-destructive techniques that are at our disposal to research the characteristic style and painting technique of Frans Hals. For this purpose, a typical Frans Hals portrait from around 1635, *Portrait of a Woman* in the Gemäldegalerie in Berlin, was researched using not only observations with the naked eye, x-rays (XR) and infrared reflectography (IRR), but also macro-x-ray fluorescence scans (MA-XRF), hyperspectral imaging also referred to as reflectance imaging spectroscopy (HI/RIS), and even neutron activated radiographs (NAR). Because the portrait is the only painting to date to have been researched with all these techniques, it provided a unique opportunity to compare and assess their potential.

The second case study focuses on distinguishing between different versions or imitations of one of Hals's best-known scenes from everyday life: his depiction of *Malle Babbe* ("Mad Meg"), the first portrait-like depiction of a woman who was mentally ill in the history of art. At the centre of this study is a painting in the storage room of the Metropolitan Museum of Art in New York that has been called an original, a forgery, and a (contemporary) imitation. By comparing it on the one hand to the famous original by Hals at the Gemäldegalerie in Berlin and on the other hand to a forgery created by Han van Meegeren (1989–1947) at the Rijksmuseum in Amsterdam, this study aims to solve the riddle of its attribution by using new techniques in combination with an in-depth analyses of primary sources.

[1] For this reason the lion share of the research done in this context is still classified, including a report (115 pages) written on the painting specifically for this lawsuit in 2017 by Violaine de Villemereuil and Anna Tummers in collaboration with Arie Wallert.

The third and largest case study concerns Hals's portraits of civic guards, his largest and most prestigious commissions. All five Haarlem civic guard portraits completed under his supervision are examined in detail, as well as a contested Amsterdam civic guard portrait, *The Meagre Company*. This last work was started by Hals and—due to a conflict with his patrons—eventually finished by another painter, Pieter Codde (1599–1678). Hals experts have long disagreed about the exact extent of Hals's and Codde's share. The purpose of this case study is to provide deeper insights into Hals's technique, style and workshop practice, and to subsequently shed new light on the attribution of *The Meagre Company*.

Because the final case study involves six large paintings and the new techniques used to study these yield enormous amounts of data, it soon became clear that data science and computational tools could greatly facilitate and enhance the analysis. It was the reason to start the Seed Money Project *21st Century Connoisseurship: Smart Tools for the Analysis of Seventeenth-Century Paintings* (2018–2022), funded by the Netherlands Institute for Conservation, Art and Science (NICAS). Building on high resolution photographs and advanced technical analyses, online computer tools are causing a turning point in the early twenty-first century. For the first time in history, it has become easier to study paintings online than offline. The aim of the project was to explore how best to (further) develop and apply a range of computational tools to attribution issues, building on the data generated by the *Frans Hals or not Frans Hals* project. This book will thus not only provide an overview of developments in Hals connoisseurship, examine how our knowledge of Hals can be expanded through the use of relatively new techniques and close reading of primary sources, but also explore the potential of a number of computer tools for enhancing the eye in attribution matters. The results of both research projects will therefore be published in both a book and an interactive website, presenting the various data visualization tools.

While the *Frans Hals or not Frans Hals* research project was underway, we were approached by on the one hand the French Ministry of Justice and on the other hand by a buyer, seller and auction house to aid with the attribution of two paintings in the style of Frans Hals that had caused confusion. The first request resulted in a joint research report written by Violaine de Villemereuil and Anna Tummers in collaboration with Arie Wallert, which is currently classified, pending an upcoming lawsuit in France (see note 1). The second request resulted in a report that formed the basis of an article published in *Burlington Magazine* in late 2019, which is also included in this book: A Recent Riddle: The Story of the Two Fisherboys (Chap. 5). It gave us the opportunity to test the relevance of our integrated approach (combining in-depth art historical study with new technical analysis and data visualization tools) in solving current attribution issues.

References

Noce, V.: Sotheby's settles claim against London dealer Mark Weiss in Frans Hals case (1 Apr 2019). The Art Newspaper (2019). https://www.theartnewspaper.com/2019/04/01/sothebys-settles-claim-against-london-dealer-mark-weiss-in-frans-hals-case

Siegal, N.: Investors must pay back Sotheby's over forged Frans Hals, Court finds (11 Dec 2019). The New York Times (2019). https://www.nytimes.com/2019/12/11/arts/design/frans-hals-fake-sothebys.html

Sotheby's declares "Frans Hals" work a forgery (6 Oct 2016). BBC News (2016). https://www.bbc.com/news/entertainment-arts-37574411

Part I
The Eye of the Expert

Chapter 2
Frans Hals Connoisseurship

Anna Tummers

Abstract Of all seventeenth-century Dutch painters, Frans Hals is the most controversial in as far as the exact scope of his oeuvre is concerned. The most prominent twentieth-century Hals experts, Harvard Professor Seymour Slive (1920–2014) and his German peer Professor Claus Grimm (born 1940) disagreed about no less than a third of Hals's oeuvre. Moreover, some of the fiercest public debates and legal battles about attributions in this field concern paintings in the style of Frans Hals. This chapter provides an overview of the evolving insights in this field, both in theory and in practice.

2.1 Connoisseurship, the Humanities and Some Recent Insights from Cognitive Psychology

Assessing a painting's characteristic qualities with the aim of determining who made it and when, known as 'connoisseurship', is one of the most difficult tasks art historians have taken upon themselves, as briefly mentioned in the introduction.[1] It involves estimating variabilities that can be tantalizingly difficult to determine. How much consistency can one expect in an artist's inventions, style and technique, choice of materials and workshop practice? Did the artist use one particular style that gradually developed over time or, instead, different manners at the same time? To what extent did he/she involve workshop assistants and was he/she consistent in doing so (or not doing so)? Even when secure evidence is scarce or missing, the art

[1] Some parts of this chapter were first published in (Tummers and Erdmann 2022) and adjusted somewhat to best suit the present publication (notably parts of 2.1, 2.3, 2.5 and 2.8). I would like to thank Ellis Dullaart and John Bezold for their thoughtful comments on an earlier version of this chapter.

A. Tummers (✉)
Ghent University, Ghent, Belgium
e-mail: Anna.Tummers@UGent.be

A. Tummers, R. G. Erdmann, *Frans Hals or not Frans Hals*, Cultural Heritage Science, https://doi.org/10.1007/978-3-031-59489-2_2

expert has to form a mental image –consciously or subconsciously– of what is characteristic of the artist in order to make a decision. Evidently, the validity of an attribution hinges on the correctness of the expert's assumptions.

Moreover, the determination of authorship also usually involves assessing the quality of the art work. Although key to the artist's goals and to the reception of the art work, the question how to define (high) quality is notoriously hard to tackle from an academic perspective. For old master paintings, given the guild regulations and possible workshop assistance, the question is not just what level of quality can be expected of work by the master's hand but also –and perhaps more importantly– what level of quality the master demanded in the paintings he/she deemed worthy of carrying his/her name (Tummers 2011, pp. 81–112). Here again, the issue of consistency is relevant: how much variation did the master allow in paintings produced in his/her workshop? And did the artist –as many old masters did– consciously produce works of different quality levels, that were priced accordingly?

In short, judging a picture is far from simple, no matter how swiftly the judgment is sometimes made. It involves myriad questions that touch on different academic disciplines, including art history and materials science. Moreover, the inherent complexity of the task entails a risk. As the psychologist and Nobel laureate Daniel Kahneman has shown, the human brain works with two different systems: a quick sub-conscious way of assessing (often referred to as 'intuitive') and a slower, conscious thought-process (Kahneman 2011). In daily life, the brain tends to simplify, thereby delegating the mental process to a sub-conscious part of our brain. When confronted with complex questions, the brain commonly substitutes a complex question with an easier one (Kahneman 2011). Though in many cases effective and efficient -and occasionally even better than conscious decisions- our intuitive tendency to simplify can also lead to dangerous biases and oversights in the decision-making process (Gladwell 2005, 48ff, pp. 263–264 (example of a racial bias); see also Dijksterhuis 2007).

Given these insights into the workings of the brain (Gladwell 2005; Kahneman 2011), it is not surprising that there have long been discussions about the nature of connoisseurship among art experts. Throughout the twentieth century there have been two opposing views as to what should have the most weight in the attribution process: the connoisseur's intuition, that is, the sudden insight that the connoisseur experiences without fully grasping its origin (Kahneman's 'system 1'), or rational, communicable arguments (Kahneman's 'system 2'; Tummers 2011, pp. 30–60). Although both aspects can be considered part of the same decision-making process, their different nature has long caused tension in both theory and practice.

The complexity of the decision-making process and the danger of oversimplifications may also explain the fierce criticism connoisseurs have often received. Already in the eighteenth century, the French scholar and theologist l'Abbé Du Bos dismissed 'the art of predicting the author of a painting by recognizing the master's hand' as 'the most faulty of all the arts, apart from medicine' (Du Bos 1993/1719, p. 296). In the twentieth century, the attribution of paintings was reviled as subjective and intuitive, and as tainted by the market (Chapman and Weststeijn 2019,

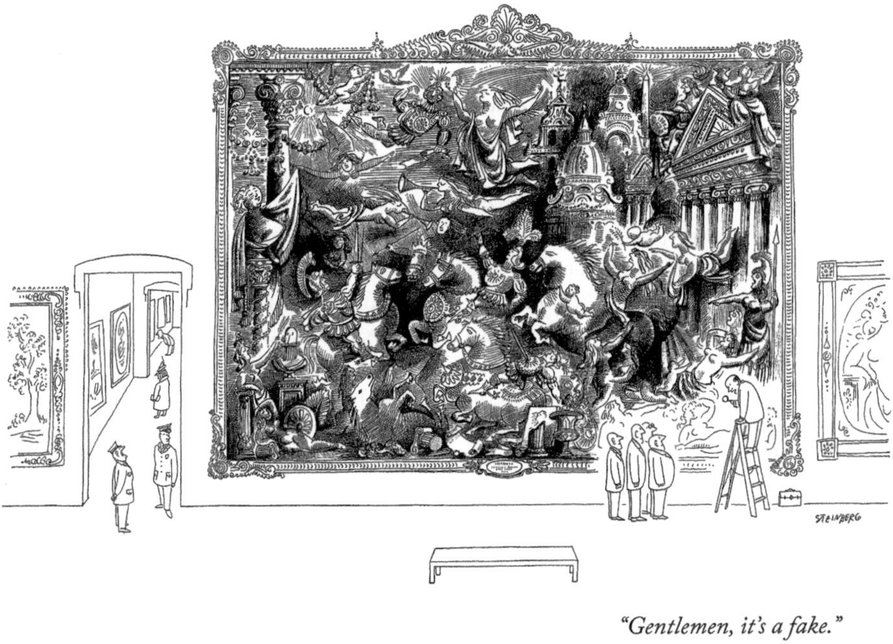

Fig. 2.1 Saul Steinberg, 'Gentlemen, it's a fake!', cartoon published in *The New Yorker*, 6 May 1950

pp. 10–15). Consequently, the term 'connoisseur' has acquired negative connotations, conjuring up the image of a presumptuous, outdated and inadequate judge of pictures –an attractive target for ridicule (Fig. 2.1).

As a result, academic art history repeatedly attempted to avoid connoisseurship, claiming it would not be quite 'theoretical' and 'scholarly' enough to be worthy of serious academic attention (Martin 1904; Muthesius 2013). Avoiding connoisseurship did not solve the issue, however; it merely left academics vulnerable to the reproach that their discipline lacked a firm foundation (Pächt 1999/1986, pp. 66–67). For the history of art cannot be written without a basic classification of who created what and when. Therefore, other art experts chose the opposite strategy and attempted to mend the situation by creating a more 'objective' and 'scientific' connoisseurship. In the Netherlands the conservator Maurits van Dantzig in particular set out to develop a concrete (verifiable) method to attribute paintings based on rational arguments that could be checked (Van Dantzig 1937, 1973).[2] Despite his efforts, however, art historians remained divided. Even the prestigious Rembrandt Research Project (the largest and most advanced research project dedicated to sorting out the oeuvre of one single painter, pioneering several advanced scientific

[2]The first one to devise concrete method to attribute paintings was the Italian art expert Giovanni Morelli. Although Van Dantzig did not refer back to him, his method seems indebted to his well-known Italian predecessor (see Tummers 2011, 30 ff).

techniques) did not believe that the intuitive component could or should be taken out of the decision making process (Bruyn et al. 1982–1989, vol. 1 (1982), XVII).

Meanwhile, another blow to the connoisseur's reputation came from the field of philosophy. In the 1950s and 1960s Arthur Koestler and Alfred Lessing argued that it made no aesthetic difference whether a painting is forged or not. Therefore, the person who pays a large sum of money for an original but would have no interest in a reproduction or imitation which he could not tell from the original, or worse, who prefers an aesthetically inferior original over an excellent forgery, is said to be at best confused and at worst a snob (Koestler 1955; Lessing 1965). It raised the question why connoisseurs should bother to tell originals and forgeries apart at all.

Although Koestler's and Lessing's claims were effectively refuted by the philosophers Nelson Goodman and Denis Dutton in the late 1960s and 1970s (see below), the fact that the validity of connoisseurship was questioned in this way, is telling. It is hard to imagine that a similar claim would be made for any other field of study (i.e. that it would make no difference if an expert's analysis and appreciation is based on an authentic or a forged piece of evidence). For example, should one value real and counterfeit money in the same manner if one cannot tell the difference? Should historians interpret and appreciate real and forged historical artefacts such as Hitler's diary or pieces of the dead sea scrolls in the same way if they cannot tell these apart? It is the emphasis on the *aesthetic* properties of course that makes the difference here. Yet the question was if the aesthetic properties could be separated entirely from any cultural or historical context.[3]

Goodman argued that since the exercise, training and development of our powers of discriminating among works of art are plainly aesthetic activities, the aesthetic properties of a picture include not only those found by looking at it but also those that determine how it is to be looked at (Goodman 1969, pp. 111–112). In his view, the knowledge that a picture is an original and not a copy, imitation or forgery is a critical and valid factor in our response to it. Indeed, the impact of such knowledge has recently been confirmed by neuroscientific research (see Huang et al. 2011; Wolz and Carbon 2014).

Denis Dutton made a similar point but arrived at it differently. He stated that all visual art is necessarily performative, as it represents an achievement within a certain cultural and historical context. It is this achievement that determines its value as an art work and makes it relevant to art history. Therefore, if our understanding of this achievement alters drastically when a work of art is exposed as a forgery, in his view it is no longer the same object, in so far as its position as a work of art is concerned (Dutton 1979, p. 314).

Although philosophers thus underscored the importance of connoisseurship, many academic art historians stayed clear of in-depth visual analysis and moved instead towards contextual and historical approaches in the 1970s and 1980s.

[3] According to Goodman the idea that one should strip oneself of all the vestments of knowledge and experience when encountering a work of art derives from the Tingle-Immersion Theory which was developed around 1800 and has since then become part of the fabric of what Goodman calls our 'common nonsense' (Goodman 1983, p. 102).

Iconography, social history and socio-economic perspectives gained ground, causing art historians to rely heavily on verbal and contextual evidence rather than on their eyes. In an effort to change this, Harvard Professor and drawings expert Henri Zerner wrote an engaging essay on connoisseurship's bad reputation in 1987. "Ours is a logocentric culture", he stated: "We trust the written document much more than our visual understanding of an image. This must be changed and we must attend to visual clues if we want to get something out of our visual legacy" (Zerner 1987, p. 290).

While major research projects dedicated to individual artists such as Rembrandt, Rubens and Van Gogh greatly expanded our visual understanding of these masters at the end of the twentieth century, the advances in connoisseurship hardly impacted the academic curriculum. In 2009, Paul Craddock sharply observed: "the subject of authenticity does not seem to be seriously studied or taught to prospective art historians/curators, much less to materials scientists [...] an honorable exception being the centre for study of forgery with its own museum at the University of Salerno" (Craddock 2009, p. 6). This lack of academic attention is disconcerting and yet somewhat understandable. Connoisseurship and authentication skills require arduous practice including extensive first-hand observation and in-depth study of important art works, copies and imitations, which not all universities can provide.

Nevertheless, the twenty-first century witnessed a renewed academic interest in connoisseurship as well as a theoretical refinement in thinking about issues of authenticity. Prominent academics who had not themselves dedicated their lives to sorting out the oeuvres of artists started to underscore the importance of this specific type of visual knowledge. David Freedberg eloquently argued that it was not just fundamental to art history but also potentially a 'core discipline in the humanities' as connoisseurship shared its 'evidential paradigm' with other types of scholarly detective work involving the interpretation of clues, symptoms and pictorial marks (Freedberg 2006; see also Ginzburg and Davin 1980). Stephanie Dickey stated that the continuing value of connoisseurship could be claimed both on theoretical and practical grounds: "Broad historical theories that build on works of art as evidence fall like a house of cards if assumptions about the authenticity of those works prove incorrect" (Dickey 2015, p. 5). In 2019 the Dutch art historical yearbook was even dedicated entirely to connoisseurship, which is framed as the 'history of visual knowledge since the Renaissance'. According to Chapman and Weststeijn, connoisseurship is now widely understood as 'an essential and ever-evolving art-historical method'. Moreover, there is an 'enhanced rigor, [an] interdisciplinary reliance on materials science and neuroscience, and [a] new theoretical awareness' that represent a departure from the past (Chapman and Weststeijn 2019, p. 7).

This chapter takes a closer look at the development of Frans Hals connoisseurship throughout the twentieth century. The main line of enquiry focuses on what types of criteria were used to determine whether or not paintings were created by Frans Hals and how these could be improved to effectively face current and future challenges.

2.2 Frans Hals: One of the Most Contested Dutch Painters

Of all seventeenth-century Dutch painters, Frans Hals (1582/83–1666) is the most controversial in as far as the exact scope of his oeuvre is concerned. The last major retrospective of Hals's work in London, Haarlem and Washington in 1989/1990 triggered a particularly fierce debate about the question whether or not the paintings on display were indeed painted by Hals. The two main twentieth-century Hals experts, Harvard Professor Seymour Slive (1920–2014) and his German peer Professor Claus Grimm (born 1940) disagreed about no less than a third of Hals's oeuvre. Since then, no other scholar has taken up the challenge of publishing a new Frans Hals oeuvre catalogue. The first large Hals exhibition in 1937 similarly caused sharp disagreements about the authenticity of the paintings on display (see below, Sects. 2.4 and 2.5). Moreover, some of the fiercest legal battles about attributions in this field also concern paintings in the style of Frans Hals (see below, Sects. 2.3 and 2.8).

It is noteworthy that in Hals's case the attribution debates did not just focus on so-called 'Monday morning' paintings (pictures of relatively poor quality that are believed to be by a well-known artist) and on other works of dubious quality.[4] Even some of the most famous and highly praised paintings that are commonly attributed to Hals were the topic of controversies, including the *Lute Player* at the Louvre in Paris, the *Laughing Boy* at the Royal Cabinet of Paintings the Mauritshuis in The Hague, *Jasper Schade* at the Narodni Gallery in Prague and *The Regentesses* at the Frans Hals Museum in Haarlem (Figs. 2.2, 2.3, 2.4 and 2.5). Hals expert Maurits van Dantzig attributed the *Laughing Boy* in 1937 to an unknown seventeenth- century master, dismissed the *Lute Player* as a copy, and even claimed that the portrait of *Jasper Schade* was a forgery, as we will see (see below, Sect. 2.4; Van Dantzig 1937, pp. 72–73, 60–61 and 107–108). He did consider *The Regentesses* an original work by Hals. The latter painting was de-attributed by Claus Grimm in 1969 on account of the thickness and consistency of the paint as well the execution of certain details such as the sharp contrasts between the heads of the women and the background, and the rather dry and harsh definition of their eyes, nostrils and mouths (Grimm 1972/1968, no. 167; see also Grimm 1990/1989, pp. 245–247).

This last painting is still somewhat contested today, while –at the same time– it is praised by many as one of Hals's most innovative and beloved pictures. In his younger years Hals followed tradition and painted women with a keen eye for detail and few sharp contrasts in their faces, whereas in this work he turned his back on tradition, and used a very bold, loose painting style instead (see Tummers 2013a, b, p. 40). Seymour Slive famously praised the loose strokes that capture the delicate skin of the elderly hand of the right-most woman, stating that this one hand told him more about the fragility of life than a warehouse full of Vanitas still lifes; in his view

[4] For an example of the use of this term ('Monday morning'), see the interesting discussion on possible variations in quality in Rembrandt's paintings by Anthony Bailey (1994, pp. 71–76): 'Monday Mornings'.

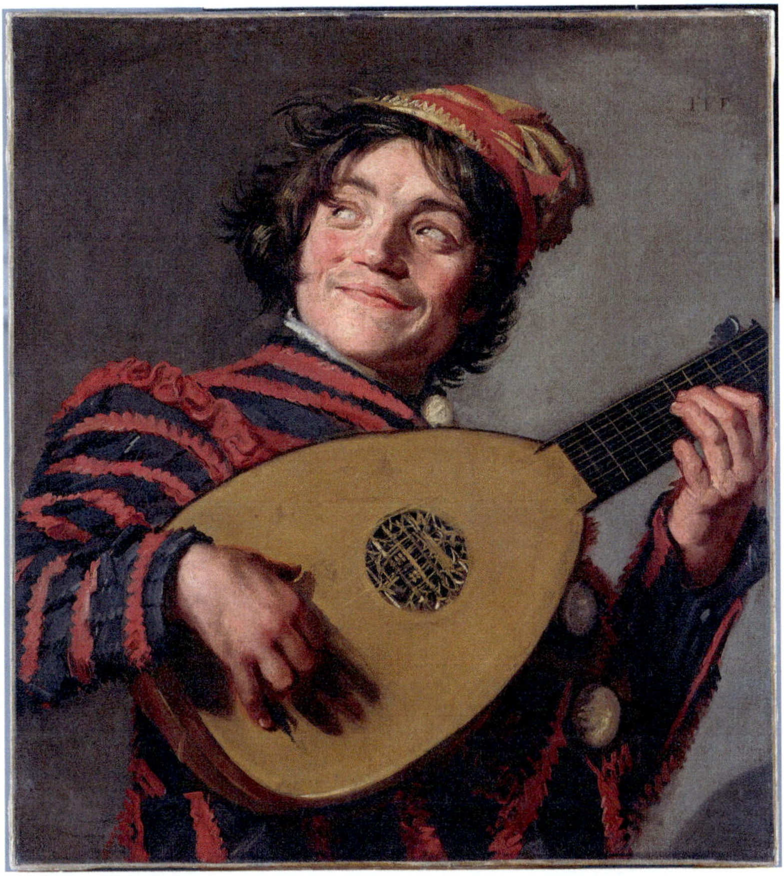

Fig. 2.2 Frans Hals, *Lute Player*, c. 1623, oil on canvas, 70 × 62 cm. (Musée du Louvre, Paris)

the painting was "one of the most penetrating portraits ever painted"(see Slive (ed.) 1989–1990, p. 368). Indeed, it has been one of his most admired works since Hals's rediscovery at the end of the nineteenth-century.[5] The famous American painter James McNeill Whistler (1834–1903) reportedly even became so enraptured by it that he actually caressed one of the women depicted on the cheek.[6]

[5]After his rediscovery, Hals was praised for his modernity and this painting was one of the ultimate examples of this, see for example the colophon in the Belgian journal L'Art modern (1883, pp. 302–303): 'Frans Hals est un moderne. Son esthéthique, son coloris, son dessin, ses procédés, appartiennent à notre époque.' And specifically about Hals's group portraits of regents and regentesses of the Old Men's Home: 'Rien dans ces deux stupéfiantes compositions ne se rattache à l'art d'autrefois.'

[6]This incident was described by his traveling companion, the German painter Georg Sauter (1866–1937), see (Robins and Pennell 1908, p. 285) (a biography written with consent from and the aid of the artist himself). See also (Jowell 1989-1990, note 156, p. 76).

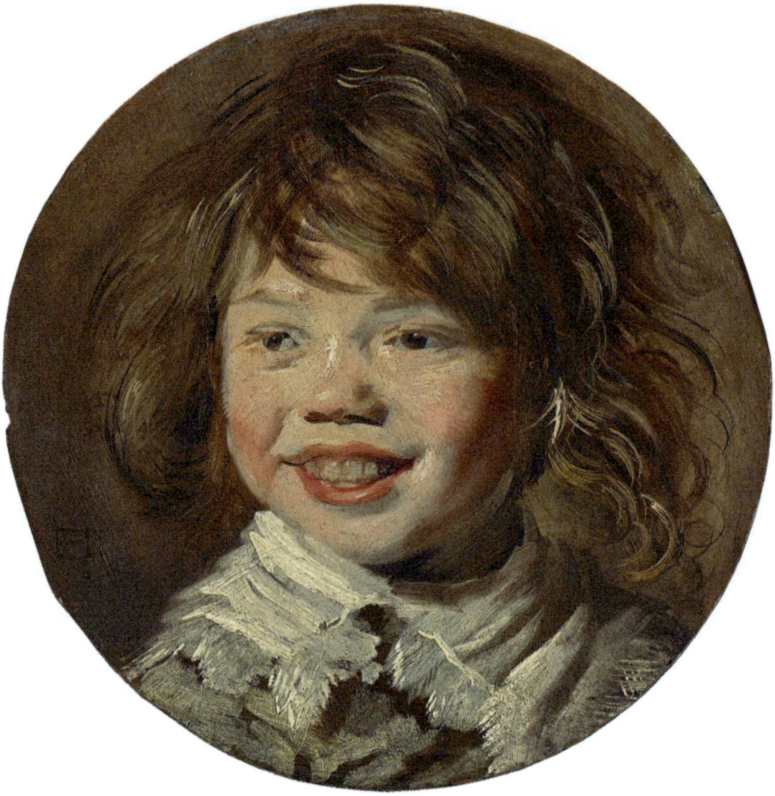

Fig. 2.3 Frans Hals, *Laughing Boy*, c. 1625, oil on panel, diameter: 30.45 cm. (Royal Picture Gallery Mauritshuis, The Hague)

The widely differing opinions on the exact scope of Hals's oeuvre raise the question of how decisions about the authenticity of paintings in his style are made, and how this should be done. Despite the fierce and often public controversies over the attribution of specific pictures, the criteria used by connoisseurs have hardly been systematically examined. What follows, therefore, is first a critical analysis of how Hals connoisseurship has developed since the early twentieth century. What criteria have been used in the process? What underlying assumptions do these criteria reveal? And what can we conclude about their relative validity?

2.3 The Eye Versus Chemistry: An Early Controversy

No incident illustrates the deep distrust of an early twentieth century art expert towards chemical evidence better than a curious booklet of 89 pages with the title *Real or fake? Eye or Chemistry?*, dated 1925 (Hofstede de Groot 1925). It is written

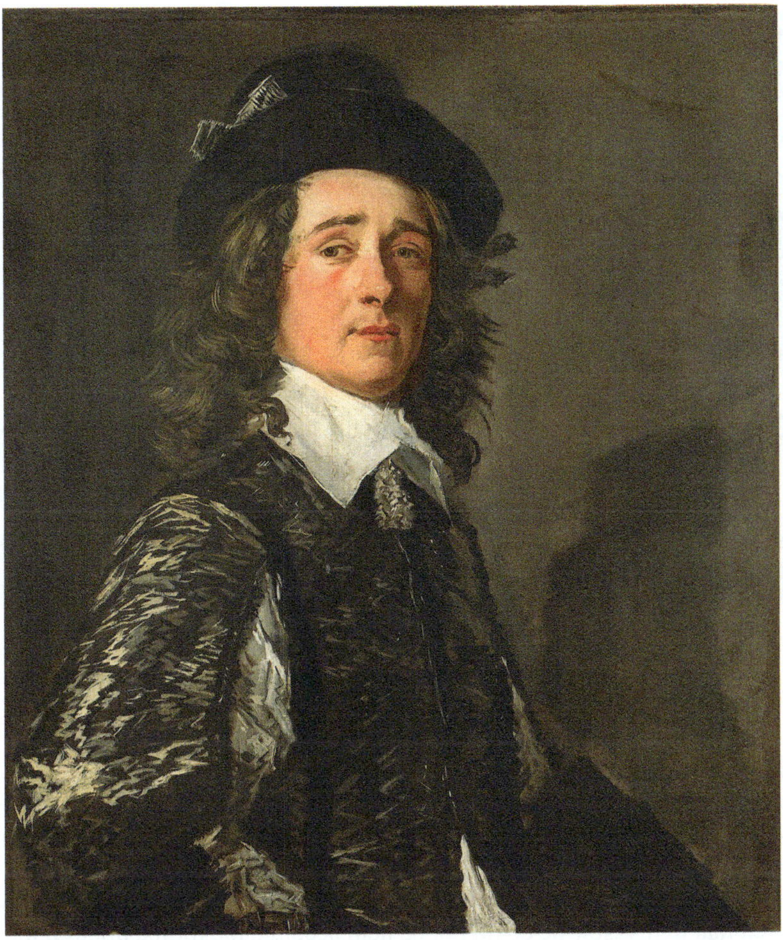

Fig. 2.4 Frans Hals, *Jasper Schade*, 1645, oil on canvas, 80 × 67.5 cm. (National Gallery, Prague)

by Cornelis Hofstede de Groot, then one of the leading art historians and the author of a ten-volume survey book on Dutch seventeenth-century painting (Hofstede de Groot 1907–1928). He published the booklet in response to the lawsuit *Fred. Muller & Co. vs H.A. de Haas*, the first court case in the Netherlands in which chemical evidence was brought to bear in an attribution matter. The bone of contention was the attribution of a small painting: the *Laughing Cavalier* (Fig. 2.6). Cornelis Hofstede de Groot had recognized it as an authentic Frans Hals (1582/83–1566) in 1923 and provided a certificate of authenticity. Both the certificate and painting had subsequently come into the possession of a certain H.A. de Haas, who had sold it via the auction house Frederik Muller & Co. to a private collector for fl. 50.000,- (at

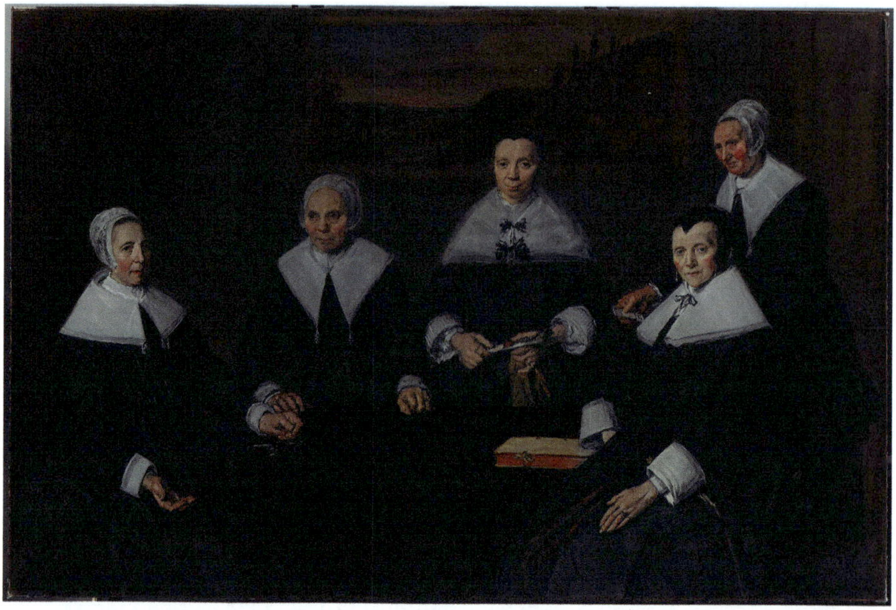

Fig. 2.5 Frans Hals, *Regentesses of the Old Men's Alms House*, c. 1664, oil on canvas, 170.5 × 249.5 cm. (Frans Hals Museum, Haarlem)

the time the equivalent of fourteen years' salary for the average man[7]). A few months afterwards, however, the buyer demanded to be reimbursed claiming that the painting was in fact a forgery. The auction house looked into the matter, agreed with the buyer, reimbursed him, and subsequently asked Hofstede de Groot to cover a third of the damages suffered (circa fl.16.666,-). Hofstede de Groot refused to do so. He indicated that in his view he could not be held accountable for the prices fetched by paintings he had authenticated, and he offered to research the painting anew. After a second inspection, however, he concluded once again that in his view the painting was by Frans Hals (see Hofstede de Groot 1925, pp. 7–8).

The auction house then subpoenaed the seller, Mr. de Haas on the 9th of December 1923. Muller and Co. demanded that the purchase contract be annulled and that the purchase amount be reimbursed including interest as well as their litigation expenses. The burden of proof that the auction house presented was substantial. The painting had been researched by a team of experts who had jointly written a report: Sir Charles Holmes, director of the National Gallery in London, Prof. dr. Wilhelm Martin, director of the Royal Cabinet of Paintings the Mauritshuis in the Hague and Prof. dr. F.E.C. Scheller, chair of Inorganic Chemistry at the Technical College of

[7] The comparison with the annual salary of an average worker comes from (Lopez 2008a, p. 46).

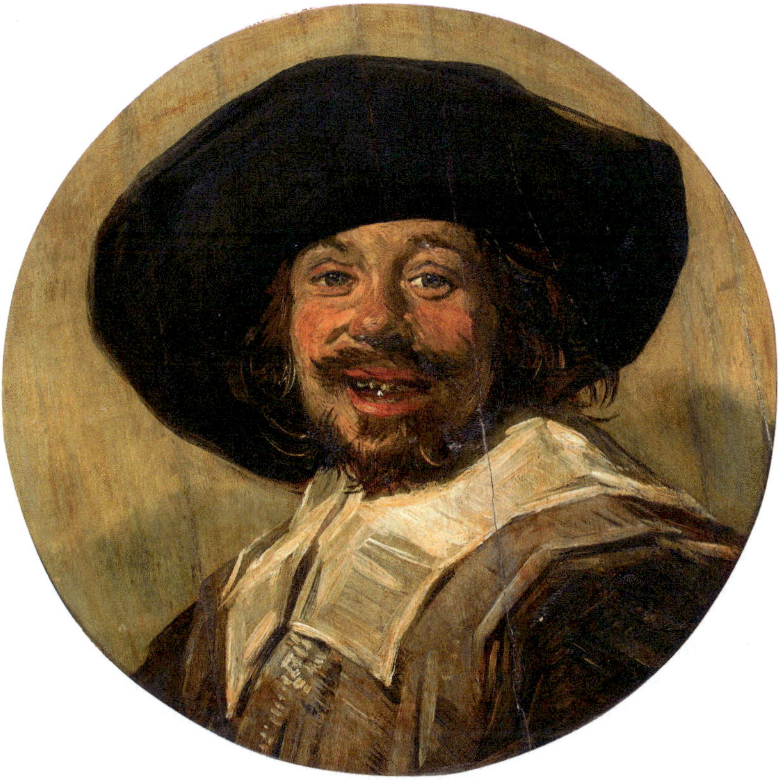

Fig. 2.6 *Laughing Cavalier*, Hals forgery created before 1923, diameter: 36 cm. (Private collection, credits obtained through the Hoogsteder Museum Foundation)

Delft (nowadays known as Delft University of Technology).[8] Especially the materials science part of the investigation was disconcerting. Only the first test had yielded a positive result: the paint layer did not dissolve when treated with the usual 96% alcohol solution, which agrees with what one would expect of a seventeenth-century painting. However, when touched lightly with a cotton ball soaked in water, the paint became soft; with a soft brush and water the paint layer could even be entirely removed. Moreover, the researchers found artificial ultramarine in several locations throughout the painting, a pigment that had only been discovered in 1826. Furthermore, the researchers observed cobalt blue (through the microscope) in several locations in the background, a pigment that was not manufactured

[8] *Rapport van het onderzoek naar de echtheid van een aan Frans Hals toegeschreven schilderij*, prepared by Sir Charles Holmes, director of The National Gallery in London, Prof. Dr. Wilhelm Martin, director of the Royal Cabinet of Paintings the Mauritshuis in The Hague, and Prof. Dr. F.E.C. Scheffer, professor of inorganic chemistry at the Technical College of Delft (Technische Hogeschool te Delft, nowadays known as Delft University of Technology), reproduced as an appendix in (Hofstede de Groot 1925, pp. 74 ff.).

commercially until the early nineteenth century.[9] A chemical analysis of the white used in the painting identified it as zinc white, a pigment that has only been available since 1781. Also, a radiography of the picture revealed two nails below the paint layer that had been hammered into the picture from the front; these were machine-made and could therefore not have been produced before the nineteenth century. The conclusion was obvious: the painting could not be by Frans Hals or one of his contemporaries; it was made by a modern forger or imitator.

The lawsuit received much press coverage and constituted a serious blow to the reputation of Cornelis Hofstede de Groot, a leading connoisseur of Dutch seventeenth-century painting. After an impressive career as deputy director of the Royal Picture Gallery the Mauritshuis (1891–1896) and director of the Print Room at the Rijksmuseum (1896–1898) and dozens of prestigious publications, he lived as an independent art historian mostly from the certificates of authenticity that he provided (Ekkart 1979). When the court case had dragged on for one and a half years, he suddenly brought it to a halt, presumably in an attempt to prevent further damage to his reputation. Before the judge could reach a verdict, Hofstede de Groot purchased the contested painting for the full amount of fl. 50.000,-, which made the law suit redundant. He then defended his point of view in his publication *Real or Fake?* In a nutshell, he argued that his expert eye should outweigh the chemical evidence presented in court. It is a position one can hardly imagine taking nowadays, and therefore an interesting benchmark in our study of the development of Hals connoisseurship. For what criteria did Cornelis Hofstede de Groot use to substantiate his attribution? And how did he come to dismiss 'chemistry' so radically?

Although Hofstede de Groot hardly defines criteria for assessing paintings, the way in which he attacks his opponents is revealing. Hofstede de Groot mainly directs his arrows at Professor Wilhelm Martin. The latter allegedly appealed too often to his 'sense of style' (*stijlgevoel*): "In an earlier polemic against Rembrandt scholars, Prof. Martin declared with great self-confidence, that by possessing a sense of style he noticed things, which must have escaped his opponents because they did not have this sense of style, or at least to a lesser degree than he did. I then very clearly pointed out to him that what he considered to be a sense of style had nothing to do with it, and I must also state here that the fact that the highly learned gentleman does not recognize the hand of Frans Hals in every brushstroke in this painting proves very clearly that he does not possess the very least feeling for the style of Frans Hals either."[10]

---

[9] According to the experts consulted for the court case, this was around 1820 to 1830 (Hofstede de Groot 1925, p. 84); today it is assumed to be 1807.

[10] "In een vroegere polemiek tegen de Rembrandtvorschers heeft Prof. Martin met groote zelfbewustheid verklaard, dat hij door het bezit van stijlgevoel dingen opmerkte, die zijn tegenstanders moesten ontgaan omdat zij dit stijlgevoel niet of althans in mindere mate dan hij hadden. Ik heb hem toen op zeer duidelijke wijze er op gewezen, dat wat hij voor stijlgevoel hield, niets daarmede te maken had en ik moet ook hier verklaren, dat het feit, dat de hooggeleerde heer in dit schilderij niet in iederen toets de hand van Frans Hals herkent, zonneklaar bewijst, dat hij ook voor den stijl

What is striking in this reasoning is that Hofstede de Groot does not reject 'sense of style' as a criterion. To him it was apparently perfectly natural for connoisseurs to have a 'feeling' for a painter's characteristic style, allowing them to see more and judge better than a layman. That 'feeling' (*gevoel*) seems to have been for Hofstede de Groot a mainly intuitive insight, which in principle needed little explanation. Circumstances did, of course, compel him to refute the arguments of his opponents. He therefore framed his booklet as *negatio*: as a denial of the contrary, of the expert opinions that had been used against him in court.

In doing so, he did not shy away from technical evidence. In particular, the discovery of modern nails in the picture and the solvability of the paint layer seem to have worried him. Therefore, he had he had gone to seek redress from the painter and restorer who had asked him to assess the painting in 1923: Theo van Wijngaarden (1874–1952), nowadays better known as the mentor and business associate of master forger Han van Meegeren (1889–1947). Van Wijngaarden immediately admitted that he had hammered modern nails into the painting. These would not be situated underneath the paint layer, however; he claimed that he himself had covered their heads with tiny retouches. He also provided an explanation for the solvability of the paint layer: he was in the possession of a product –invented by himself but kept a secret– that could render any old master painting in oil solvable in water, which he demonstrated on the spot on another seemingly old painting. Ignorant of the massive swindling for which Van Wijngaarden would later become known, Hofstede de Groot did not raise further questions. He mentioned Van Wijngaarden in good faith in his publication and indicated that the restorer was willing to demonstrate his product on any old master painting (Hofstede de Groot 1925, p. 13).

Having thus 'refuted' the chemical evidence, Hofstede de Groot then proceeded to counter Martin's style analysis. In the expert report, Martin had indicated that he recognized a certain similarity to Hals' oeuvre (notably elements that seemed to have been copied from Hals's famous *Merry Drinker* at the Rijksmuseum, Fig. 2.7), but that he did not encounter the distinctive characteristics of Hals' own hand, which he had described in rather broad terms as a 'playfulness of spirit' (*dartelheid van geest*), a 'secure hand' (*zekerheid van voordracht*), a 'virtuoso manner of painting' (*gave schilderwijze*) and 'a light-hearted mobility in head and body' (*luchtige bewegelijkheid in hoofd en lichaam*) (Hofstede de Groot 1925, pp. 82–83). Several elements in particular deviated from what Martin would have expected of Hals: the stockiness of the shoulder area in relation to the head, the rough indication of the left cheek and neck which did not show the underlying structure, the coarse definition of the hair roots, the way in which the mouth and teeth were depicted and the light reflection on the lower lip.

According to Hofstede de Groot, however, a 'secure hand' was a rather 'subjective feeling'. Although this remark seems to imply that he believed that Martin's criteria were perhaps not objective or clear enough, he merely objects to Martin's

van Frans Hals niet het allergeringste gevoel bezit"(Hofstede de Groot 1925, pp. 28–29). On the debate about Rembrandt attributions see (ibid., p. 12).

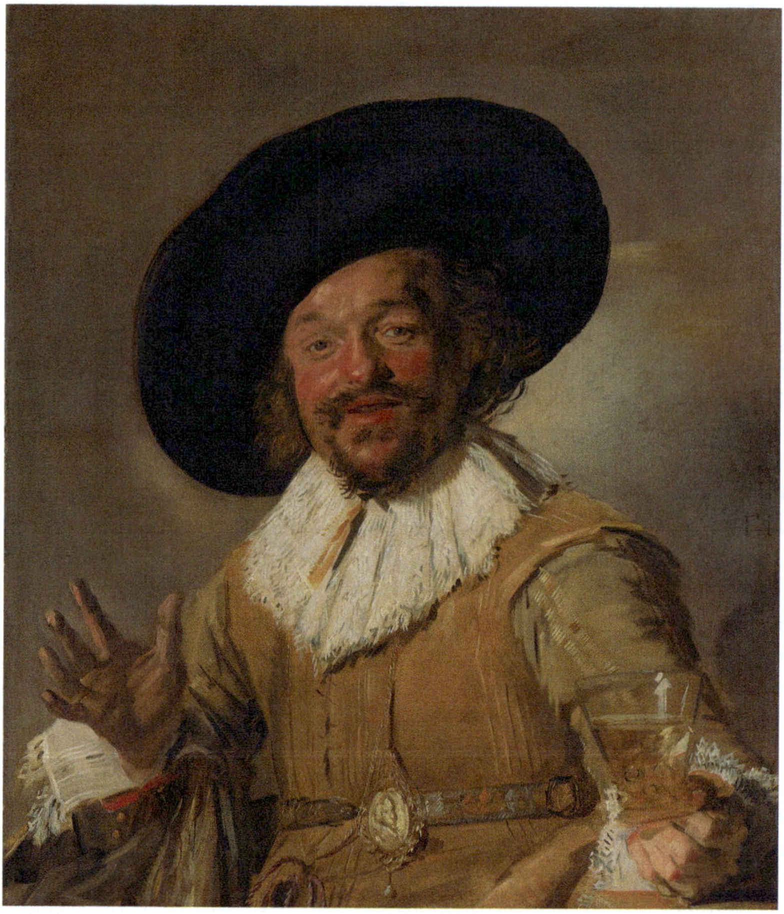

Fig. 2.7 Frans Hals, *A Militiaman Holding a Berkemeyer, Known as the 'Merry Drinker'*, c. 1628–1630, oil on canvas, 81 cm × 66.5 cm. (Rijksmuseum, Amsterdam)

application of the criterion. Hofstede de Groot argues that the disputed picture does in fact show 'a secure hand'. He also believes that 'playfulness of spirit' is a valid criterion, but he claims that it does not apply to all Hals's works. Hals's late regent group portraits, for example, are far from playful in his view, thereby touching indirectly on the ambivalence of the term. For did it allude to a certain playfulness in the subject depicted or rather in handling of the brush? Moreover, in Hofstede de Groot's view, the coarse brushwork was not unusual for Hals, and neither was the stockiness of the shoulder area; he provides no fewer than 20 comparative examples to substantiate his claims (Hofstede de Groot 1925, pp. 20–21). Ironically, one of the reference works he uses has the same provenance as the contested *Laughing Cavalier*: it is a picture of an *Amused Smoker* that Theo van Wijngaarden had also asked him to assess in 1923 and that Hofstede de Groot had liked so much that he

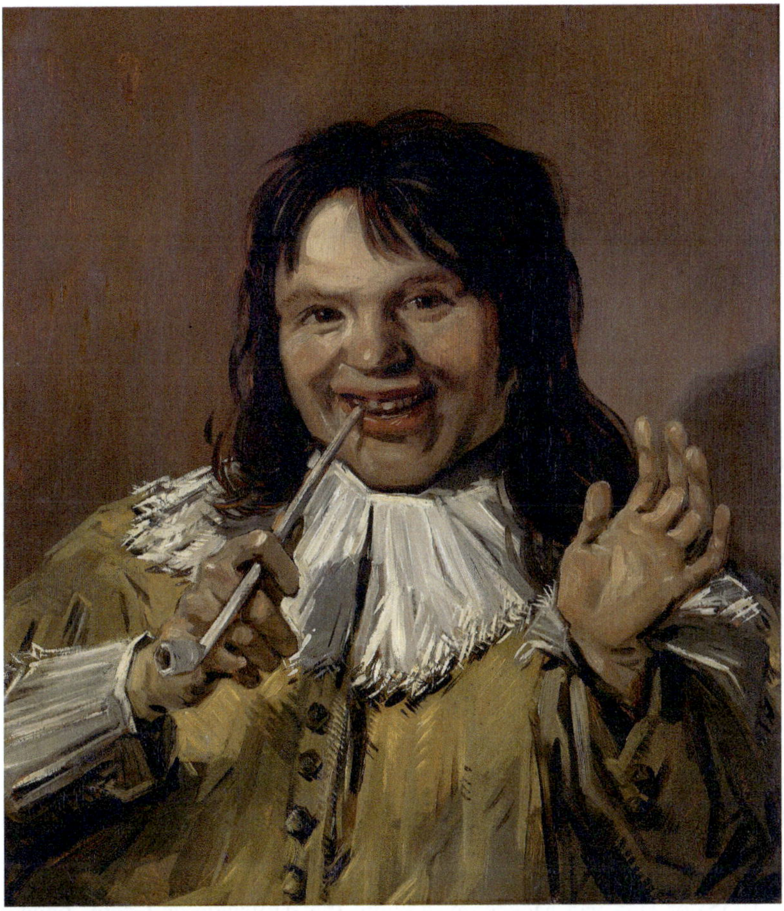

Fig. 2.8 *Amused Smoker,* forgery created before 1923, oil on panel, 57.5 × 49 cm. (Groninger Museum, Groningen)

had purchased it for himself (Fig. 2.8).[11] The similarities were not coincidental: the picture is also a forgery, presumably by someone from Theo van Wijngaarden's workshop, possibly Han van Meegeren (Kraaijpoel and Van Wijnen 1996, p. 49; Lopez 2008a, p. 42).

Many of Hofstede de Groot's reference works have been de-attributed since then. He thus did not just lack clear criteria to distinguish between an authentic Hals and an imitation, but also a clear frame of reference, which makes his mistaken attribution somewhat understandable. In Wilhelm Valentiner's 1923 Hals oeuvre catalogue, for example, a selection of 322 paintings is presented as by the master without

[11] See receipt, RKD—Netherlands Institute for Art History, The Hague, Archive Cornelis Hofstede de Groot, access no. 0356, inv. no. 78RKD/HDG archives, inventory 78. See also (Lopez 2008a, p. 45).

much explanation (which is about 25% more than today's most positive estimate, see Slive 2014) (Valentiner 1923). It brings to mind how broadly Hals's oeuvre was defined at the beginning of the twentieth century and how much the paintings ascribed to Hals's hand varied in quality. The fact that Hofstede de Groot stuck to his attribution even when confronted with ample counter arguments is also less surprising than it might seem at first glance. It is reminiscent of the *sunk cost fallacy*, the phenomenon described by Daniel Kahneman and Amos Tversky that investors who lose money several times on the same project tend to repeat their mistake and - against their better judgment - invest in the same project again (Kahneman and Tversky 1979; see also Kahneman 2011). After all, Hofstede de Groot had invested in his mistake both literally (with his purchase) and figuratively (with his reputation).

While the unclear frame of reference gave Hofstede de Groot some leeway for new attributions, the chemical analyses did not leave room for doubt. If the painting had indeed been made with modern materials, it could not possibly be by Hals. Like an alley cat, and perhaps against his better judgment, Hofstede de Groot opted for the frontal attack: the samples taken would not be from the original paint layer but exclusively from later retouches. The Professor in chemistry had not understood what exactly he had been researching. In the introduction he explained that his defense was directed mostly at Professor Martin, not at Prof. F.E.C. Scheffer "for one cannot argue with a chemist about art. In painting the eye has to hold the highest authority, just like the ear does in matters of music. Here not the tuning fork; there not the test tube."[12]

The fact that Hofstede de Groot did not bother to have the paint layer that he believed to be original tested by a chemist, gives the impression that he must have at least suspected something was wrong. For the outside world, his booklet did not put an end to speculations about the status of the *Laughing Cavalier*. Shortly after its publication, a rumor spread that the picture was a forgery by Leo Nardus (1868–1955) or Han van Meegeren, a claim that Hofstede de Groot -once again-denied firmly and publicly. In an interview with the newspaper *Het Vaderland* of 10th of June 1926, he exclaimed: "they should wish they could paint like that!"[13] Over the following decades, scientific evidence was no longer dismissed so radically by art historians. It would still take a long time, however, before technical research became an integral part of advanced Hals attributions.

[12] "[...] omdat men over kunst nu eenmaal niet met een chemicus kan redetwisten. In zake schilderkunst moet het oog de hoogste instantie zijn, evenals in muziek het oor. Hier niet de stemvork, daar niet de reageerbuis" (Hofstede de Groot 1925, p. 5).

[13] "[Z]ij zouden willen dat zij zoo konden schilderen!" (Het Vaderland 1926), cited in (Lopez 2008a, p. 48 and pp. 64–65). On the attribution of this painting to Han van Meegeren, see (Lopez 2008a; Lopez 2008b; Lammertse et al. 2011, Fig. 38, p. 49 and p. 74). Previously, Arthur attributed the painting to the 'workshop of Theo van Wijngaarden' (see Wheelock 1995, pp. 271–275).

2.4 The First Retrospective Exhibition of Hals's Works: Largely Forgeries?

The first retrospective exhibition of Frans Hals's paintings took place in 1937: from July 1 to September 30 at the Frans Hals Museum in Haarlem.[14] "After Rembrandt, Jan Steen and Vermeer, it is finally Frans Hals's turn," wrote museum director Gratama enthusiastically in the introduction to the exhibition catalogue (Gratama and Van Rijckevorsel 1937). To present as complete an overview as possible of Frans Hals's oeuvre, the museum had brought together 116 paintings from all over the world. No small feat. However, this achievement was overshadowed in the final weeks of the exhibition by negative reports in the press. In these, the authenticity of the paintings on display was openly questioned. According to some scholars, a few pieces in the show were not painted by Frans Hals; others even believed that the majority of the paintings on display were not by his hand.

On August 18, 1937, one J.H. de Bois signals in the newspaper *Haarlems Dagblad* that there are "objections to be made" against "a number" of the exhibited paintings, pointing as an example to a small *Merry Company*, which appears to be based on Hals's famous *Merrymakers at Shrovetide* at the Metropolitan Museum of Art in New York (Figs. 2.9 and 2.10). Doubtful pieces should not, in his opinion, have been included in the exhibition: "Paintings with question marks are unpleasant things to the lay public, interesting only to the art scholars, who fight over them, and the owners who can speculate in them. A most delightful sport, but for which, in our opinion, a public museum is not the appropriate place, especially if the object in question is not museum property."[15]

Then, on August 29, a much more extensive and critical piece appears in the *Telegraaf* newspaper with the headline: 'Doubtful Hals paintings exhibited in Haarlem' (De Telegraaf 1937). An unnamed expert who, according to the newspaper, enjoys great authority both in the Netherlands and abroad, is quoted extensively. He makes short shrift of a number of paintings. For example, he says of a *Portrait of a Lady* (Fig. 2.11): "That young, coquettish woman, with her hand under her chin, has a pose - even leaning obliquely on the arm of the chair! - that Hals would never have painted. That pose is really nineteenth century, and already shows that we are dealing with an imitation - very clever, by the way - which could be about 100 years old, but also somewhat younger. [...] I would further like to draw your attention to [...] the sleeves, especially the left one. Is that Hals's way of

[14] An earlier exhibition did not aim to be a retrospective, but nevertheless included 50 works (Valentiner 1935).

[15] "Schilderijen met vraagteekens zijn voor het leekenpubliek onaangename dingen, alleen interessant voor de kunstgeleerden, die er over strijden, en de bezitters die er in speculeeren kunnen. Een alleraardigsten sport, maar waarvoor, naar ons inzicht, een openbaar museum niet de aangewezen plaats is, zeker niet zoo het object in quaestie niet tot het museumbezit behoort" (De Bois 1937).

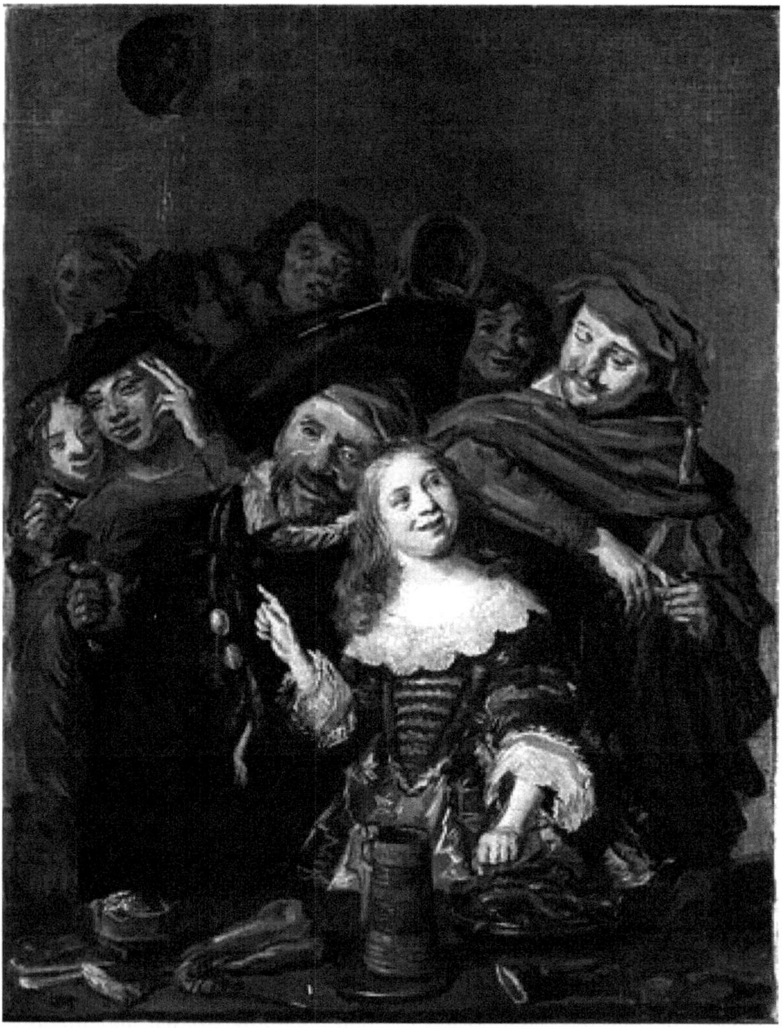

Fig. 2.9 Style of Frans Hals, *Merry Company*, c. 1616–1620, oil on canvas, 67.5 × 51.5 cm. (Current location is unknown. Photography Collection RKD—Netherlands Institute for Art History)

painting? And what is on her head? Hair? Feathers? And where did Hals ever use such" colors"? I admit that the work as a whole is cleverly done, but a Hals??"[16]

[16] "Die jonge, kokette vrouw, met haar handje onder het kinnetje, heeft een houding—en nog wel schuin leunend op de arm van de stoel!—die Hals nooit zou geschilderd hebben. Die houding is echt 19e eeuwsch, en toont reeds aan dat we met een—overigens heel handige—imitatie te doen hebben die wellicht reeds circa 100 jaren oud, maar ook iets jonger kan zijn. [...] Ik wijs u voorts [...] op de mouwen, vooral de linker. Is dat Hals' schilderwerk? En wat bevindt zich op haar hoofd? Haar? Veeren!? En wáár gebruikte Hals zulke "kleurtjes"? Ik geef toe dat het geheel allerhandigst werk is, doch Hals??" (De Telegraaf 1937). This painting was restored recently by Martin Bijl,

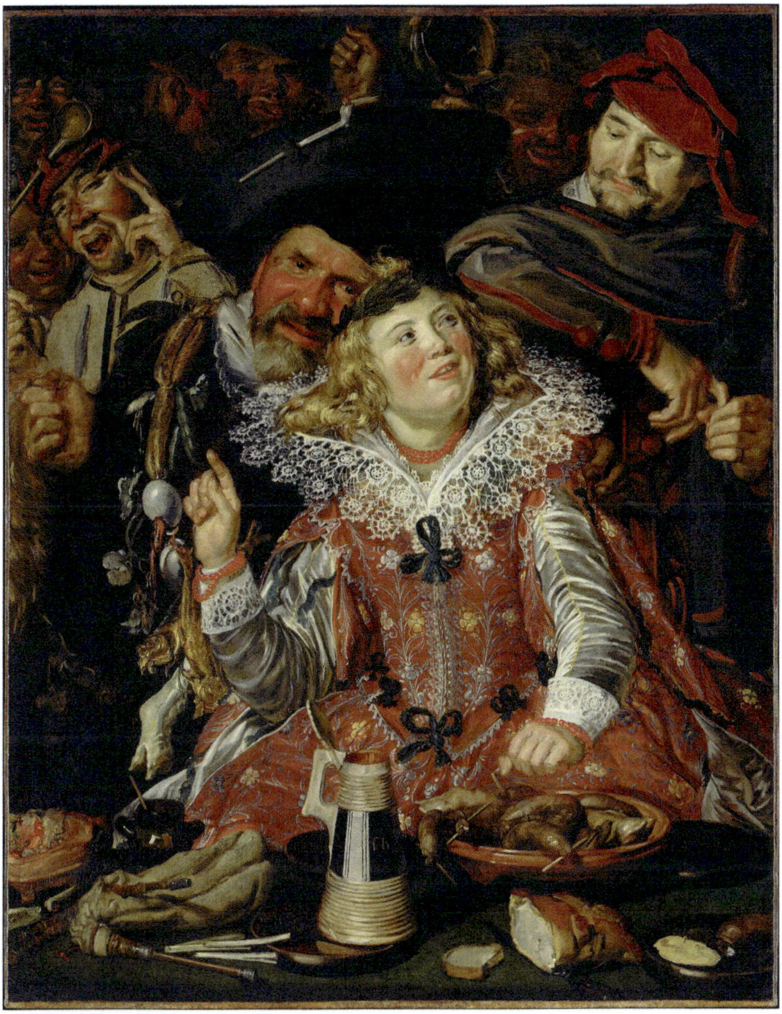

Fig. 2.10 Frans Hals, *Merrymakers at Shrovetide*, c. 1616–1617, oil on canvas, 131.4 × 99.7 cm. (Metropolitan Museum of Art, New York)

A man's portrait is of French origin, according to the same expert, a *Lute Player* (cat. no. 42) not by Hals at all, and a *Rommelpot Player* (cat. no. 24; Fig. 2.12) he recognizes as a copy by a son of Hals after a lost original by the master himself. From a portrait of Judith Leyster (cat. no. 9; Fig. 2.13) Hals is said to have painted

who kindly informed me that he believes that the work does date from the seventeenth century and might be unfinished.

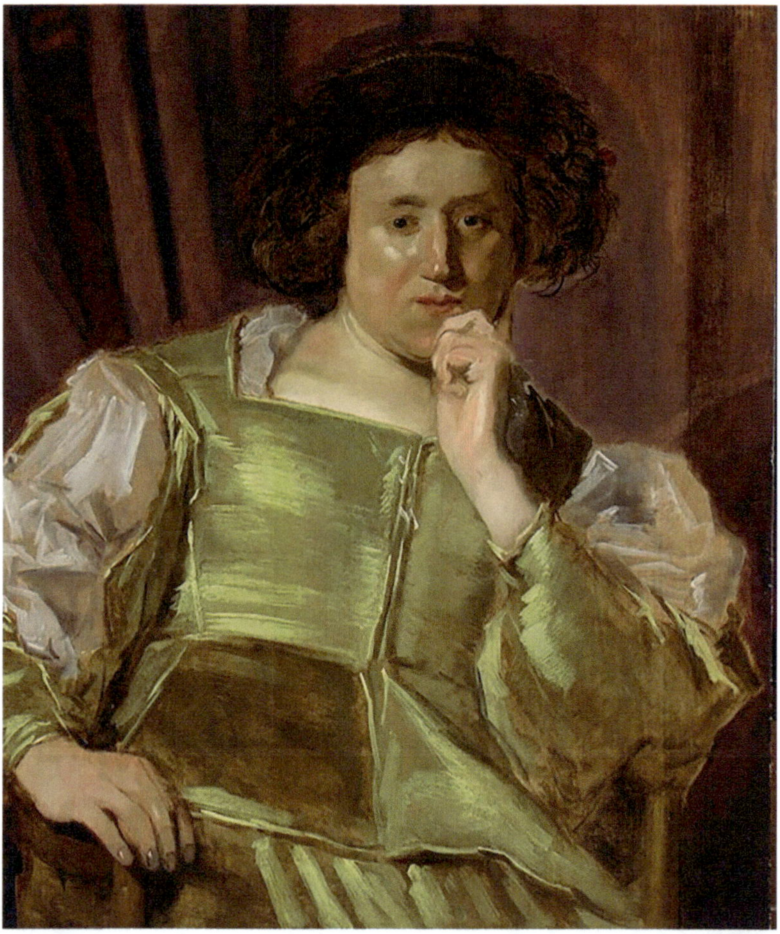

Fig. 2.11 Frans Hals, *Portrait of a Lady*, c. 1650–1652, oil on canvas, 78.5 × 62 cm. (Private collection)

at most the face (nowadays this painting is considered a self-portrait by Leyster[17]); from another portrait everything but the face (cat. no. 27; Fig. 2.14).

In total –according to the article– between 30 and 33 paintings were "unworthy" of Hals's name: these paintings had either been made in Hals's studio, possibly by one of his five sons, or had a "dubious provenance" (*vreemde herkomst*). Part of the purpose of the newspaper article was to assuage the unease prevailing among the public; rumors were circulating that half of the paintings in the exhibition were forgeries.

[17] It is now in the collection of the National Gallery of Art in Washington D.C. (acc. no. 1949.6.1), see https://www.nga.gov/collection/art-object-page.37003.html

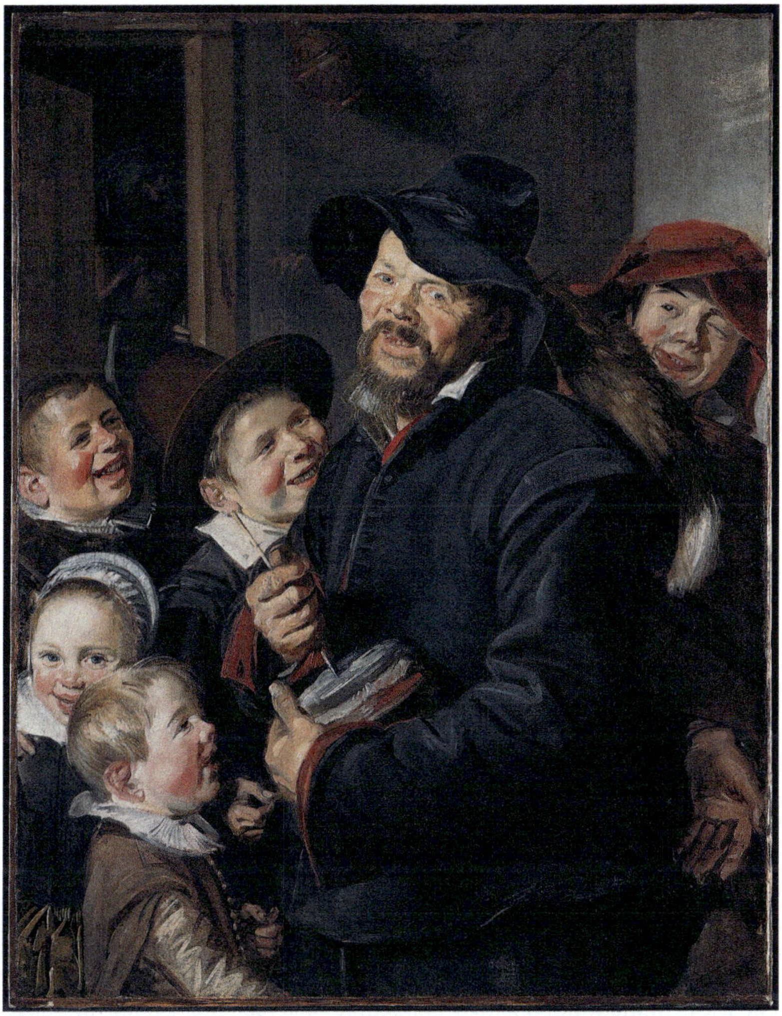

Fig. 2.12 Follower of Frans Hals, *The Rommel-Pot Player*, c. 1625–1635, oil on panel, 39.1 × 30.5 cm. (Art Institute of Chicago, Chicago)

Although the expert is not mentioned by name in the paper, Gratama, the director of the Frans Hals Museum, recognizes him immediately from his comments: Abraham Bredius (1855–1946), a well-known specialist of seventeenth-century painting, who had been director of the Mauritshuis for 20 years (1989–1909) and had since moved to Montecarlo. To confirm his suspicions, Gratama sends him a letter on September 14, 1937: "*De Telegraaf* published an extensive article on August the 29th about questionable paintings in the Frans Hals exhibition, in which an anonymous art expert is cited. What is reported therein corresponds so strongly with the opinions expressed by you during your visit to the exhibition that the

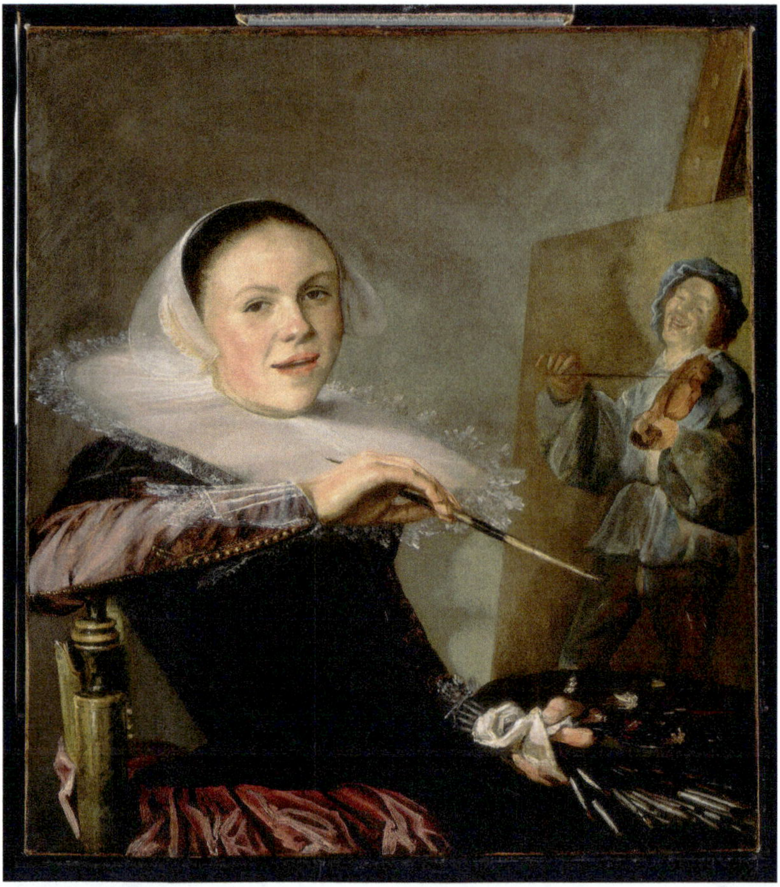

Fig. 2.13 Judith Leyster, *Self Portrait,* c. 1630, oil on canvas, 74.6 × 87.6 cm. (National Gallery of Art, Washington)

question arises in my mind whether you could have been the expert interviewed by *De Telegraaf*." Gratama calls the article "particularly unpleasant"; he regrets that "when differences of opinion between experts arise, the general public is startled and made distrustful by an ineffective publication, while it is unable to understand the issue properly."[18]

[18] In Dutch: "bijzonder onaangenaam" and "wanneer meeningsverschillen tusschen de kenners rijzen, door ondoelmatige publicatie het groote publiek wordt opgeschrikt en tot wantrouwen gebracht, terwijl het niet in staat is de questie zuiver te verstaan." Letter from Mr. Gratama to Mr. Bredius, dated 14 Sept 1937, Noord-Hollands Archief, Haarlem, archive number 1374, inventory number 89. With thanks to Mariken Kamp, who researched archival documents on the 1937 exhibition as part of Elmer Kolfin's BA seminar on Frans Hals connoisseurship at the University of Amsterdam (2012).

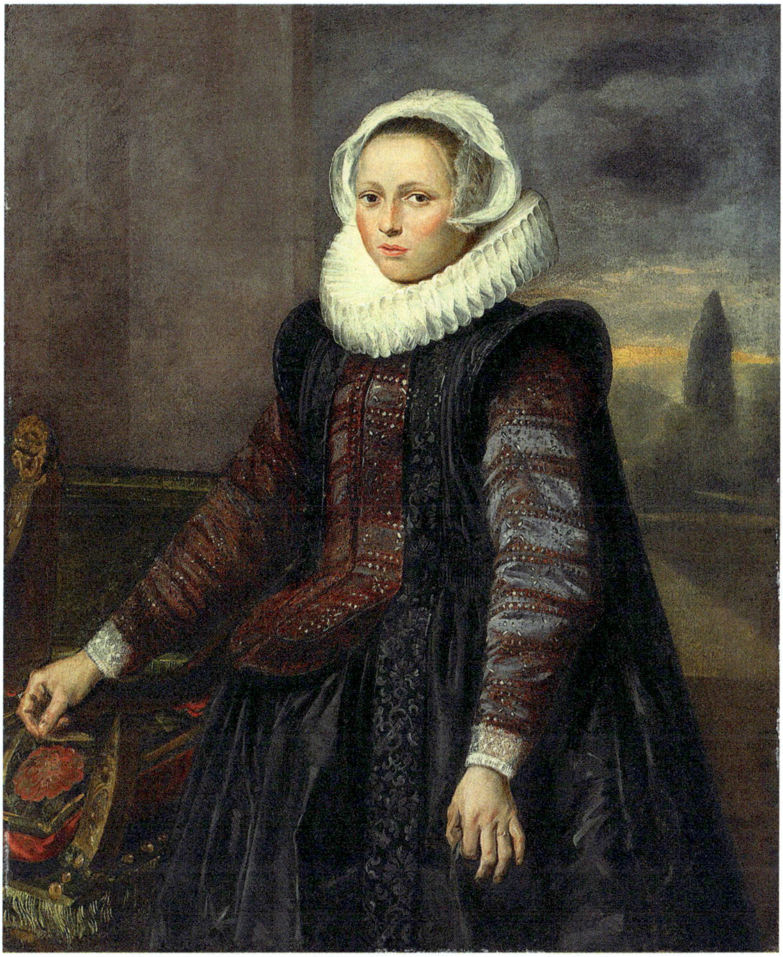

Fig. 2.14 Surroundings of Frans Hals, *Portrait of a Woman*, c. 1626–1628, oil on canvas, 112 × 91 cm. (Private collection, ©2016 Christie's Images Limited)

Three days later, on September 17, 1937, Bredius replies: it was him indeed. He had heard such disagreements about the works in the exhibition and had been approached from all sides; he could not keep his opinion under wraps when he was asked about this and put on the spot. But he had also praised the exhibition: "In the meantime, I continue to congratulate you on having given that wonderful exhibition to the world."[19] Somewhat reassured by the kind words about the exhibition as a whole, Gratama then tries to get him to express his admiration for all the beauty in

[19] Letter from Mr. Bredius to Mr. Gratama, dated 17 Sept 1937, Noord-Hollands Archief, Haarlem, archive number 1374, inventory number 89.

the exhibition publicly as well, preferably in *De Telegraaf*.[20] In vain: Bredius does not comply.

In the months that follow, it becomes clear that Bredius is not the greatest critic of the Frans Hals exhibition, as his letter to Gratama already suggested. Before the end of the year, a razor-sharp book by the painter and restorer Maurits van Dantzig (1903–1960) is published, *Frans Hals: Echt of Onecht (Frans Hals: Genuine or Fake)*, in which he downgrades most of the paintings in the exhibition: of the 116 works shown, only 33 are undisputed originals in his view. The others are forgeries, copies, pictures by other seventeenth-century painters and paintings that were partly by Hals's hand, according to him; only a few works he cannot classify (Van Dantzig 1937). Possibly, Van Dantzig's insights had prompted the rumors circulating as early as August about the large quantity of forgeries in the exhibition.

Before delving into Van Dantzig's views on Hals attributions, however, it is important to first understand what selection criteria Gratama uses. In the exhibition catalogue, Gratama describes what he believes characterizes Hals as a painter and makes him important to the development of Dutch art. He calls Hals the "most Dutch of all 17th-century painters"; Hals was a painter who grew up in an "academic atmosphere" [i.e.: among painters such as Karel van Mander (1548–1606), Hendrick Goltzius (1558–1617) and Cornelis van Haarlem (1562–1638) who had founded an early academy—at] but remained uninfluenced by this and became a pioneer of Dutch art (Gratama and Van Rijckevorsel 1937, p. 17). Gratama characterizes Hals's innovations as follows: "In a striking way he portrayed the distinctive milieu of our Dutch Golden Age. Sparkling with life, betraying his exuberant, Southern nature, his work is characterized as that of a very spontaneous painter and striking realist, unequaled in his technical ability. [...] And how he wielded the brush! Not in the smooth, even way of his predecessors, but in a very individual, revolutionary way, here with a rough dab, there with a longer brushstroke, bringing out the most characteristic [in his topic]. In his way of working he shows himself to be a true impressionist, who knew how to represent the momentary in an unparalleled way."[21] Even more than in his portraits, Hals's characteristic innovation was to be seen in his free work: "His genre pieces reveal even more strongly the master as an impressionist, when he gives snapshot-like impressions of a Rommelpot player surrounded by screaming children, of fisherboys, of singers and drinkers. [..] Hals has left us immortal *instantanés* in that genre, which have never been equaled.

[20] Letter from Mr. Gratama to Mr. Bredius, dated 20 Sept 1937, Noord-Hollands Archief, Haarlem, archive number 1374, inventory number 89.

[21] "Op frappante wijze heeft hij het karakteristieke Hollandsche milieu van onze Gouden Eeuw uitgebeeld. Sprankelend van leven, zijn uitbundigen, Zuidelijken aard verradend, kenmerkt zich zijn werk als dat van een zeer spontaan schilder en treffend realist, ongeëvenaard in zijn technisch kunnen. [...] En hoe voerde hij het penseel! Niet op de gladde, gelijkmatige manier zijner voorgangers, maar op een hem zeer eigen, revolutionaire wijze, hier met een zet, daar met een streek of sliert, het meest karakteristieke naar voren brengend. In zijn wijze van werken toont hij zich een echte impressionist, die het momenteele op ongeëvenaarde wijze wist [weer] te geven" (Gratama and Van Rijckevorsel 1937, pp. 19–20).

Perhaps therein lies his greatest strength."[22] According to Gratama, Hals' spontaneous manner of painting was at the basis of a crucial development in art: "The spirit of Frans Hals, the spirit of simple, pure naturalness has always lived on. [...] The line, which begins with Frans Hals, leads over Delacroix past Courbet and Manet. It is the line of trusting on one's intuitions, uninfluenced by any compulsion."[23]

Hals was - according to Gratama - thus the opposite of an academic painter, which was a common way of thinking at the time. Professor Wilhelm Martin, in his 1935 survey work *Frans Hals and his time*, even calls him an "anti-academician" (Martin 1935, p. 322 ff). Moreover, Gratama sees Hals as an impressionist-avant-la-lettre, a champion of simple naturalness, with an entirely unique, revolutionary type of brushwork that enabled him to depict a sudden moment. In his paintings, according to Gratama, Hals expressed not only his own "feeling," but also the "spirit" of his time, which would subsequently echo in the spontaneous brushwork of modern painters such as Courbet and Manet. The idea that artists were a kind of seers who expressed the "spirit" of their time ("*Zeitgeist*") was also prevalent at the time; it was based on the ideas of the influential German philosopher Georg Hegel (1770–1831).

How exactly Hals's spontaneity and painting style can be recognized in his works and how the selection of paintings for the exhibition came about is barely explained in the exhibition catalogue. At the beginning of the catalogue section it is stated that as far as possible a chronological order has been maintained according to the insights of Prof. Valentiner. Whether the latter was also involved in the selection of works remains unclear, however; he is not thanked for this in the preface. In the catalogue entries, brief mention is made of the relevant literature on each painting (by Moes 1909; Hofstede de Groot 1907–1928, vol. 3 (1910); Bode 1914; Valentiner 1936; among others); so the literature on Hals was certainly consulted. In addition, seven new paintings are included, that had not yet been discussed in the literature (cat. nos. 8, 11, 12, 79, 104 and 111).

Only when the exhibition received bad press did the director explain his working methods in more detail, first in a newspaper article and then in a report on the exhibition written for the Haarlem mayor and deputy mayors. In a letter to the editor in the newspaper *Nieuwe Haarlemsche Courant* of November 26, 1937, he states that "when making the selection of works for the exhibition, as much as possible was done to bring together works that were mentioned in the standard works about Frans

[22] "Zijn genrestukken openbaren nog sterker den meester als impressionist, wanneer hij momentopnamen gaf van een rommelpotspeler omringd door joelende kinderen, van visschersknapen, van zangers en drinkers. [..] Onsterfelijke instantanés heeft Hals ons in dat genre nagelaten, welke nooit zijn geëvenaard. Misschien ligt daarin wel zijn grootste kracht" (Gratama and Van Rijckevorsel 1937, p. 25).

[23] "De geest van Frans Hals, de geest van eenvoudige, zuivere natuurlijkheid is steeds blijven voortleven. [...] De lijn, die bij Frans Hals begint, voert over Delacroix langs Courbet en Manet. Het is de lijn van het zich intuïtief laten gaan op zijn gevoel, onbeïnvloed door eenigen dwang" (Gratama and Van Rijckevorsel 1937, p. 26).

Hals." To which he adds, somewhat indignantly: "In my opinion, a better guarantee for the authenticity of the exhibited works could not be given."[24]

In his report for the mayor and deputy mayors dated January 17, 1938, he adds that the purpose of the exhibition was to give the most complete overview of Frans Hals's oeuvre. He also indicates that he had selected the works himself on the basis of the existing literature and recent discoveries. Leading up to the exhibition, he had traveled to Germany, Austria, Hungary, Czechoslovakia, Belgium, France and England to view the works in person. A trip to America had not been possible, and for this reason he had enlisted the help of Hals expert Dr. W.R. Valentiner, and the art dealers Mr. N. Katz of Dieren and H. Schaeffer of New York. Thanks in part to them, it was possible to obtain 36 American loans. To what extent these gentlemen also gave substantive advice regarding the loans is not entirely clear.

The newspaper had suggested that the insights of Hals experts like Valentiner had not been taken into account in the preparation of the exhibition, and that the exhibition contained all sorts of attributions that would not have met with his approval. However, two works that received much criticism in the newspaper—the Little *Merry Company* (see Fig. 2.9 above) and the *Rommelpot Player* (Fig. 2.12)—were not new attributions, but drew on the literature. Gratama was therefore rather incensed by the criticism: "In general, the newspapers have been very frivolous in their judgments of the works exhibited at the Frans Hal exhibition, where more respect for the authority of the great art experts would have been desirable."[25]

The suggestive newspaper articles about forgeries and questionable works in the exhibition had certainly done the reputation of himself, the museum and the experts involved no good, and it is understandable that Gratama would have preferred more deference. At the same time, the usual reverence for the connoisseur's judgement also seems to be part of the problem. In the aftermath of the exhibition the great experts were clearly not in agreement. Moreover, it was far from clear on what knowledge or insights the connoisseurs' judgments were based; their benchmarks, standards and frame of reference were hardly made explicit. In fact, they judged mainly intuitively and were not accustomed to explaining their attributions. As the reliability of their judgments was doubted, however, the adequacy of their methods was also called into question.

---

[24] Letter from Mr. Gratama to the editor in chief of the *Nieuwe Haarlemsche Courant*, dated 26 Nov 1937, Noord-Hollands Archief, Haarlem, arch. no. 1374, inv. no. 91.

[25] With thanks to Hannie Koomen who gave me a copy of Gratama's report, and with her team inventoried and described much of the museum archives. To my knowledge, this report has not been researched before.

2.5 Pictology and the Search for Objective Criteria

The lack of clear criteria for attributing and de-attributing paintings was a thorn in Maurits van Dantzig's side. In the introduction of his book *Frans Hals: Echt of onecht (Frans Hals: Genuine or Fake)* he sharply criticizes art experts 'who have the habit of answering every question relating to the value of an artwork with the Yes! Or No! of their so-called aesthetic feeling. They assume that they are moved by works of art, while non-art works leave them cold or repel them. [...] However, they pertinently refuse to make *conscious* the experiences on which their judgment of beauty rests (for does not every judgment rest on experiences?)".[26] The subtitle of his book leaves no doubt about his main bone of contention: *on the occasion of the Frans Hals exhibition in 1937*. Van Dantzig had seen more exhibitions in which a large part of the attributions were, in his opinion, incorrect, but the "Haarlem exhibition surpassed in this respect all its predecessors of recent years"; he perceives a "disastrous influence [...] of such "shows" "—because they "inevitably also cloud and spoil the sense of art of the impressionable public", and therefore speaks of a "serious cultural danger" (Van Dantzig 1937, p. 2).

In his view clear and verifiable criteria were needed to determine if a painting was an original, copy, imitation, forgery or other type of work. He developed a new method, which he would later call 'pictology' (Van Dantzig 1947, 1973). On the basis of his own observations of the well-documented and securely attributed core oeuvre of Frans Hals (mainly his well-known large group portraits made for semi public display: three regent group portraits and five civic guard portraits), he made a list of 44 traits that he deemed characteristic of the artist.[27] He subsequently applied the criteria to the 116 works on display in the exhibition and reached his devastating conclusion: only 33 were authentic works by Frans Hals in his view, 5 doubtful, 42 were wrongly attributed and 36 paintings were even forgeries in his opinion.

Instead of broad remarks about the "spirit" of Hals or his time, Van Dantzig's observations are mostly very concrete and detailed. He observes a certain unity of action in the group portraits. The figures depicted are all engaged in something and their postures are in tune with one another. At least as important is the role of the viewer. The latter is involved in the painting, counterbalancing the composition, according to Van Dantzig: "A work by Hals without an observer is *au fond* an unbalanced painting. He begs, as if it were, for direct contact with you; without you the work is incomplete." He subsequently elaborates on how Hals engages the viewer with his figures, choosing casual, natural poses in which all limbs make angles with

[26] In Dutch: "[kenners] die gewoon zijn, elke vraag naar de kunstwaarde van eenig werk beantwoord te achten met het Ja! Of Neen! van hun zgn. Aesthetisch sentiment. Zij nemen daarbij aan, dat zij geroerd worden door kunstwerken, terwijl niet-kunstwerken hun koud laten of afstooten. [..] Zij weigeren evenwel pertinent de ervaringen, waarop hun schoonheids-oordeel berust (want berust niet ieder oordeel op ervaringen?) *bewust* te maken" (Van Dantzig 1937, p. 2).

[27] These large group portraits never left the city of Haarlem; they became municipal property and are all on display at the Frans Hals Museum in Haarlem.

each other and with the picture plane. "This is not accidental [...] a parallel arrangement makes the model a backdrop that you pass by. An angular arrangement, on the other hand, breaks up the picture plane and establishes the connection with you."[28]

He also noticed that with Hals, the background has the character of an enclosure placed closely behind the model; even when the model is depicted outside, almost always he or she is placed in front of dense foliage that acts as a kind of backdrop. Only besides the figures does the landscape extend into the distance. Other pertinent observations concern the anatomy of the figures (their heads are relatively small), his brushstrokes (Hals consistently blends the transitions between the sharpest light and shadow areas somewhat, which creates a suggestion of depth; he also blends his final touches often wet-in-wet with the foreground and background), the difference he makes between main and secondary elements in his pictures (the main elements usually being heads, hands and protruding parts), and his rapid style of painting (Hals draws and paints, as if it were, simultaneously, applying color and form in one go; for example, the little touch of paint with which Hals indicates the highlight on a nose, itself has a form that corresponds to the curve of the nose).

Van Dantzig's own experience as a painter and restorer enables him to even speculate about Hals's use of binding materials: this must have been an emulsion (a mixture of two or more liquids that are normally immiscible such as oil and egg yoke), according to Van Dantzig. He believed that Hals adjusted such a mixture at will to influence the drying speed of various passages.[29] Pigments that usually dry slowly seem to have been painted over quickly, without showing cracks or wrinkles in the paint film caused by underpaint that has not fully dried. On the other hand, subtle layers of glazes and fluidly blended passages betray the use of oil in his opinion. With this observation, Van Dantzig is the first to suggest that Hals had not merely used oil paint as a binder—an observation that was confirmed nearly 80 years later during the restoration of Hals's *Regents of the St Elisabeth Hospital* (1641) when a black brushstroke in the underpainting was discovered that had nearly completely dissolved in the past due to an aqueous binding medium.[30] Whether or not Hals did indeed use emulsions for the upper layers in his paintings—Van Dantzig

[28] "Dat is niet toevallig [...] een evenwijdige opstelling maakt het model tot een coulisse waar ge langs gaat. Een hoekige opzet daarentegen verbreekt het vlak en brengt de verbinding met u tot stand" (Van Dantzig 1937, p. 7).

[29] Emulsions must have played an important role at the time but still have not been researched sufficiently, as Patrick Dietemann argued in his recent lecture "The Whole is More Than the Sum of its Parts—Colloid Chemistry and Microstructure of Paints" at the Gordon Research Conference Science for Cultural Heritage, 10–15 July 2022 in Les Diablerets, Switzerland (12 July 2022).

[30] As presented by Herman van Putten in his lecture 'The restoration of Hals's Regents of St Elisabeth's Hospital', at the conference Frans Hals: A Survey or Current Research, 8–9 Jan 2023 (9 Jan 2023). The research was carried out with Liesbeth Abraham, Mireille te Marvelde and Annelies van Loon, and the lecture will be published in the forthcoming conference proceeding (scheduled to be published in 2024).

suspected that this was particularly the case in the black and red parts—deserves further investigation.[31]

That an observation like the one above can be substantiated or refuted by technical research was exactly what Van Dantzig had in mind: he wanted to base his attributions on concrete and verifiable criteria.[32] The starting point for him were measurable and more or less fixed characteristics of Hals's painting style: "Just as one [...] can recognize each person by very specific characteristics, so can one distinguish the works of one hand from those of another."[33] In his view, two aspects were key when attributing paintings to Hals: "the greatness of the artist Hals" and "the specific features [...] that distinguish him from others".[34] He believed that a great artist exhibited a relatively high degree of consistency: his characteristic traits could be recognized both in the whole and in the details of his works, in which important and subordinate passages stood in a clear relationship to each other. Later he would add "spontaneity" as a criterion. In his view, there was a certain tension underlying every old master painting, caused by the artist's urge to create and the difficulty of depicting objects recognizably. Great masters were able to overcome the resulting challenges with ease, resulting in a high degree of spontaneity in their brushwork, which copies and forgeries usually lacked, according to Van Dantzig (see also below; Van Dantzig 1947, pp. 58–90, 1973, pp. 5–11; see also Tummers 2011, pp. 33–39).

What is striking about Van Dantzig's method is its static character; except for a few remarks about Hals's earliest and last work, Van Dantzig does not discuss Hals's development as an artist. [In Hals's earliest civic guard piece from 1616, according to Van Dantzig, his colors are still a bit too harsh (Fig. 2.15); while his very late regent pieces from about 1664 (Fig. 2.4) stand out for their extremely sparse detailing]. The format and type of painting –e.g. a preliminary study, design for a print, commissioned portrait, or free work– are also completely ignored; they do not affect

[31] Since the 1930s the Doerner Institute in Germany has done research on emulsions. Here, among other things, the construction of Hals's painting of a young woman known as *La Bohémienne* (Louvre) was researched and reconstructed using tempera as the bottom layer, witness a letter to the editor by a certain J.PH. Aussems, published in the Dutch newspaper *NRC Handelsblad* in March 1990 ('alla prima') (Aussems 1990). [A clipping is present in the restoration archive of the Frans Hals Museum which mentions the month of publication but not the exact date.] Grimm also speculates on the use of emulsions because of the surface structure of the paint stroke, with deeper middle section (Grimm 1990/1989, p. 70). Hendriks and Levy-van Halm mention that the Laboratoire de Recherche des Musées de France has identified an emulsion in the priming layer of two portraits attributed to Hals, the portraits of Paulus van Berensteyn and Catharina van der Eem, as evidenced by a 1988 report (Hendriks et al. 1991, note 59).

[32] "Niemand zou zich echter meer verheugen dan de schrijver van dit werkje, indien zijn argumentatie voor weerlegging op redelijke gronden vatbaar zou blijken" (Van Dantzig 1937, p. 2).

[33] "Zooals men [..] ieder mensch aan zeer specifieke verschijnselen herkennen kan, zoo kan men ook de werken van de eene hand onderscheiden van die van een andere hand" (Van Dantzig 1937, p. 5).

[34] "de grootheid van den kunstenaar Hals"; "de specifieke eigenschappen van Hals, die hem van anderen onderscheiden" (Van Dantzig 1937, p. 5).

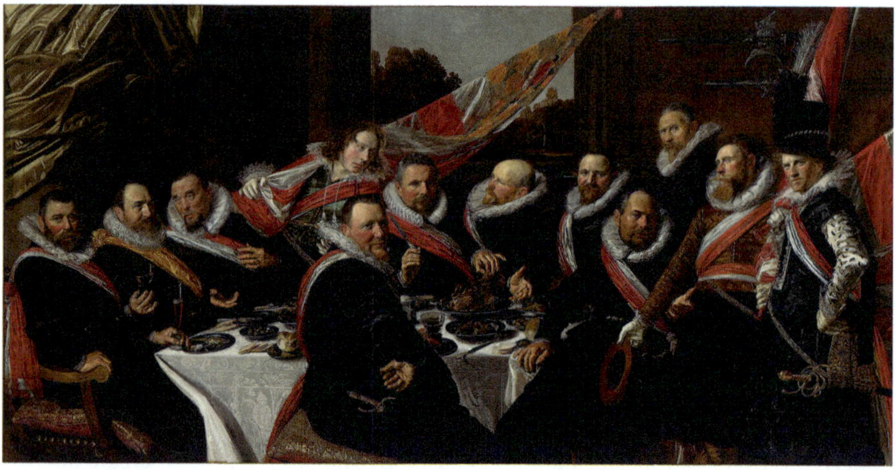

Fig. 2.15 Frans Hals, *Banquet of the Officers of the St. George Civic Guard*, 1616, oil on canvas, 175 × 324 cm. (Frans Hals Museum, Haarlem)

the criteria used.[35] Van Dantzig focuses on constant factors: in his view, the hand of the painter reflected his characteristic traits in quite a literal sense. For example, he notices that the figures in Hals's paintings appear full of life (*sprankelend*); they seem to have "dashing, witty personalities" (*zwierig-statige, geestige persoonlijkheden*). He cannot imagine, however, that the citizens of Haarlem, including merchants, brewers and mayors, would differ much from the Haarlem citizens with similar functions of his own time, and concludes that Hals must have imparted something of his own character in each of his portraits: "Hals creates for himself a genus of bubbly, vibrant people."[36] It is an aspect of Van Dantzig's method that he would later develop further for the benefit of the business world, where pictology was used for some time to analyze the character of employees through drawings by their hand (Van Dantzig 1973, p. VIII).

Van Dantzig also elaborated his method further for art attributions. A major forgery scandal immediately after the end of World War II prompted him to explain his method in more detail in the book *Johannes Vermeer, The Disciples at Emmaus and the Critics* (Van Dantzig 1947). The painting in the title, *The Disciples at Emmaus* (Fig. 2.16) had been discovered by Abraham Bredius in 1937 and was considered one of the most important works by Johannes Vermeer, until it was exposed as a forgery by Han Van Meegeren shortly after the war during a high-profile court case (see Lammertse et al. 2011; Tummers 2011, p. 23 ff; Van den Brandhof 1979). Van Dantzig claimed he had recognized the painting as a forgery already before Van

[35] Only when studying Van Gogh did Van Dantzig formulate partially different criteria for different stylistic periods (see Van Dantzig 1973, pp. 55–57).

[36] In Dutch: "Hals schept zich een geslacht van sprankelend-levende menschen" (Van Dantzig 1937, pp. 21–22).

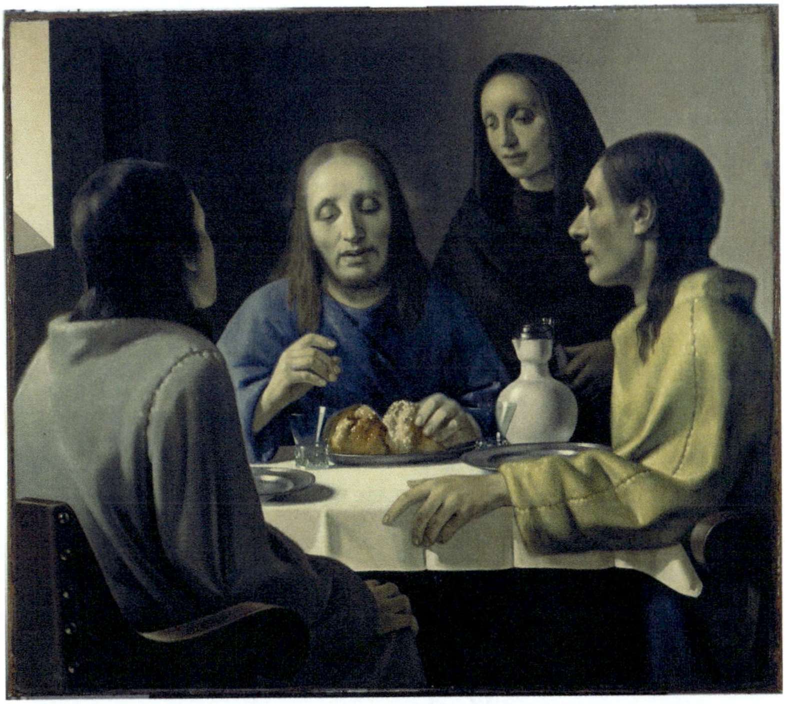

Fig. 2.16 Han van Meegeren, *Christ and his Disciples at Emmaus*, c. 1937, oil on canvas, 118 × 130.5 cm. (Museum Boijmans van Beuningen, Rotterdam)

Meegeren was arrested, and used the occasion to demonstrate and expand his attribution method (see Tummers 2011, p. 26 and note 10, p. 253). Moreover, after Van Dantzig's death, some of his students completed what he had initiated: they systematically described his approach in the book *Pictology: an analytical method for attribution and evaluation of pictures* (Van Dantzig 1973).

In short, Van Dantzig's method did not only get a name (first mentioned in the 1947 publication), but became increasingly systematic, and even began to look a bit like a formula. According to Van Dantzig, a list of some 100 characteristic features was needed to make successful attributions. Assuming that these characteristics were correctly defined, a new work could be attributed to the painter with certainty if at least 75% of the characteristics described could be seen in it. With less than 50% corresponding characteristics, the work was definitely *not* by the same hand. With more than 50% but less than 75% similar characteristics, further research was needed to be able to give a definitive answer. He also made spontaneity measurable by clear drawings (Fig. 2.17) and this became one of his most important criteria for judging artistic quality (and thus for distinguishing originals from copies and forgeries).

Although Van Dantzig himself does not refer to him in any way, his method is strongly reminiscent of the astute Italian art expert Giovanni Morelli

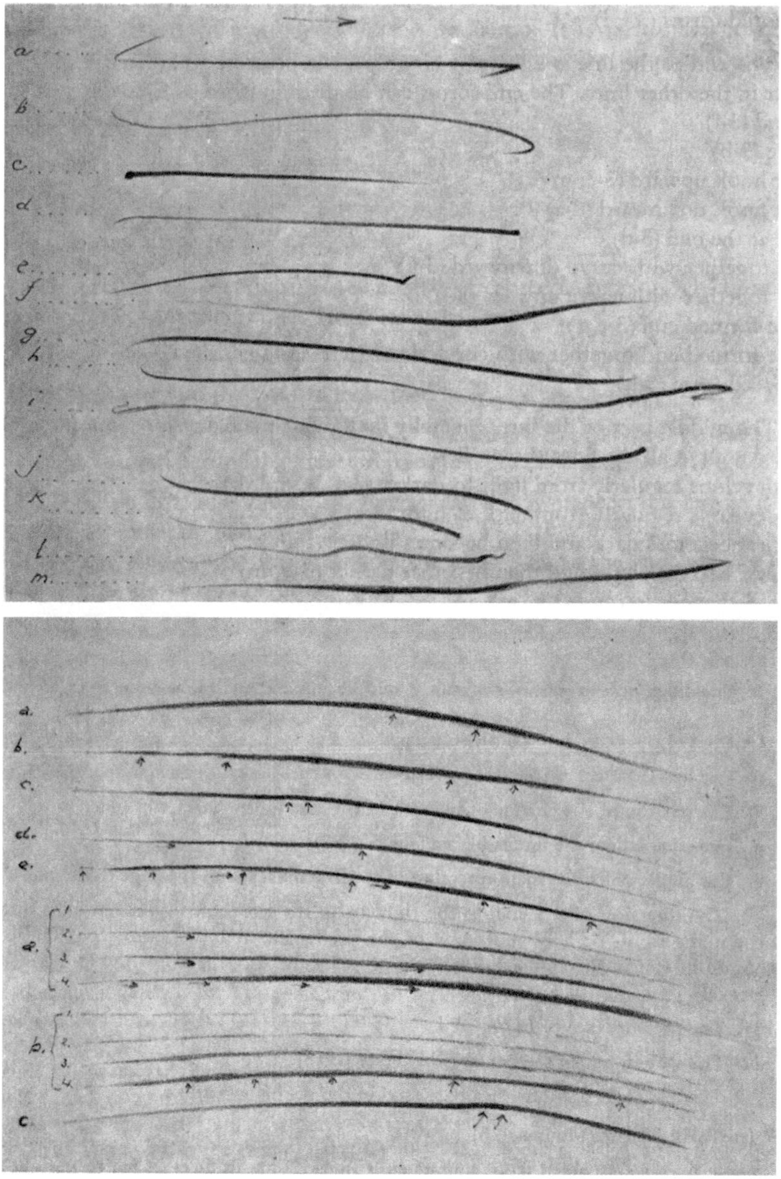

Fig. 2.17 Examples of (a) spontaneous line and (b) copied lines, in: Van Dantzig, M.M.: Pictology: an analytical method for attribution and evaluation of pictures. Brill, Leiden (1973)

(1816–1891). The latter became very famous in the late nineteenth century with his razor-sharp pleas for rational and verifiable standards in the attribution of paintings, undermining many attributions in public museums. Morelli's eye for detail and characteristic patterns that painters repeated –probably unconsciously– in the depiction of, say, an eye, a hand or a nose inspired Freud in developing his famous psycho-analysis, and must have been an important inspiration for Van Dantzig as well (Ginzburg and Davin 1980). According to Van Dantzig's pupil Storm van Leeuwen, his teacher looked particularly at features that emerged from characteristic movements of an artist: unconscious regularities.[37] A clear difference from—or rather an addition to—Morelli's method is Van Dantzig's emphasis on quantity: the importance of a very extensive list of characteristics, which could then be tested.

In the case of Frans Hals, Van Dantzig's list of characteristic features grew over the years: from 44 to a whopping 146 features (see Van Dantzig 1937, pp. 5–37; 1973, p. 78 ff). Although in many cases the characteristics became more precise, sometimes the nuance also disappeared: for example, Van Dantzig's interesting remark about the drying speed of Hals' paint (characteristic no. 42) no longer appears in the 1973 list of Hals's characteristics. There is also a subtle, yet significant difference in the way Van Dantzig's characteristics are applied. The 1937 application shows that he had an eye for possibly unfinished passages; for example, he stated that the head of the third figure from the left in Hals's earliest civic guard piece was not entirely completed (sn.pub/ohdtkx; Fig. 2.15). He also regularly detected additions by studio assistants in the secondary work, for example in the wedding portrait of Isaac Massa and Beatrix van der Laan (Fig. 2.18): "The view into the distance is too clumsily executed and the architecture is also uncertain. They are both by other hands. Presumably this somewhat unusual composition is the result of a wish of the clients who were portrayed and who may have wanted their possessions immortalized together with themselves. Also, the foliage of the trees is too weak for Hals. The brushstrokes are not adjusted to the perspective, but instead parallel in their placement, and many separate leaves are repeated over and over again."[38] The 1973 comprehensive analysis, however, does not address the possibility of unfinished passages and studio assistance; the analysis focuses purely on Hals's characteristic inventions and handwriting. Even when the list of characteristics is applied –by way of example– to specific portraits, the possibility of studio assistance is not considered.

In 1937, Van Dantzig checked all 116 exhibited paintings against his list of characteristics, which led him to his shocking conclusions. No less than 36 pictures were forgeries in his view. Moreover, he gave several well-known paintings a new

[37] "In Van Dantzig's view especially these features are important; they have been brought about by movements which the artist made unconsciously and they were more or less constant throughout his life, even when his style changed radically."(Storm van Leeuwen 1979, p. 58). Despite the great similarities with Morelli's approach, the latter is not mentioned in this passage, see also (ibid., note 10, p. 90) on the difference with Morelli.

[38] "Het verschiet is te peuterig en ook de architectuur is onzeker. Zij beide zijn van andere hand. Vermoedelijk is geheel deze eenigszins ongewone compositie het gevolg van een wensch der geportretteerde opdrachtgevers die hun bezittingen tegelijkertijd met henzelf vereeuwigd wilden zien. Ook het gebladerte der bomen is te zwak voor Hals. het heeft geen perspectivisch verloopende, doch evenwijdige penseelstreken, terwijl de te vele aparte blaadjes zich steeds eender herhalen" (Van Dantzig 1937, p. 41).

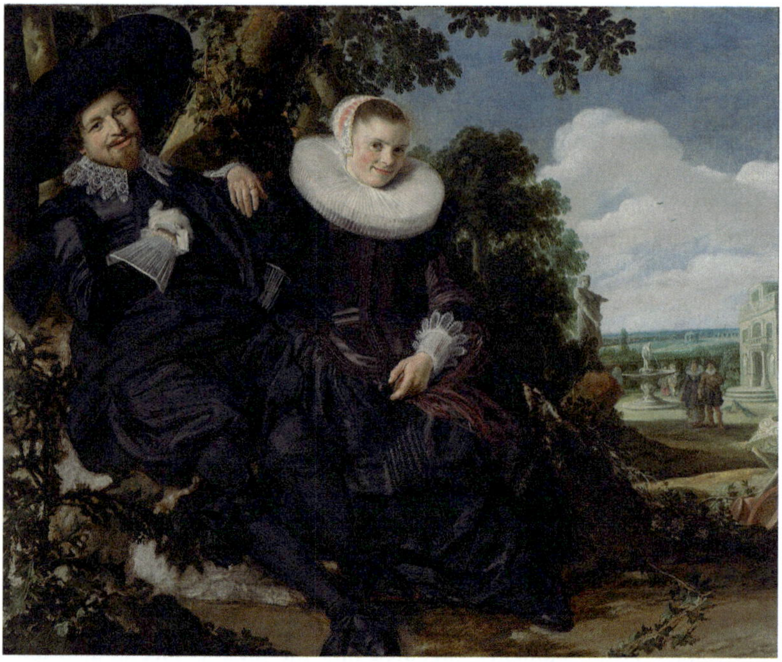

Fig. 2.18 Frans Hals, *Portrait of a Couple, Probably Isaac Abrahamsz Massa and Beatrix van der Laen*, c. 1622, oil on canvas, 140 cm × 166.5 cm. (Rijksmuseum, Amsterdam)

attribution. He believed that the portrait of *Zaffius* in the Frans Hals Museum (Fig. 2.19) was a copy, a piece that has since been alternately identified as either an original or a copy (see Grimm 1990/1989, no. 1, pp. 43, 271; Hendriks et al. 1991; Slive 1970–1974, vol. 3, p. 1; Van Thiel 1993; Köhler et al. 2006, cat. no. 186, p. 491 ff). He also considered the famous *Lute Player* in the Louvre (Fig. 2.1) to be a copy, an opinion that has not been supported since: "Not one economic, accurate touch can be seen in the entire piece," Van Dantzig stated, while at the same time observing: "The pose of the figure and the expression of the head are too strong for such a weak painter."[39] More shocking still was his opinion about the famous portrait of *Jasper Schade*, known - both then and now - as one of Hals' best individual portraits (Fig. 2.3) (see Tummers (ed.) 2013a, b, p. 88 ff). According to Van Dantzig, it was a forgery and he counted it among a group of 11 that were probably made by the same hand. Among others things, he finds it peculiar that he cannot identify the fabric of Jasper Schade's clothing, he believes that the head is placed too high on the shoulders, and he misses a light accent in the hat and folds in the drapery. Moreover, he characterizes the forger's character as "reserved-defensive" and that of Hals as

[39] "Op het gehele stuk komt niet één puntige, rake toets voor", aldus Van Dantzig, terwijl hij tegelijkertijd observeert: "De houding van de figuur en de uitdrukking van den kop zijn te sterk voor een zoo zwak schilder" (Van Dantzig 1937, pp. 60–61).

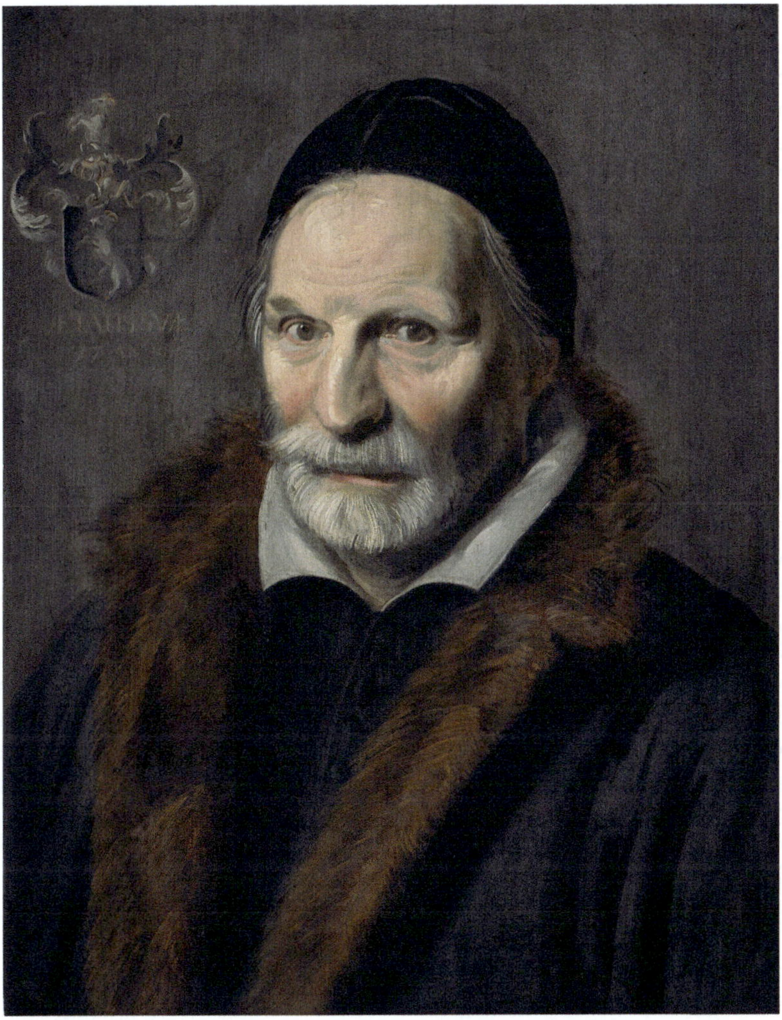

Fig. 2.19 Frans Hals, *Jacob Hendricksz Zaffius*, 1611, oil on panel, 54.5 × 41 cm. (Frans Hals Museum, Haarlem)

"accommodating-contact seeking".[40] Given the great importance of character analysis in pictology, this last argument seems particularly revealing: the facial expression—described in the literature as 'as proud as a peacock'—clearly did not match Van Dantzig's image of Hals (Slive (ed.) 1989–1990, cat. no. 62). His description further suggests that the painting must have been covered under a thick layer of discolored varnish: "The whole is covered with an artificial, flat brown layer" (Van

[40] In Dutch: 'terughoudend-afwerend' versus 'tegemoetkomend-contact zoekend' (Van Dantzig 1937, pp. 105–111, esp. pp. 107–108 and 110).

Dantzig 1937, pp. 105–111, esp. p. 108). The *Laughing Boy* that is in the Mauritshuis collection today (Fig. 2.2) was also de-attributed: according to Van Dantzig, it was made by another seventeenth-century painter (Gratama and Van Rijckevorsel 1937, cat. no. 22; Van Dantzig 1937, cat. no. 55).

Van Dantzig's optimism about making "objective" statements using his method and his conclusions—especially the large number of forgeries he identified—met with disbelief and repugnance from other specialists. According to his pupil Storm van Leeuwen, virtually the entire Dutch art world turned against Van Dantzig, while he remained virtually unknown abroad (Storm van Leeuwen 1979, p. 57). Gratama, in his report to the mayor and deputy mayors of Haarlem, indicates that Van Dantzig's book was laughed at in art historical circles. He briefly dismisses the book as implausible, citing the example of *Jasper Schade*. The low esteem that academics had for Van Dantzig –an outsider who had not trained as an art historian– is particularly clear in a comment from Wilhelm Vogelsang (1875–1954), the first university professor in Art History in the Netherlands. In a copy of Van Dantzig's book that he donated to the library of Utrecht University, he wrote: "Donated as an example of trashy literature."[41]

Despite the many negative reactions, Van Dantzig does seem to have struck a chord. Although his selection was too rigid and severe for many, he paved the way for a more precise and reasoned approach to defining Hals's oeuvre. Van Dantzig's pupil Storm van Leeuwen carefully retested his method in 1977 as far as its application to Hals's oeuvre was concerned, and in a well-researched article came to somewhat different conclusions and standards than his teacher. Among other things, he further elaborates on the characteristic proportions of Hals' figures, compares the brushstrokes characteristic of Hals with those of Jan Steen, identifies a preference for triangular arrangements in Hals's compositions, and points to characteristic features of so-called "residual forms," the negative spaces around the depicted objects and figures (a criterion that seems rather dated today).[42] Of particular interest are the two points of improvement he suggests. According to Van Dantzig, each characteristic was exactly equally important; however, Storm van Leeuwen emphasizes that in many cases the absence of certain characteristics may be important, while their presence may be virtually meaningless, and vice versa. How this could be included in the analysis in a quantitative way, he did not know. Storm van Leeuwen also felt that Van Dantzig had been too quick to jump to the conclusion "false"; in his opinion, this type of conclusion was not the domain of a pictologist alone and should always be substantiated with chemical research (Storm van Leeuwen 1979, pp. 88–89). By the time Storm van Leeuwen analyzed and tested his teacher's method, however, research into the exact scope of Hals' oeuvre had already taken new turns, thanks to the work of two foreign researchers who would become the greatest Hals scholars of the twentieth century: Harvard Professor Seymour Slive (1920–2014), and German scholar Dr. Claus Grimm (born 1940).

[41] "Overgedragen als een monster van prullige literatuur." (Storm van Leeuwen 1979, note 3, p. 89).

[42] In Dutch: "restvormen"(Storm van Leeuwen 1979, p. 65 ff).

2.6 Slive, Grimm and the 1989/1990 Controversy

The second retrospective of Frans Hals's oeuvre in 1962 differs considerably from its 1937 predecessor. Of the 116 paintings on view in 1937, only 46 return in 1962. The new exhibition consists of 76 paintings and is accompanied by a well thought-out catalogue, compiled by the relatively young Harvard professor Seymour Slive (1920–2014). Slive is working on a large-scale catalogue of Frans Hals's oeuvre at the time and uses his research for the exhibition publication (Slive 1970–1974). He discusses each painting in the show in an individual entry, analyzing it critically and relating it to broader developments in art history and iconography. Unlike the 1937 retrospective, this new presentation does not create controversy about what Hals may or may not have painted. On the contrary, reactions are full of praise, and the exhibition catalogue becomes a model for similar shows.[43]

It was not until 1989 that a new controversy arose, and Slive's vision clashed with opposite insights from his 20 years younger colleague Claus Grimm. The latter - inspired by the 1962 retrospective exhibition - began to work on an oeuvre catalogue of Frans Hals, which he completed as a Ph.D. dissertation in 1968, and published in 1972 (Grimm 1968, 1972). In it, Grimm arrived at a different understanding of Hals's oeuvre than Seymour Slive, whose three-volume retrospective appeared between 1970 and 1974. Although the different views had been clear on paper for some time, they initially generated little discussion. Both specialists further elaborated their views on Hals in the following decades, and –each separately– came to a renewed definition of Hals's oeuvre in 1989.

Slive presented his vision in the third major retrospective of the oeuvre of Frans Hals in Washington, London and Haarlem in 1989/1990, which consisted of 86 works. He had been even more selective in his choices than in 1962. For example, he had now labeled the *Malle Babbe* in the Metropolitan—which he still held in 1962 to be an autograph work by Frans Hals—a "version of [the Berlin] *Malle Babbe*" by an anonymous hand (Figs. 2.20 and 2.21) (Slive (ed.) 1989–1990, cat. no. 37, Fig. 37c; Slive and Hees 1962, cat. no. 31, pp. 50–51; see also below Chap. 4). The exhibition was accompanied by a hefty 439-page catalogue with very extensive commentaries by Slive on each individual painting. It also contained innovative essays by a variety of specialists, covering such topics as Hals's Haarlem clientele, his reception in the nineteenth century and his painting technique, as well as a complete overview of all the archival sources on Hals. It was a milestone in the study of the master, against which earlier catalogues stood in stark contrast (for comparison, in 1937 and 1962, the catalogues—excluding images—covered 56 and 79 pages, respectively).

However, the extensive scholarly studies accompanying the exhibition could not prevent the presentation from becoming the focal point of a particularly fierce battle

[43] Eric Jan Sluijter, 'Frans Hals in the 21st Century', lecture on 1 Sept 2016, Bavokerk, Haarlem. https://www.ericjansluijter.nl/wp-content/uploads/2016/08/Frans-Hals-lezing-1-9-2016-tekst-def.pdf

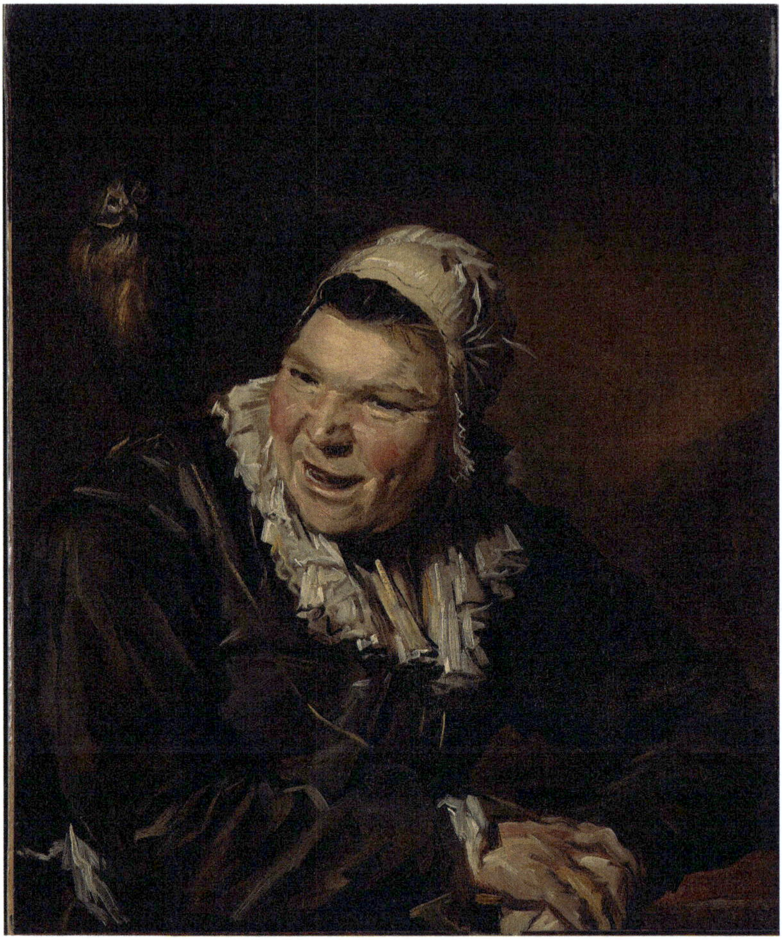

Fig. 2.20 Style of Frans Hals, *Malle Babbe,* c. 1625–1650, oil on canvas, 74.9 × 61 cm. (Metropolitan Museum of Art, New York)

over the definition of Hals's oeuvre. It seems no coincidence that it was yet another retrospective of Hals's work that gave rise to controversy. After all, such a retrospective provides an excellent opportunity to compare the various paintings. Moreover, precisely during the run of the exhibition, Claus Grimm's new oeuvre catalogue of Frans Hals appeared, which differed even more strongly from Slive's view than his earlier catalogue. While according to Slive the total, surviving oeuvre of Frans Hals covered 222 paintings (Slive 1970–1974), Grimm made a smaller selection. Initially (in 1972) he still accepted 168 works; in 1989 he further reduced that number to 145 paintings, including six works that Slive—in turn—did not consider to be original Halses and four that Slive was not aware of (Grimm 1972, 1990/1989, p. 292 for

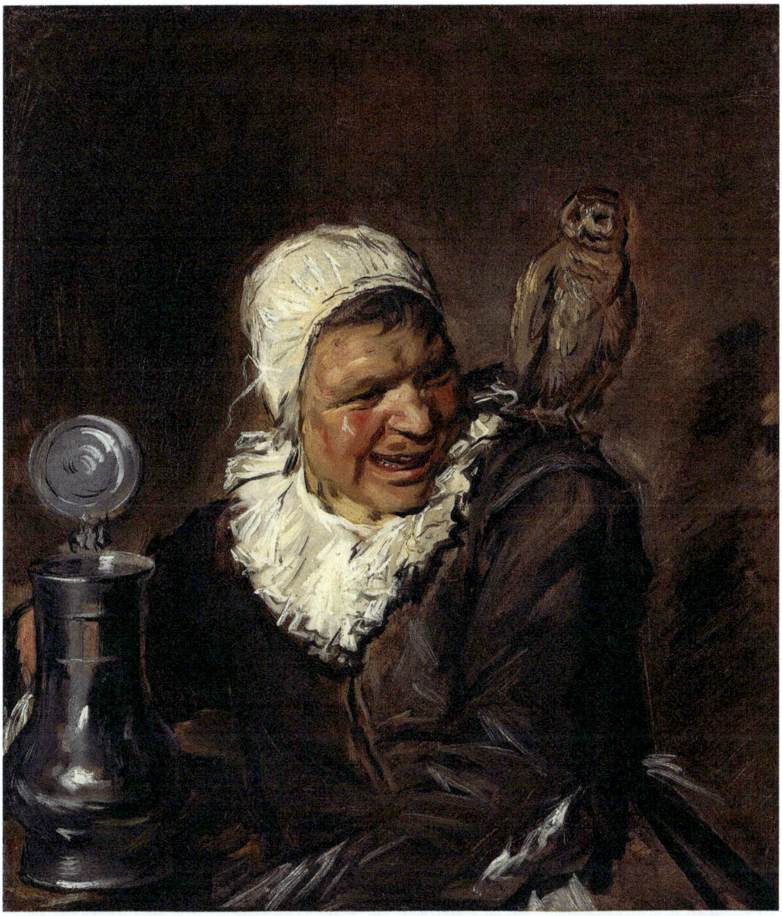

Fig. 2.21 Frans Hals, *Malle Babbe*, c. 1640–1646, oil on canvas, 78.5 × 66.2 cm. (Gemäldegalerie, Staatliche Museen zu Berlin)

correspondence).[44] Of the 68 paintings that hung in the London exhibition, as many as 28 were not painted by Hals, according to Grimm. It led to a veritable explosion of media speculation about possible forgeries and misattributions.

For example, the Dutch weekly *Vrij Nederland* devoted a cover and editorial to the controversy just before the exhibition was to open in Haarlem on May 12th, 1990 (Fig. 2.22). Fueled by the disagreement between Slive and Grimm, reviewer Ella Reitsma came up with de-attributions of her own, which she explained in eight pages: 'Frans Hals? It can be seen with the naked eye. A guide to visiting the great

[44] Six paintings of these had been written off by Slive (1970–1974, D44, D52, D54, D66, D72 and D74). These include the relatively smoothly painted *Portrait of a Man* from the Kremer collection, which Grimm dates c. 1637–1640 (Van de Ploeg et al. 2008, cat. no. 14, pp. 66–69).

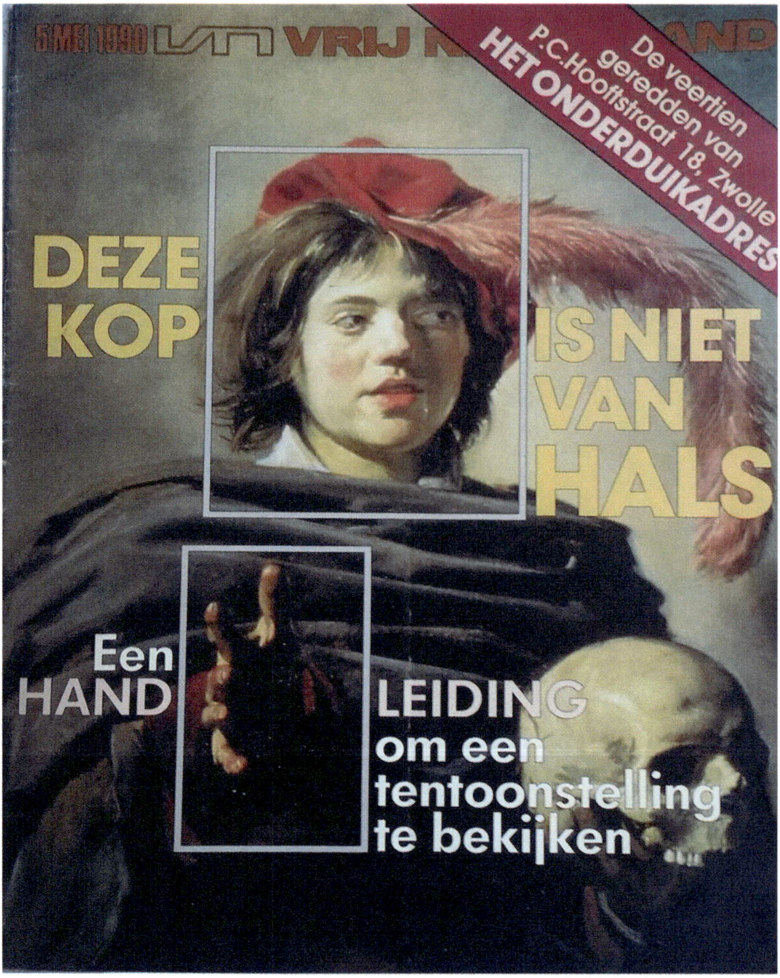

Fig. 2.22 Front page, *Vrij Nederland*, 5 May 1990

Haarlem exhibition' (Reitsma 1990). According to her, it could be observed 'with the naked eye' and 'without prior art-historical knowledge' that at most 30 of the 68 works were by one hand, by Frans Hals. In particular, the innovative and freely painted paintings were suspect in her opinion: "In the seventeenth century the urge to be original and unique did not exist at all", she explained (a view that actually clashes with seventeenth century art theory, see Chap. 6).[45] Almost all the genre paintings in the exhibition seemed to her to be nineteenth-century for this reason, including such well-known and undisputed masterpieces as the *Young Man with*

[45] "In de zeventiende eeuw bestond de drang om origineel en uniek te zijn helemaal niet" (Reitsma 1990, p. 17).

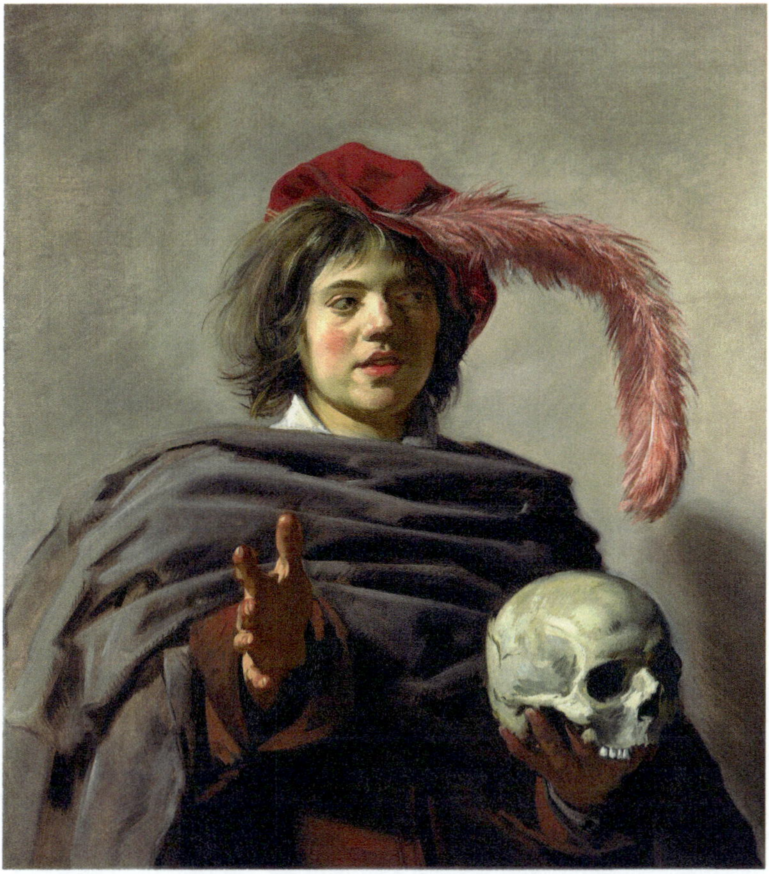

Fig. 2.23 Frans Hals, *Young man holding a Skull (Vanitas)*, 1626–1628, oil on canvas, 92.2 × 80.2 cm. (The National Gallery, London)

Skull in the National Gallery in London (Fig. 2.23) (Reitsma 1990, p. 18).[46] She also pointed out the possibility that Hals's painting sons could have collaborated in his style and on his works.

It was one of the sharpest attacks on the exhibition, but it was certainly not the only one. Ad Blom, president of the Art Historical Society (*Kunsthistorische Vereniging*) in Amsterdam was interviewed by several newspapers and also took a stab at the attributions in the exhibition: in his opinion, eight paintings in the show

[46] A salient detail, by the way, is that the then director of the Frans Hals Museum, Derk Snoep, is Reitsma's ex-husband and the father of her two sons, which may explain her rather personal conclusion: "I would therefore advise the director of the Frans Hals Museum to go and have a look [at the paintings] every morning together with a group of trained viewers before the public comes in" ["Ik raad de directeur van het Frans Hals Museum dan ook aan om met een groepje getrainde kijkers elke ochtend voordat het publiek binnenstroomt te gaan kijken."]

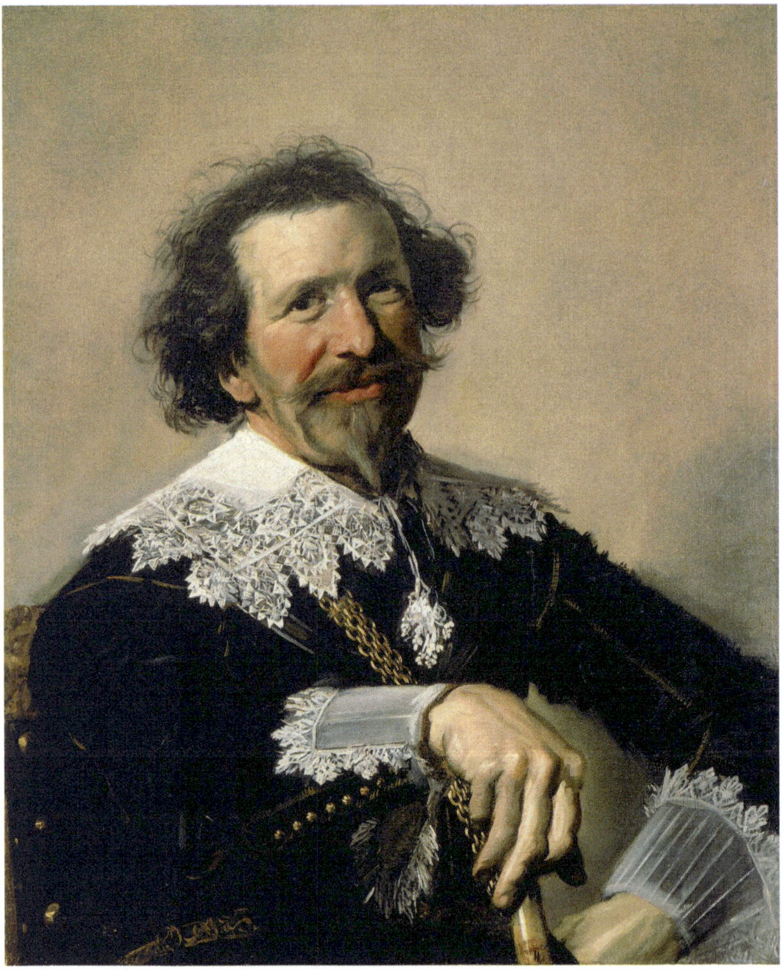

Fig. 2.24 Frans Hals, *Pieter van den Broecke*, c. 1633, oil on canvas, 71.2 × 61 cm. (Kenwood House, London)

were "truly unacceptable," including Hals's famous portrait *Pieter van den Broecke* which had been reproduced in print shortly after it had been created and was accepted by all Hals experts (Fig. 2.24).[47] He gives no arguments for this particular de-attribution, and appears to have revised his opinion about this painting in a conversation with Derk Snoep, director of the Frans Hals Museum, a few days later. In a letter to the latter, dated February 5, 1990, he mentions only seven

[47] Ad Blom mentions cat. Nos. 36, 44, 53, 65, 66, 67 and 84 in an article of his in the newspaper *NRC Handelsblad* of January 31, 1990 (Blom 1990b); Earlier, he referred to eight unacceptable pieces in the newspaper *Volkskrant* of January 26, 1990 (Blom 1990a).

pieces –unfortunately unspecified– that he finds "totally unacceptable".[48] The remaining de-attributions are presumably based on Grimm's views, to which Blom refers frequently; arguments hardly appear in the papers. About the man's portrait in Boston (Fig. 2.25), for example, Blom says, "That's no Hals, you see that, don't you? Look at that greased hair!"[49]

Amid the media storm, a surprisingly nuanced piece by historian Roelof van Gelder also appeared in the newspaper *NRC Handelsblad* on May 10th, 1990: 'No final word on Hals'. He points out how difficult it is to determine the exact scope of Hals's oeuvre: "Did Frans Hals paint in one style all his life? Was he never in a hurry, never having a bad day? If so, then the number of 'authentic paintings by Hals is smaller than some assume. But perhaps the painter evolved, applied different styles, and also occasionally made mistakes. In that case, many more paintings qualify for the label 'authentic'."[50]

Van Gelder hits the nail on the head. Defining the oeuvre of Hals is no easy task. Any selection necessarily rests on assumptions; for how much consistency and variety can the expert expect? The widely divergent views of Slive and Grimm are in fact based on opposing presuppositions.

Slive starts with a broad, synthetic view on Hals and his historical context. Hals was –in his view– an innovative painter who was mainly appreciated within Haarlem in his own time and later became famous for his dynamic compositions, his lifelike figures, and the suggestion of a sudden moment in his paintings, the so-called "snapshot" effect. His earliest group portrait, the civic guard piece of 1616 announced the Golden Age of Dutch painting like a "cannon shot", according to Slive, so radically did it differ from the earlier, less vivid group portraits of civic guards (compare Figs. 2.15 and 2.26; see also below Chap. 6). The overall quality of the figures depicted, and the suggestion of inner life, are not exactly easy to put into words, but this is no reason for Slive not to attempt this as, for example, in his discussion of the various portraits of *Malle Babbe (Mad Meg)* (Figs. 2.21, 2.22 and 2.27):

> Other portraits of Malle Babbe are known. Among the existing variants, the one at the Metropolitan Museum [. . .] comes closest to the standard set by the Berlin picture, but when seen next to the latter, betrays another hand. With all its spontaneity, the pictorial organization of the Berlin painting remains clearly though out. Malle Babbe's sharply turned head makes a strong counter-movement to the emphatic diagonal thrust of the design, which is

48 Letter from Ad Blom to Derk Snoep, dated 5 Feb 1990, doc. no. 198, Frans Hals exhibition, copy in the archive compiled by Ella Hendriks, Frans Hals Museum. In it, Van der Blom refers to no less than five newspaper articles in which he is quoted in response to the exhibition (which appeared in the newspapers *Volkskrant*, *NRC Handelsblad*, *Telegraaf*, and *Haarlems Dagblad*). He also advises Snoep to start a research project.

49 "Dat is geen Hals, dat zie je toch? Kijk naar dat gesmeerde haar!" (Schenke 31 Jan 1990).

50 "Heeft Frans Hals zijn hele leven in één trant geschilderd? Had hij nooit haast, nooit een slechte dag? Als dat zo is, dan is het aantal 'authentieke Halsen' kleiner dan sommigen veronderstellen. Maar misschien heeft de schilder zich ontwikkeld, heeft hij verschillende stijlen toegepast en heeft hij ook wel eens een steekje laten vallen. In dat geval komen veel meer schilderijen voor het predikaat 'authentiek' in aanmerking." (Van Gelder 1990).

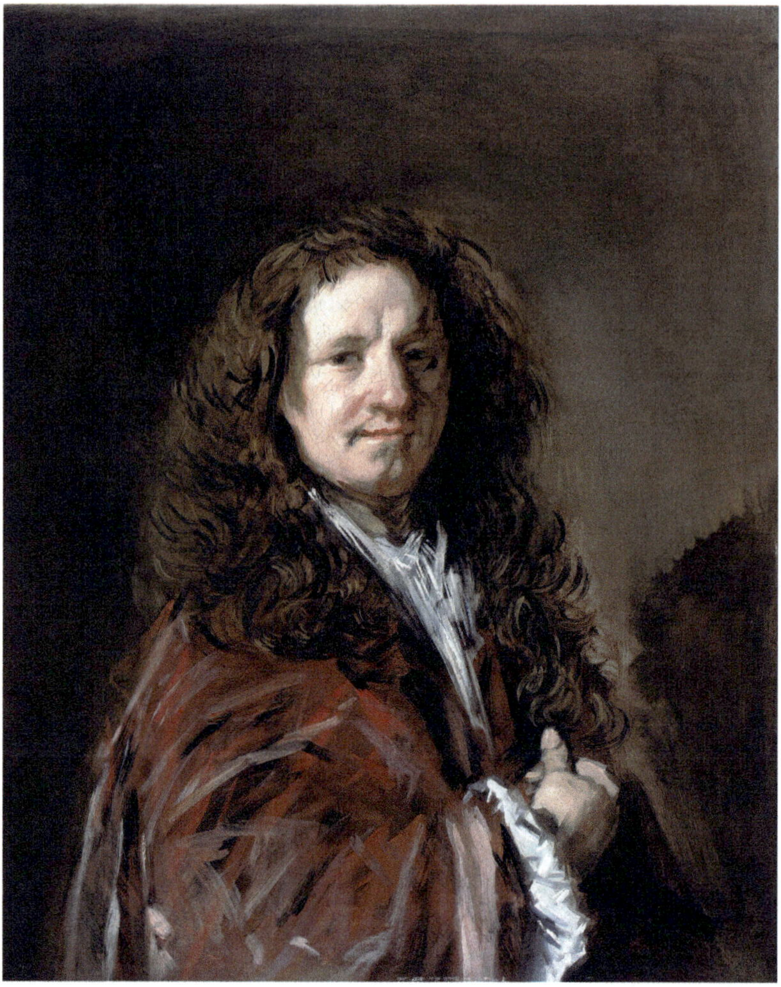

Fig. 2.25 Frans Hals, *Portrait of a man*, c. 1665, oil on canvas, 85.8 × 67 cm. (Museum of Fine Arts, Boston)

subtly reinforced by the direction of the energetic brushwork on her collar, cap, sleeve and apron. In the New York painting the tautness has slacked considerably, the detached strokes are hardly part of the dominant rhythm established by the movement of the figure and we miss the decisive accents which give a convincing roundness to the forms even when they are suggested with a single touch of light or dark paint. [...] A more recent effort to rival Hals's peerless original was made by the Dutch forger Han Van Meegeren (1889–1947), best known for his Vermeer fakes. [...] And, as always when we turn to works by the master's followers, copyists and forgers, something else is missing; there is not a trace of the psychological penetration found in Hals's works. Admittedly, judging this aspect of his legacy will always remain partially subjective, but who would argue that Van Meegeren's gross drinker touches depths of the human spirit? (Slive (ed.) 1989–1990, cat. no. 37, pp. 238–239).

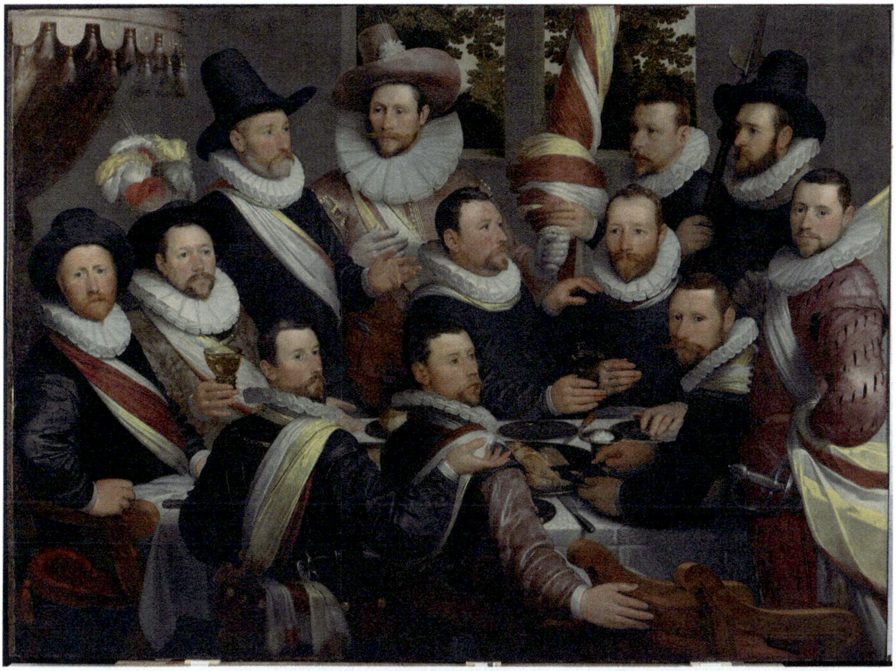

Fig. 2.26 Cornelis van Haarlem, *Banquet of the Officers and Subalterns of the St. Calivermen Civic Guard*, 1599, oil on canvas, 169 × 223.5 cm. (Frans Hals Museum, Haarlem)

In addition to his visual analysis of Hals's painting style, Slive also takes a variety of prints and historical sources into consideration, and accepts a certain variation in quality. For example, he points at a number of passages that are "mechanical and even akward" in Hals's paintings of evangelists, such as Luke's sleeve and Mathew's hands and huge book (Figs. 2.28 and 2.29), but these do not impact the overall attribution: "Hals, *au found* a portraitist, with characteristically free and impulsive brushwork, concentrated on the heads: Luke convincingly lost in his thought and Matthew studying a text under the admiring gaze of his little angel." (Slive (ed.) 1989–1990, cat. nos. 22 and 23, pp. 196–197). In three paintings he believes he recognizes another hand in the background landscape: that of landscape specialist Pieter de Molijn: *Isaac Massa* (1626, Art Gallery of Ontario, Toronto, cat. no. 21; Fig. 2.30), *Family Portrait* (c. 1635, Cincinnati Art Museum, Cincinnati, cat. no. 49; Fig. 2.31) and *Family Group in Landscape* (c. 1648, Thyssen-Bornemisza Museum, Madrid, cat. no. 67; Fig. 2.32).

Grimm's approach is in fact complementary to Slive's. He starts from a very careful study of the paintings themselves in all their details, incorporating as many conservation insights as possible regarding damages, later changes and additions. On this basis, he arrives at a different chronology than Slive and a much more limited selection. Documentary information such as early reproductive prints carry relatively little weight; his visual analysis is decisive. For example, according to

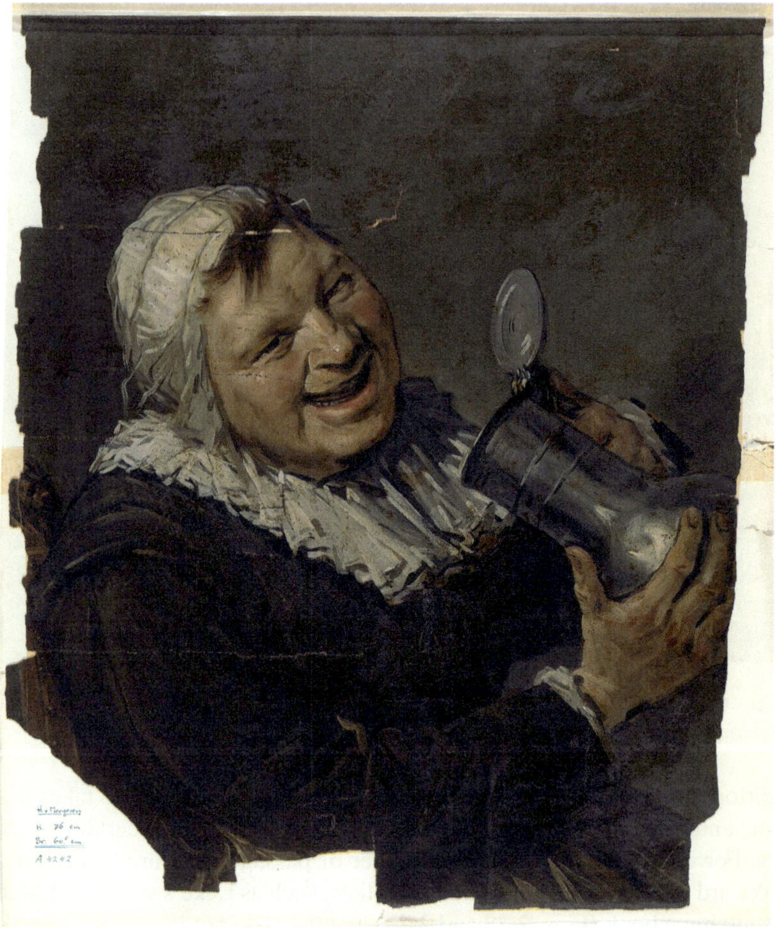

Fig. 2.27 Han van Meegeren, *Malle Babbe,* 1930–1940, oil on canvas, 76 × 60 cm. (Rijksmuseum, Amsterdam)

Grimm, the signed and dated portrait of *Conrad Viëtor (FH 1644)* is not by Hals himself, but "stiff studio work" (Fig. 2.33) (Grimm 1990/1989, p. 46). The painting, however, was reproduced in print by Jonas Suyderhoef (1614–1686), who belonged to Hals's extended family (his brother was married to Hals's niece) and described it as an original, witness his inscription: '*Frans Hals pinxit*' ('painted by Frans Hals') (Fig. 2.34).

Grimm is focused on distinguishing Hals's personal share and his brushwork from that of studio assistants, pupils and imitators. In doing so, he works as objectively as possible, as a kind of "graphologist or detective," reconstructing Hals's oeuvre as best he can (Grimm 1990/1989, note 26, pp. 267–268). In his view, Hals is an "independent, keen observer of human nature and at the same time a virtuoso craftsman with paint and brush, who mastered the technique to the finest

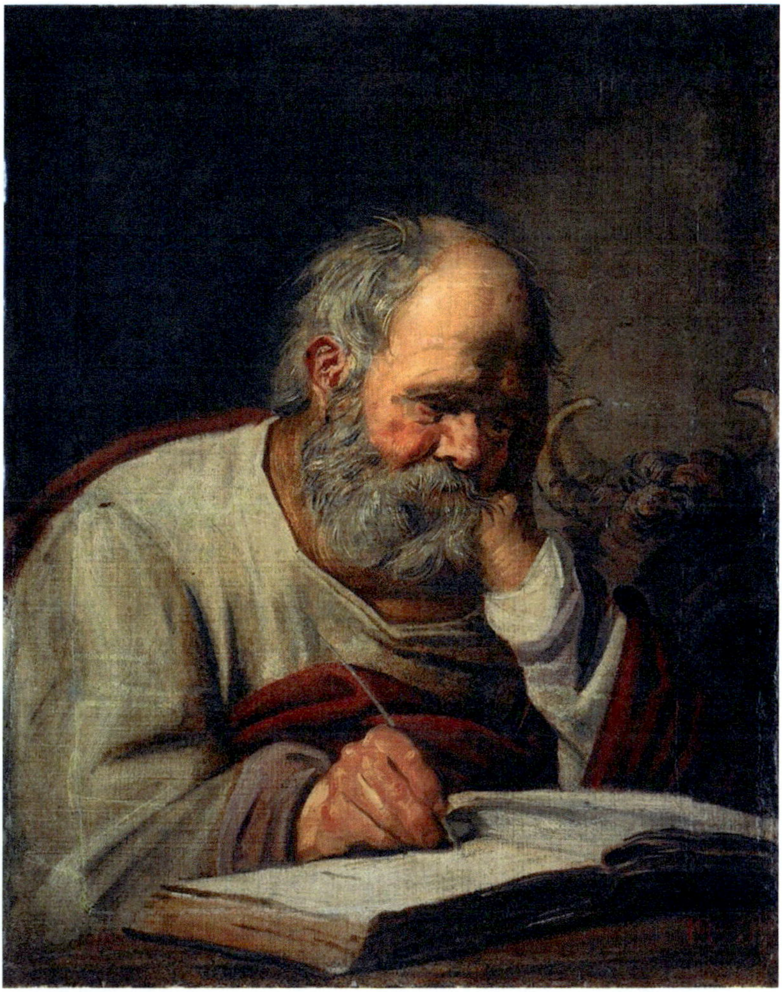

Fig. 2.28 Frans Hals, *Saint Luke,* c. 1625, oil on canvas, 70 × 55.2 cm. (Museum of Western and Oriental Art, Odessa)

detail"(Grimm 1990/1989, p. 9). Grimm is less forgiving than Slive regarding weaker passages and expects greater consistency (although he does accept the aforementioned evangelists as autograph works by Hals). He maps out Hals' stylistic development in 15 successive phases of one or several years each, designated by the letters 'A' through 'O' (Grimm 1990/1989, p. 177 ff). Certain types of pictures he removes entirely or almost entirely from the oeuvre as Slive had defined it: all the fishermen's children, the family portraits –except for the Van Campen family portrait which has been preserved in parts (Grimm 1990/1989, cat. Nos. 11, 12 and

Fig. 2.29 Frans Hals, *Saint Matthew*, c. 1625, oil on canvas, 70 × 55 cm. (Museum of Western and Oriental Art, Odessa)

13)[51]– and all small-scale portraits except the portrait of *Jean de la Chambre* in the National Gallery in London (Fig. 2.35). Furthermore, he distinguishes two different hands in the paintings that he recognizes as studio work, an anonymous hand (A) and a hand (B) that he suspects to be Johannes Hals (c. 1620–1654), one of Frans Hals's sons.

[51] The different surviving parts of the painting are currently in the Toledo Museum of Art, The Museum of Fine Arts in Brussels and in a Belgian private Collection and were recently reunited in an exhibition (see Nichols et al. 2018).

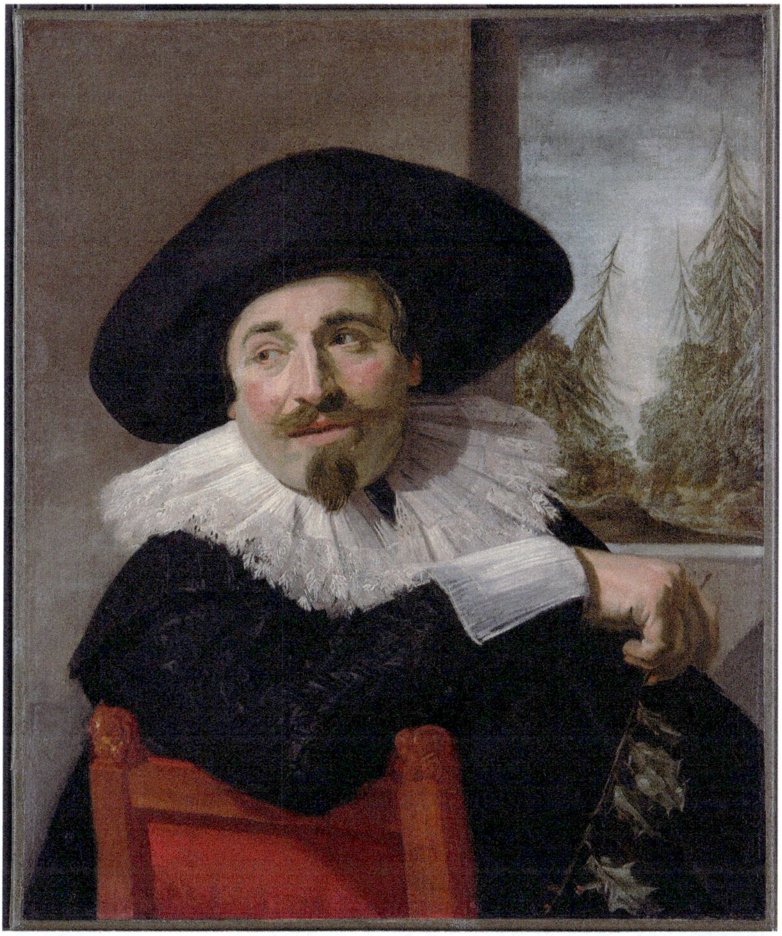

Fig. 2.30 Frans Hals, *Isaac Abrahamsz. Massa*, 1626, oil on canvas, 79.7 × 65.1 cm. (Art Gallery of Ontario, Toronto)

Slive and Grimm's divergent views of Hals's oeuvre raise questions about their presuppositions. For how consistent was Hals in the quality he delivered as a painter? And did he develop his style in a logical sequence or did he alternate between different styles depending on the type of work? Perhaps the most pressing question is whether we can assume that a "Hals" was commonly painted solely by the master himself. Although Grimm's updated chronology received acclaim when it was published in 1972, and his explicit and therefore verifiable observations were praised, the question also arose as to whether his definition of Hals's oeuvre was too narrow. For example, Rudi Ekkart, in his review of Grimm's oeuvre catalogue, argued that the weaker passages Grimm identified in the family portraits, which were reason for him to remove the portraits from Frans Hals's oeuvre, might be better interpreted as contributions by studio assistants in larger commissions by the

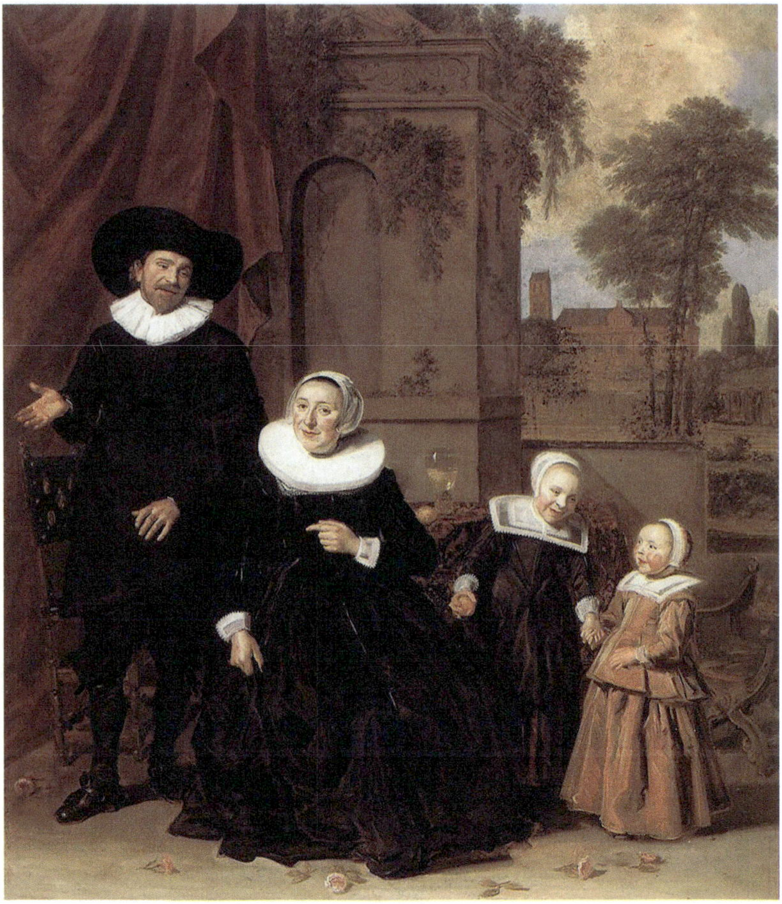

Fig. 2.31 Frans Hals, Portrait of a Dutch family, c. 1635, oil on canvas, 111.8 × 89.9 cm. (Cincinnati Art Museum, Cincinnati)

master: "Certainly in large paintings such as [...] family portraits, some of the weaknesses would in my opinion indicate such assistance rather than a work created solely by an apprentice."[52]

Slive's speculation about Hals's use of landscape specialist Pieter de Molijn is also interesting in this context. An archival document dated November 6, 1656, discovered in the 1950s by H.E. Hees suggests that it was not uncommon for Hals to employ such a specialist. It mentions portraits of the grandparents of the Haarlem painters Nicolaes and Jan (de) Kemp, which had been painted by Hals, with contributions from the landscape and genre specialist Willem Buytewech (1591/92–1624):

[52] "Zeker bij grote schilderijen als [..] familieportretten zouden sommige zwakheden mijns inziens veeleer wijzen op zulk een assistentie dan op vervaardiging door een leerling alleen" (Ekkart 1973).

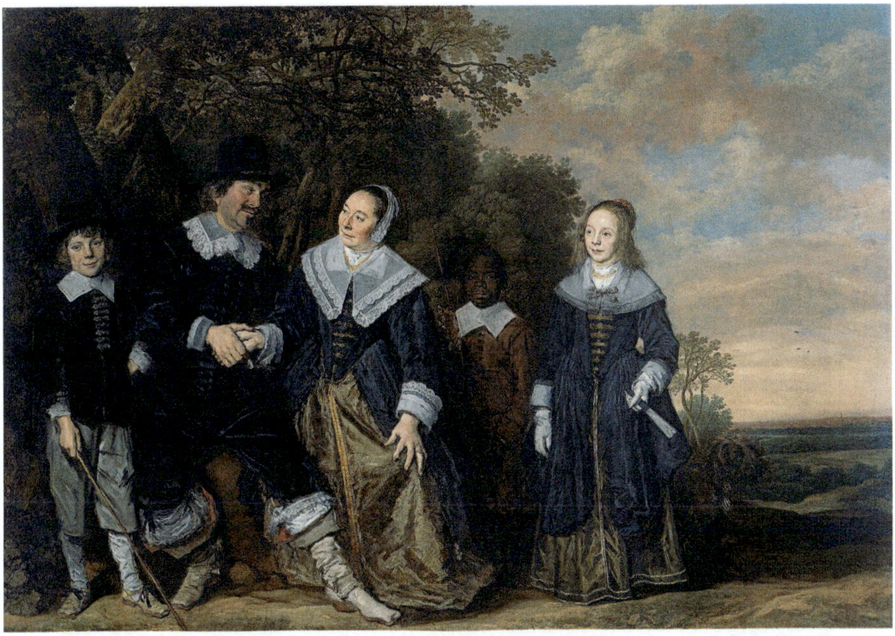

Fig. 2.32 Frans Hals, *Family Group in a Landscape*, c. 1645–1648, oil on canvas, 202 × 285 cm. (Museo Nacional Thyssen-Bornemisza, Madrid)

"[the] portraits are made and painted by Mr. Franchoys Hals the elder and the surroundings by Buytewech, also called Witty William".[53]

Although the contradictions between Slive and Grimm lead mostly to speculation about later imitations and forgeries in the media, the differences between their views are actually more subtle. Rather, the main bones of contention were the distinction between different types of seventeenth-century paintings and the consistency that could be expected in style and quality.

2.7 Technical Investigation of Frans Hals Paintings

Because of the 1990 media storm, the director of the Frans Hals Museum decides to commission a technical investigation of the paintings while the exhibition is still running. The Dutch Ministry of Culture (WVC) grants a subsidy of fl. 23,000 for this purpose. Three research questions are central. First, the question whether

[53] "conterfeijtsels sijn gemaeckt en geschildert bij Mr. Franchoys Hals den oude en het comparquement bij Buytewegh, ofte anders genaemt Geestige Willem" (Van Hees 1959, p. 37). The word "comparquement" is difficult to translate; it could indicate a painted cartouche, a frame or a background (perk?), according to Van Hees.

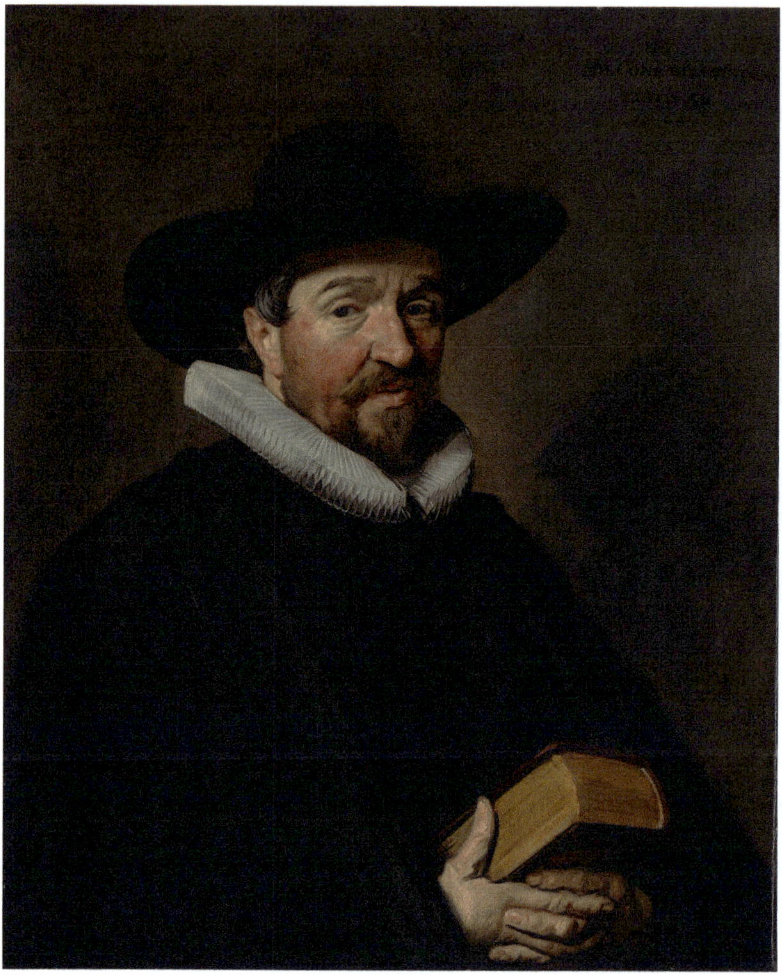

Fig. 2.33 Frans Hals, *Conradus Viëtor,* 1644, oil on canvas, 82.6 × 66 cm. (The Leiden Collection, Leiden)

eighteenth-century or later copies or forgeries can be distinguished in the exhibition. Next, the question of whether the exhibited works show similarities in material and technique that indicate a specific studio practice. And finally, the question of whether or not it is possible to determine if a painting is autograph, a studio product or a seventeenth-century copy in the style of Hals. The research is conducted by conservator Ella Hendriks and art historian Koos Levy-Halm in collaboration with Professor J.R.J. van Asperen de Boer. They recorded their findings in an unpublished report, which they completed in February 1991: *Report Concerning a*

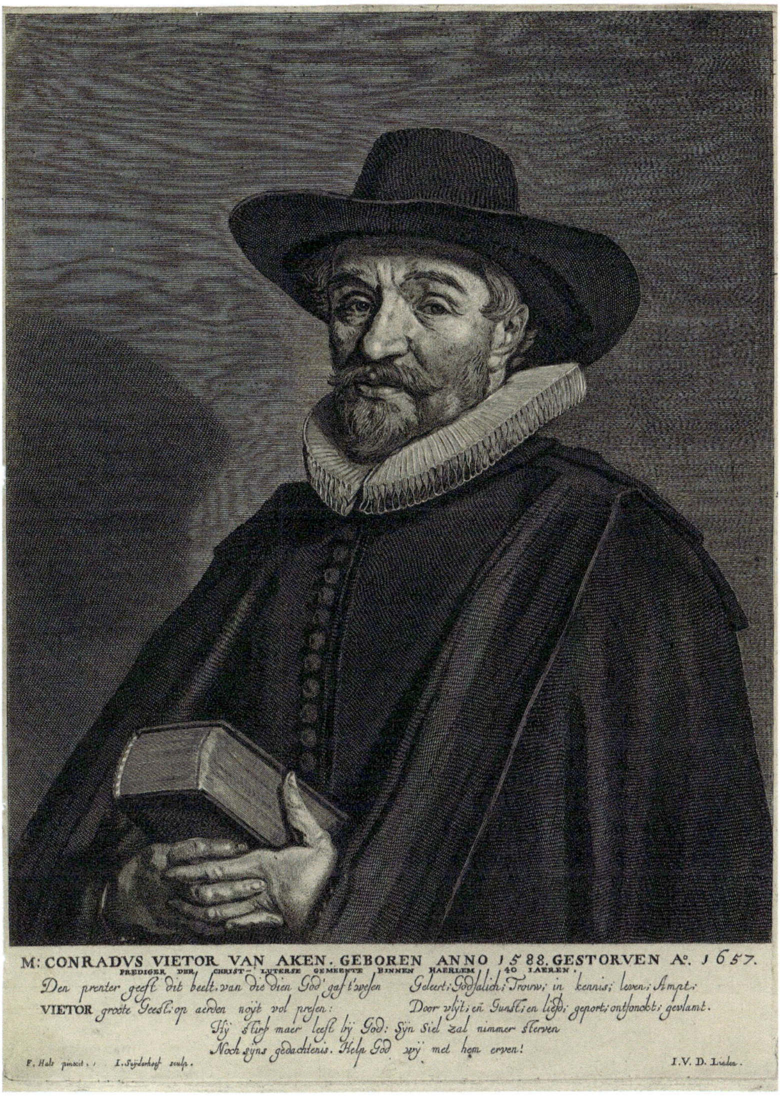

Fig. 2.34 Jonas Suyderhoef, after Frans Hals, *Conradus Viëtor van Aken*, c. 1657, engraving. (National Gallery of Art, Washington D.C.)

Preliminary Technical Investigation of Paintings Exhibited During the Frans Hals Exhibition, Held From May 11 to July 22 1990 in the Frans Hals Museum, Haarlem.[54]

The first question was answered most succinctly. The wild speculations in the media that all kinds of nineteenth-century forgeries and imitations would be on

[54] A copy is available at the library of the RKD—Netherlands Institute for Art History, The Hague.

Fig. 2.35 Frans Hals, *Jean de la Chambre at the Age of 33*, c. 1638, oil on panel, 20.6 × 16.8 cm. (The National Gallery, Londen)

display in the exhibition are not confirmed. The researchers find no indication that pictures in the show were modern pieces: "The condition, material and technique of all the paintings examined are consistent with those one would expect to find in a seventeenth-century Dutch painting" (Hendriks et al. 1991, p. 49).[55] Of the 86 paintings in Slive's catalogue, 59 could be examined in the exhibition, mostly in the galleries and outside opening hours; six works were taken out of their frames. The research focused on the entire construction of the pieces, from priming to finishing. Observations were made with the naked eye, the (stereo) microscope, and various

[55] Early eighteenth-century materials and techniques do not differ enough to be distinguishable, according to Hendriks and Levy-van Halm.

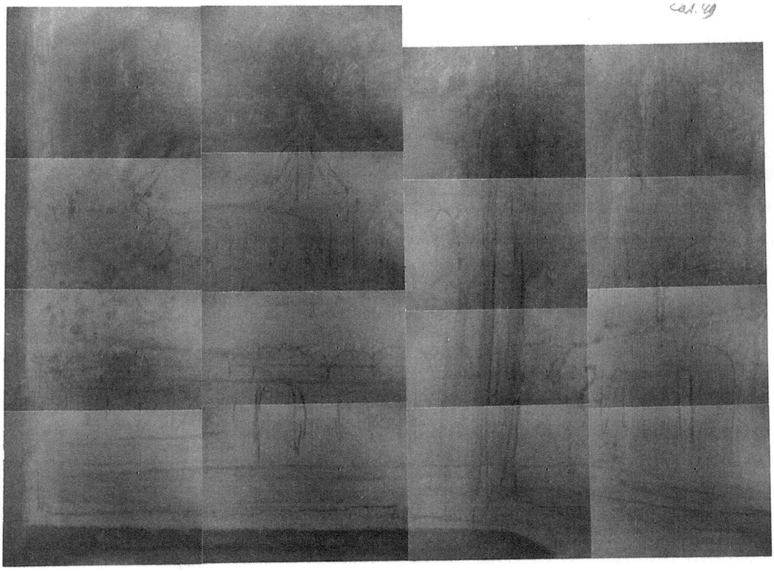

Fig. 2.36 Infrared Reflectogram (IRR) created by Van Asperen de Boer, showing the background in Frans Hals, *Portrait of a Dutch Family* (c. 1635, Art Museum, Cincinnati). (Archive of the Frans Hals Museum)

types of photographs were used (macro, detail, and micro). Furthermore, radiographs were made whenever possible; samples were taken in some cases and infrared reflectograms (IRR) were constructed of a small number of pieces.

The other two questions require more extensive answers. Indeed, there appears to be some consistency in material and technique that indicates a specific workshop practice and a broader Haarlem tradition. The canvases and their thread densities seem characteristic of Haarlem, especially the exceptionally large one-piece canvases.[56] Also, the paintings examined show a spontaneous painting technique from beginning to end. The composition appears to have been sketched directly on the primed support, using painted lines and areas of underpaint. This last stage appears cursory in his genre paintings. By contrast, in the Family Portrait from Cincinnati, the researchers found a very elaborate underdrawing beneath the background landscape (Fig. 2.36), which seems to confirm Slive's theory that the background was done by another hand (Van Asperen de Boer 1991, p. 47). Furthermore, the portraits show many small changes made during the working process. These suggest a creative searching, which one would not expect in later copies. Usually the background is applied first, but not always. When a change requires skin color to be applied over

[56] Regarding the priming and *imprimatura*, no immediate conclusion can be drawn; further research is needed to determine the extent to which the observed similarities are characteristic of Haarlem and/or Hals's workshop. In the examined works the *imprimatura* varies in colour from cool gray to dark flesh pink.

a dark background, allowance has been made in the thickness of the applied skin color, which could possibly be specific to Hals. Contours are adjusted in successive layers and only fixed at the last stage (Hendriks et al. 1991, p. 50).

The thickness and structure of the loose accents is also noteworthy: often the brush stroke is shallow in the middle and has two raised edges, which could possibly be explained by the binder used. The construction of black areas may also be typical of Hals: first lampblack, sometimes mixed with bone black and lead white, then bone black, sometimes mixed with umber, for the darkest shades, and lampblack mixed with lead white for the grayish highlights. Some ochre and copper also occurred, the latter was probably added as a siccative (Hendriks et al. 1991, pp. 50–51).

The last question is the most difficult to answer. The researchers contradict the widespread assumption that Hals was an *alla prima* (wet-in-wet) painter; they recognize different stages of execution in all the examined paintings, even in those that appear to have been done most quickly (see also below Chap. 4). They therefore assume that he always painted in separate stages and may have collaborated with others in the process (Hendriks et al. 1991, pp. 51–53). Too little is known about Hals's studio practice to provide a clear framework, but the researchers consider several types of collaboration likely. Catalogue numbers 33 (*Fruit and vegetable seller*, 1630) and 49 (Cincinnati family portrait) (Figs. 2.32 and 2.37) show contributions from a specialist in the surrounding still life and background landscape, respectively, as Slive suggested in the exhibition catalogue. In addition, the researchers believe that studio assistants may also have participated in a particular phase in the execution of a portrait. According to them, this seems to be the case with the portrait of *Feyntje Steenkiste* (c. 1635, Rijksmuseum, Amsterdam), a portrait attributed by Grimm to Judith Leyster, while he did recognize the pendant, the portrait of her husband, as a Hals (Figs. 2.38).[57] Technical analysis revealed a difference in their canvas supports and in the initial paint layers applied. In their finish and contour lines, however, the pendants are, on the contrary, very similar, suggesting that the pieces may have been set up by two different hands but completed by one and the same hand to achieve unity (Hendriks et al. 1991, pp. 51–52).

In short, several characteristics point to a specific studio practice. Moreover, a preliminary survey using infrared reflectography revealed some initial sketch lines using dark strokes on a light ground in flesh areas. According to the researchers, the character of these lines and the way in which they are integrated in the subsequent layers could provide criteria for the separation of hands. Furthermore, the small portraits on panel showed a relatively wide variety in the preparatory layers and paint application, which could mean that these came from different studios. All in all, the question of what distinguishes an autograph work, a studio work, a contemporary or later copy is a complex one that should be answered cautiously, according

[57] In his upcoming Hals catalogue, Grimm lists *Feyntje Steenkiste* as a painting done entirely by Frans Hals himself. See his forthcoming updated oeuvre catalogue published via the online platform RKD Studies, hosted by the Netherlands Institute for Art History (Grimm Forthcoming).

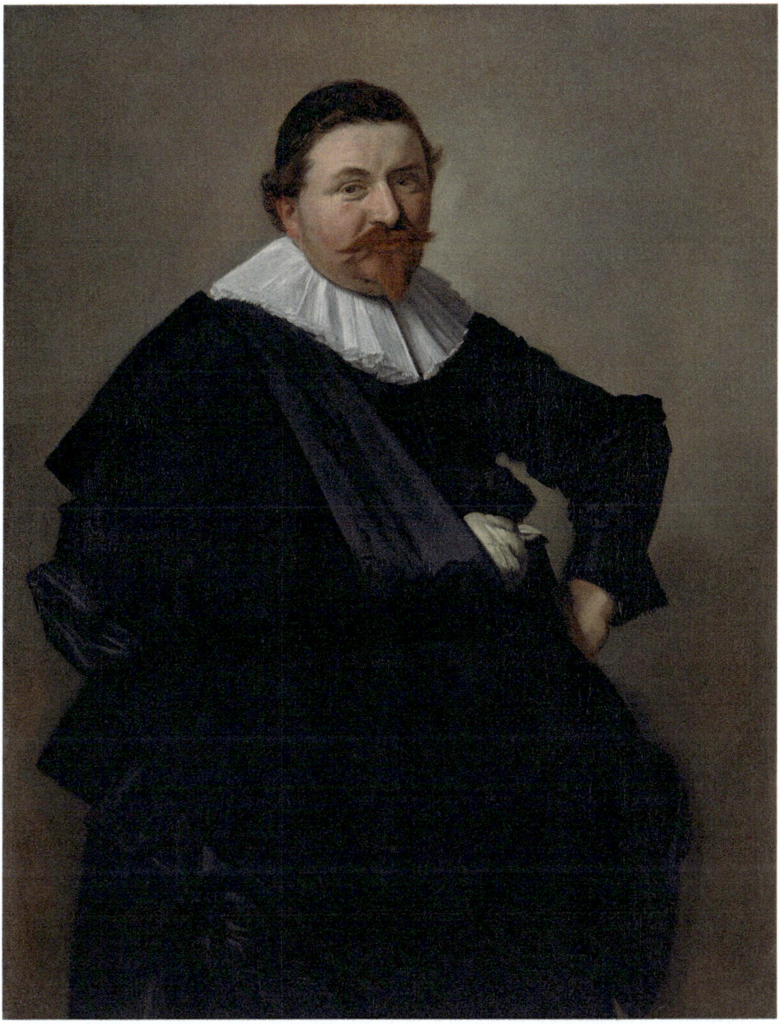

Fig. 2.37 Frans Hals, *Lucas de Clercq*, c. 1635, oil on canvas, 121.6 × 91.5 cm. (Rijksmuseum, Amsterdam)

to the researchers.[58] They advocate further technical research, particularly binding medium analysis, dendrochronology, infrared reflectography, and paint analysis. Further research into seventeenth-century art theory and sources on studio practice is also needed to put the technical findings in context.

[58] Technical examination of *Zaffius* (cat. no. 1) produced evidence that it might be copy, as Van Dantzig had suggested (see Hendriks et al. 1991, pp. 52–53). Grimm, however, still recognizes this painting as an autograph work (see Grimm forthcoming).

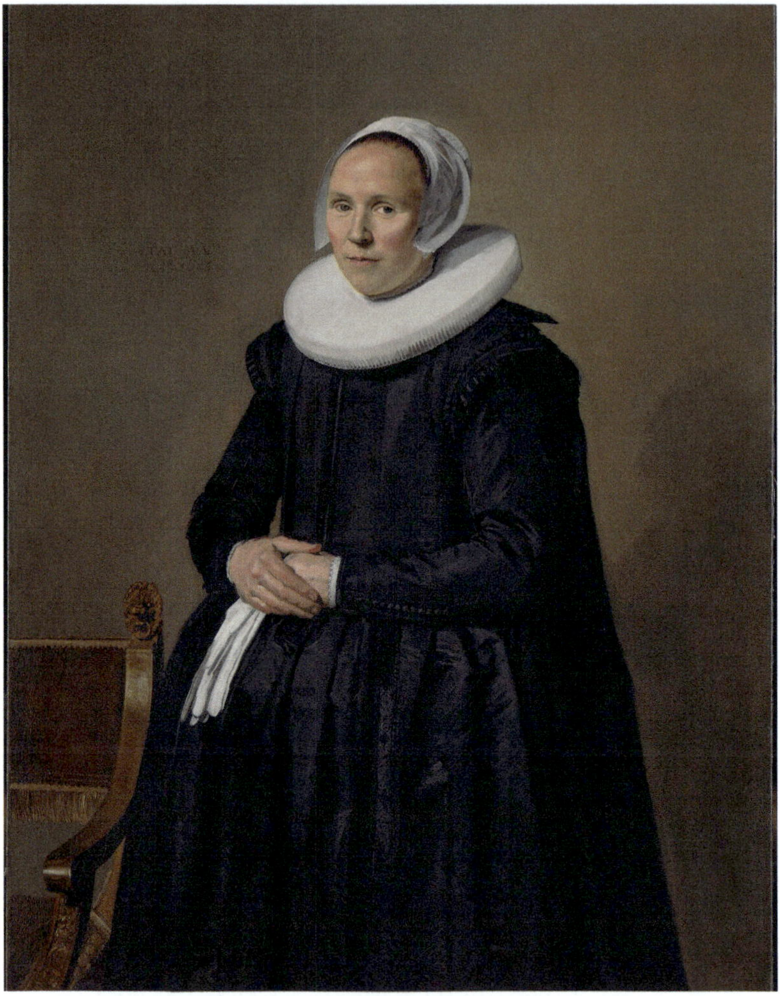

Fig. 2.38 Frans Hals, *Feyntje van Steenkiste*, c. 1635, oil on canvas, 121.9 cm × 91.5 cm. (Rijksmuseum, Amsterdam)

The careful, erudite report by Hendriks, Levy-van Halm and Van Asperen de Boer constitutes a major step forward in Hals connoisseurship. For the first time, technical research is integrated in a systematic way in the attribution process. Building on some of the technical research done in preparation for the 1989–1990 exhibition, the authors explore the use of different scientific techniques specifically for Hals attributions.[59] It is reminiscent of the large research project dedicated to sorting out the oeuvre of Rembrandt: the Rembrandt Research Project (RRP),

[59] Notably the study by Hendriks and Groen in (Slive (ed.) 1989–1990).

founded in 1968, and indeed the report refers repeatedly to comparative findings by the RRP (e.g. Hendriks et al. 1991, pp. 5, 8 and 52).[60] While the funding and scope of the Hals research project is of course much more limited than the decades long investigation by the RRP (which lasted eventually until 2014), it is nevertheless the most extensive technical examination of Hals's oeuvre to date, and therefore an important benchmark for all subsequent studies in this field.[61]

The integration of scientific techniques in Hals authenticity research echoes broader developments in the field of connoisseurship. After the Van Meegeren forgery scandal at the end of the Second World War, scholars became more cautious when attributing and dating pictures, and it became the rule rather than the exception to support attributions with rational arguments. Moreover, experts gradually started to expand their visual analysis by systematically integrating scientific techniques (see Tummers and Erdmann 2022). Whereas previous generations of art historians had been hesitant and sometimes even skeptical about incorporating chemical research, in the early 1980s the members of the RRP signal rather an excessive optimism: a relatively widespread belief that science holds the answers and could potentially replace the eye in matters of attribution (Bruyn et al. 1982–1989, vol. 1 (1982), p. XIIIff). It explains perhaps why the Frans Hals Museum announces its research project with much emphasis on its objective, technical component: while opinions on attributions are just subjective, their research is 'purely scientific'.[62]

It gradually becomes evident however, that technical evidence seldom yields conclusive answers in attribution matters. While a technical or chemical analysis *can prove* that a work is *not authentic* by demonstrating, for example, that the materials used are anachronistic, a positive attribution cannot be done without a visual analysis. For if the materials are consistent with the period, one still needs to analyze the particular style and 'handwriting' of the artist in order to determine if the attribution is correct, to differentiate between different types of workshop products, contemporary copies and imitations. Moreover, definitive proof that a work is a later imitation or forgery is exceedingly rare. The RRP identified very few potential forgeries among the more than 600 paintings they studied (e.g. Bruyn et al. 1982–1989, vol. 1 (1982), XX, C12). Likewise, the Hals research project did not expose an evident later imitation among the 59 works studied, as we have seen.

Nevertheless, the integration of scientific techniques in authenticity research allows for a much more in-depth understanding of an artist's characteristic use of materials and techniques. Consequently, the pioneering Frans Hals research project –like the bigger RRP– constitutes a major step in significantly refining the criteria used for attributions as they include various material aspects and deeper layers into the analysis, thus focusing not only on the final product but also on

[60] On the RRP see (Tummers 2011; Tummers and Erdmann 2022)

[61] See for example the recent technical study of the *Laughing Cavalier* (Packer and Roy 2021–2022).

[62] Statement by Ella Hendriks cited by John Bezold in his lecture 'The Many Frans Hals Exhibitions and Connoisseurs: From Thoré to Today' at the conference *Frans Hals*: *A Survey of Current Research*, 8–9 Jan 2023 (9 Jan 2023). The conference proceeding are scheduled to be published in 2024.

different stages of the painting process. Moreover, the lengthy and often detailed descriptions of the research makes it possible to analyze and check the observations, reference data and underlying assumptions to a much greater extent than had been possible before.

2.8 Twenty-First Century Perspectives

When the *Frans Hals of nor Frans Hals* research project just started in 2016, a new Hals controversy arose. A previously unknown painting that surfaced at auction house Christie's in Paris in 2007 was widely recognized as an original by Frans Hals (Buvelot and Ducos 2014). Shortly after its discovery, the Louvre Museum in Paris attempts to buy the painting but fails to raise the necessary five million euros. Subsequently, the panel comes into the hands of London art dealer Mark Weiss, who sells it in 2011 through Sotheby's New York for just under ten million euros ($10.8 m) to a Seattle real estate developer.

Weiss includes it in his 2011 anniversary catalogue *Facing the Past* with an entry written by Pieter Biesboer, the retired curator of the Frans Hals Museum, who also supports the attribution: "One can immediately recognize the fluid brushwork and freedom of handling of Frans Hals, so characteristic of his late work" (Biesboer 2011, p. 64). Subsequently, curators Quentin Buvelot of the Mauritshuis and Blaise Ducos of the Louvre publish the painting in February 2014 in the leading art historical journal *The Burlington Magazine*: 'A rediscovered portrait by Frans Hals.' They deem the panel a "very important addition to Hals's oeuvre" and receive support for their attribution from Hals expert Seymour Slive (Buvelot and Ducos 2014, p. 102 and note 2; Slive 2014, p. 314). The basis for the attribution are stylistic considerations, especially a comparison with three other, relatively small portraits on panel of circa 1660. The painting belongs to Hals's late period, according to the authors, as evidenced by the loose painting technique, the vivid, snap-shot-like character of the portrait and the convincing facial expression and penetrating gaze.

The painting suddenly becomes suspect when, in March 2016, French police seize a painting they suspect to be a forgery in the style of Lucas Cranach: *Venus*. Both paintings are from the same Parisian collection of Giuliano Ruffini and otherwise have no traceable provenance history.[63] Sotheby's asks permission from the buyer of the alleged Hals to have the piece technically examined by Orion Analytical and subsequently concludes that it is indeed a forgery. The auction house therefore reimburses the buyer in July 2016, which suddenly becomes headline news in early October. Pending an upcoming lawsuit, Sotheby's does not release the research

[63] The same is true of a *David with Goliath's head* in the style of Orazio Gentilleschi and a *Saint Hieronymus* in the style of Parmigianino, both of which are on display as originals in prominent museums: respectively the National Gallery in London and the Metropolitan Museum of Art in New York.

report on the picture; however, the auction house does reveal that modern materials were found in the painting.[64]

While the details of the investigation are still classified and Weiss commissions a counter-evaluation, traditional connoisseurship is already heavily criticized. Art historian Bendor Grosvenor (2016) concludes that "the system on which the art market relies in determining authenticity does not work" in his blog *Art History News*, and Sarah Cascone (2016) of Artnet News states that: "technical research succeeds where connoisseurship failed." The confusion understandably raises questions about the criteria and methods used in old master attributions. The lack of clear standards and the absence of extensive technical studies and reference material has left Hals prone to doubts and misattributions; it underscores the need for the current research projects.[65]

Moreover, the visual analysis has become both more challenging and easier in the twenty-first century, due to a paradigm shift in the humanities, myriad new technical possibilities and advanced digital tools. Within the humanities, the very notion of what constituted 'authenticity' was redefined, especially in the field of old master painting (Tummers and Erdmann 2022). As a result, art experts increasingly moved away from a simple binary perspective (either by the master or not), became more aware of the complex range of possibilities, and started introducing more nuanced categories of thought. A case in point is Claus Grimm's upcoming renewed Frans Hals oeuvre catalogue, which will be published online though the RKD Studies platform, hosted by the Netherlands Institute for Art History (RKD). The usual three attribution categories (A-by the artist, B-not by the artist, and C-uncertain) are subdivided further to allow for a broader variety of works. It makes a significant difference for the works in category A, which now includes four groups, based on the extent of the involvement of the master (see also Chaps. 4 and 6).[66]

Furthermore, new technical possibilities are revolutionizing the field. In particular, macro x-ray fluorescence scanning (MA-XRF) and hyperspectral imaging or reflectance imaging spectroscopy (HI/RIS) are currently used by pioneering teams of experts (often including conservators, conservations scientists and curators) to gain a deeper understanding of old master painting techniques and use of materials, for example in Vermeer research and in Operation Night Watch. The knowledge thus gained often helps to significantly reduce the number of possible attributions to

---

[64] See for example: https://www.bbc.co.uk/news/entertainment-arts-37574411; https://www.nytimes.com/2016/10/27/arts/design/a-dubious-old-master-unnerves-the-art-world.html

[65] As discussed in the introduction, we have contributed to an extensive research report on the painting written for the French Ministry of Justice, which is classified for now because of an upcoming lawsuit.

[66] The sub-categories will be as follows: *A1 Works that were executed in all areas by the master, or appear reworked by him throughout* (120 works)*; A2 Works by the master with contributions by other masters or workshops* (14 works)*; A3 Works with discernible differences in separate areas: combining sections that were recognizably painted by the master with contributions by presumed assistants* (65 works)*; A4 Works that were executed under Hals's supervision, probably on the basis of his compositional designs and using his templates* (133 works).

a specific master and occasionally yields unique insights that make the creation of new forgeries virtually impossible.[67]

Although these advanced research projects using techniques such as MA-XRF and HI/RIS commonly focus on just a few works of art, the digital data generated is nevertheless substantial. Consequently, one of the main challenges of twenty-first-century connoisseurship is to manage and process all the information and to effectively select the most relevant parts. While early twentieth-century connoisseurs heavily relied on their visual memory when judging attributions, their twenty-first-century equivalents face a different reality. Digital tools are increasingly facilitating one of their core tasks: making effective comparisons. Although various computer programs have been developed with more ambitious goals, namely to substitute the connoisseur's analysis of brushstrokes or overall visual assessment, these have not proven to be effective yet in practice (which must be due to the complexity of such decisions).[68] By contrast, algorithms with more modest goals (namely simply facilitating comparisons) have a powerful impact on the field (see below Chap. 7).

In short, the divergent views on Hals's oeuvre are all the more reason to think critically about the criteria used in attributions to Hals, and the criteria that should be used. What types of paintings can we expect and how do we distinguish them? Moreover, it is high time to use technical research to further enhance our understanding of Hals's painting technique and studio practice. The following four pilot case studies therefore combine a variety of research methods to generate new insights. Stylistic analyses are combined with broader art historical and archival research, as well as in-depth technical analyses using a stereo microscope, an Osiris infrared camera, a handheld XRF, MA-XRF, a hyperspectral laser scanner, and lead isotope analysis, among others (see Chaps. 3, 4, 5 & 6). Moreover, a number of digital tools, or smart tools, have been developed specifically for this study in order to facilitate the interpretation of large amounts of information and to enable a new type of visual analysis and comparison (see Chap. 7).

References

Aussems, J.P.H.: Alla Prima, Letter to the editor (March 1990). NRC Handelsblad (1990)

Bailey, A.: Responses to Rembrandt, Who Painted the Polish Rider? A Controversy Reconsidered. Timken Publishers, New York (1994)

Biesboer, P.: Entry 20 "Frans Hals, An Unknown Man". In: Facing the Past a Catalogue of Early Portraiture 1530–1780, pp. 62–65. The Weiss Gallery, London (2011)

Blom, A.: Frans Hals, zijn werk en wat ervoor doorgaat (26 Jan 1990). Volkskrant (1990a)

[67] As in the case of Van Gogh drawings researched in the context of the REVIGO Project, see https://www.vangoghmuseum.nl/en/about/knowledge-and-research/completed-research-projects/revigo

[68] As discussed in my lecture 'Future Frontiers for Heritage Research by Strengthening Collaboration Between the Humanities and Science', at the Gordon Conference Science for Cultural Heritage, Les Diablerets, Switzerland, 15 July 2022.

Blom, A.: Tentoonstellingen vervalsen de kijk op Frans Hals (31 Jan 1990). NRC Handelsblad (1990b)
Bode, W.: Frans Hals: His Life and Work, With an Essay by M. J. Binder. Photographische Gesellschaft, Berlin (1914)
Bruyn, J., Haak, B., Levie, S.H., Van Thiel, P.J.J., Van de Wetering, E.: A Corpus of Rembrandt Paintings, vols 1–3. Springer, The Hague/Boston (1982, 1986, 1989)
Buvelot, Q., Ducos, B.: A rediscovered portrait by Frans Hals. Burlingt. Mag. **156**(1331), 102–103 (2014)
Cascone, S.: Suspected $255 million Old Master forgery scandal continues to rock the art world (11 Oct 2016). Artnet News (2016). https://news.artnet.com/art-world/more-details-old-mastery-forgery-ring-695777
Chapman, H.P., Weststeijn, T.: Connoisseurship as knowledge: an introduction. In: Chapman, H.P., Weststeijn, T., Meijers, D. (eds.) Connoisseurship and the Knowledge of Art, pp. 6–41. Brill, Leiden (2019)
Craddock, P.T.: Scientific Investigation of Copies, Fakes and Forgeries. Elsevier, Oxford (2009)
De Bois, J.H.: Kunst in Haarlem en daarbuiten, Nog een paar Halszaken (18 Aug 1937). Haarlems Dagblad (1937)
Dickey, S.S.: The art of connoisseurship. In: Podedworny, C., MacDonald, B.L., Hathaway, N., McNeill, F., Holubizky, I., Spronk, R., Pocobene, G., Murray, A., Bronson, J., Dickey, S.S. (eds.) The Unvarnished Truth: Exploring the Material History of Paintings, pp. 1–8. McMaster University Museum of Art, Hamilton (2015)
Dijksterhuis, A.: Het slimme onbewuste: denken met gevoel. Prometheus/Bert Bakker, Leiden/Amsterdam (2007)
Du Bos, J.B.: Réflexions critiques sur la poésie et la painture: ut pictura poesis. Jean Mariette, Paris (1719) (cited edn. ÉNSB-A, Paris 1993)
Dutton, D.: Artistic crimes: the problem of forgery in the arts. Br. J. Aesthet. **19**(4), 302–314 (1979)
Ekkart, R.E.O.: Review of Frans Hals. Entwicklung, Werkanalyse, Gesamtkatalog. Oud. Holl. **87**(4), 252–255 (1973)
Ekkart, R.E.O.: Hofstede de Groot, Cornelis (1863-1930). In: Biografische Woordenboek van Nederland, pp. 248–249. Martinus Nijhoff, The Hague (1979)
Freedberg, D.: Why connoisseurship matters. In: Van Der Stighelen, K. (ed.) Minuscula Amicorum: Contributions on Rubens and His Colleagues in Honour of Hans Vlieghe, pp. 29–43. Brepols, Turnhout (2006)
Ginzburg, C., Davin, A.: Morelli, Freud and Sherlock Holmes: clues and scientific method. Hist. Work. J. **9**, 5–36 (1980). https://doi.org/10.1093/hwj/9.1.5
Gladwell, M.: Blink: the Power of Thinking Without Thinking. Little, Brown, New York (2005)
Goodman, N.: Languages of Art: an Approach to a Theory of Symbols. Oxford University Press, London (1969)
Goodman, N.: Art and authenticity. In: Dutton, D. (ed.) The Forger's Art: Forgery and the Philosophy of Art, pp. 93–115. University of California Press, Berkeley/Los Angeles/London (1983)
Gratama, G.D., Van Rijckevorsel, J.L.A.A.M.: Frans Hals tentoonstelling ter gelegenheid van het 75-jarig bestaan van het Gemeentelijk Museum te Haarlem op 30 juni 1937, exh. cat. De Erven F. Bohn, Haarlem (1937)
Grimm, C.: Frans Hals: Entwicklung, Werkanalyse, Gesamtkatalog. Dissertation, Munich (1968)
Grimm, C.: Frans Hals: Entwicklung, Werkanalyse, Gesamtkatalog. Gebr. Mann, Berlin (1972)
Grimm, C.: Frans Hals: The Complete Work (1989). Translated from German by Riehle, J. H.N. Abrams, New York (1990)
Grimm, C.: Frans Hals and his Workshop. RKD, The Hague (forthcoming)
Grosvenor, B.: Fakes, fakes, everywhere? (ctd.) (18 Oct 2016). Art History News (2016). https://www.arthistorynews.com/entries_by_date/2016/October/2
Hendriks, E., Levy-Van Halm, K., in collaboration with Van Asperen De Boer, J.R.J.: Report Concerning a Preliminary Technical Investigation of Paintings exhibited during the Frans Hals Exhibition, held from May 11 to July 22 1990 in the Frans Hals Museum Haarlem (1991)

Hofstede de Groot, C.: Echt of onecht? Oog of chemie? Beschouwingen naar aanleiding van het mansportret door Frans Hals uit het proces Fred. Muller & Co. contra H.A. de Haas. W.P. Van Stockum, The Hague (1925)
Hofstede de Groot, C.: Beschreibendes und kritisches Verzeichnis der Werke des hervorragendsten Holländischen Maler des XVII. Jahrhunderts, 10 vols. Paul Neff Verlag/F. Kleinberger/MacMillan, Esslingen/Paris/London (1907–1928)
Huang, M., Bridge, H., Kemp, M.J., Parker, A.J.: Human cortical activity evoked by the assignment of authenticity when viewing works of art. Front. Hum. Neurosci. **5**, 1–9 (2011). https://doi.org/10.3389/fnhum.2011.00134
Jowell, F.: The rediscovery of Frans Hals. In: Slive, S. (ed.) Frans Hals, Exh. Cat, pp. 61–86. National Gallery of Art/Royal Academy of Arts/ Frans Hals Museum, Washington/London/Haarlem (1989-1990)
Kahneman, D.: Thinking, Fast and Slow. Farrar, Straus and Giroux, New York (2011)
Kahneman, D., Tversky, A.: Prospect theory: an analysis of decision under risk. Econometrica. **47**, 263–291 (1979)
Koestler, A.: The anatomy of snobbery. Anchor. Rev. **1**, 1–25 (1955)
Köhler, N., Levy- Van Halm, K., Runia, E., Van Thiel, P., Van Thiel-Stroman, I.: Painting in Haarlem, 1500–1850: the Collection of the Frans Hals Museum. Ludion, Gent (2006)
Kraaijpoel, D., Van Wijnen, H.: Han van Meegeren 1889–1947 en zijn meesterwerk van Vermeer. Waanders, Zwolle (1996)
Lammertse, F., Garthoff, N., Van Laar, M., Wallert, A.: De Vermeers van Van Meegeren: kennerschap en de techniek van het vervalsen. Museum Boijmans Van Beuningen, Rotterdam (2011)
Le Modernisme de Frans Hals (23 Sep 1883). L'Art Modern, 301–303 (1883)
Lessing, A.: What is wrong with a forgery? J. Aesthet. Art Critic. **23**(4), 461–472 (1965). https://doi.org/10.1111/1540_6245.jaac23.4.0461
Lopez, J.: The Man Who Made Vermeers: Unvarnishing the Legend of Master Forger Han van Meegeren. Harcourt Books, Orlando (2008a)
Lopez, J.: Van Meegeren's early Vermeers. Apollo. **25**, 23–29 (2008b)
Martin, W.: Kunstwetenschap in theorie en practijk. Martinus Nijhoff, The Hague (1904)
Martin, W.: Frans Hals en zijn tijd, onze 17e eeuwsche schilderkunst in het algemeen, in hare opkomst en rondom Frans Hals. Meulenhoff, Amsterdam (1935)
Moes, E.W.: Frans Hals: sa vie et son oeuvre. Translated by De Bosschere, J. Oest, Brussels (1909)
Muthesius, S.: Towards an 'exakte Kunstwissenschaft'(?), part II: the new German art history in the nineteenth century: a summary of some problems. J. Art Hist. **9**, 1–16 (2013)
Nichols, L.W., De Belie, L., Biesboer, P.: Frans Hals Portraits. A Familiy Reunion, exh. cat. Hirmer, Munich (2018)
Pächt, O.T.: Practice of Art History: Reflections on Method (1986). Translated by Britt, D. Harvey Miller Publishers, London (1999)
Packer, L., Roy, A.: The making of the laughing cavalier. In: Packer, L. (ed.) Frans Hals: the Male Portrait, exh. cat., pp. 81–91. Philip Wilson, London (2021–2022)
Reitsma, E.: Frans Hals? Het is met het blote oog te zien. Een handleiding voor een bezoek aan de grote Haarlemse tentoonstelling (5 May 1990). Vrij Nederland, 12–18 (1990)
Robins, E., Pennell, J.: The Life of James McNeill Whistler, 2 vols. William Heinemann, London (1908)
Schenke, M.: Tussen echt en onecht: discussie laait op over werken van Frans Hals (31 Jan 1990). De Telegraaf (1990)
Slive, S.: Frans Hals, 3 vols. Phaidon, London (1970–1974)
Slive, S. (ed.): Frans Hals, exh. cat. National Gallery of Art/Royal Academy of Arts/Frans Hals Museum, Washington/London/Haarlem (1989–1990)
Slive, S.: Frans Hals, revised edn. Phaidon, London (2014)
Slive, S., Van Hees, C.A.: Frans Hals: tentoonstelling ter gelegenheid van het honderdjarig bestaan van het Gemeentemuseum te Haarlem, 1862–1962. Frans Hals Museum, Haarlem (1962)

Storm van Leeuwen, J.: The concept of Pictology and its application to works by Frans Hals. In: Jaffé, H.L.C., Storm van Leeuwen, J., Van Der Tweel, L.H. (eds.) Authentication in the Visual Arts: a Multi-Disciplinary Symposium, Amsterdam, 12th March 1977, pp. 57–92. B. M. Israël, Amsterdam (1979)

Tummers, A.: The Eye of the Connoisseur: Authenticating Paintings by Rembrandt and his Contemporaries. Amsterdam University Press/Getty, Amsterdam/Los Angeles (2011)

Tummers, A.: Frans Hals: Frans Hals: rightly admired by the greatest painters. In: Tummers, A. (ed.) Frans Hals: Eye to Eye with Rembrandt, Rubens and Titian, exh. cat, pp. 13–40. Frans Hals Museum/NAI010 Publishers, Haarlem/Rotterdam (2013a)

Tummers, A. (ed.): Frans Hals: Eye to Eye with Rembrandt, Rubens and Titian, exh. cat. Frans Hals Museum/NAI010, Haarlem/Rotterdam (2013b)

Tummers, A., Erdmann, R.G.: The eye versus chemistry? From twentieth to twenty-first century connoisseurship. In: Colombini, M.P., Degano, I., Nevin, A. (eds.) Analytical Chemistry for the Study of Paintings and the Detection of Forgeries, pp. 3–45. Springer, Cham (2022)

Twijfelachtige Hals-schilderijen geëxposeerd te Haarlem. (29 Aug 1937). De Telegraaf (1937)

Valentiner, W.R.: Frans Hals: des Meisters Gemalde in 322 abbildungen, 2nd edn. Deutsche Verlags-Anstalt, Stuttgard/Berlin/Leipzig (1923)

Valentiner, W.R.: An Exhibition of Fifty Paintings by Frans Hals, exh. cat. Detroit Institute of Arts, Detroit (1935)

Valentiner, W.R.: Frans Hals Paintings in America. F.F. Sherman, Westport (1936)

Van Asperen de Boer, J.R.J.: Infrared Reflectography of Frans Hals Paintings. In: Hendriks, E., Levy-Van Halm, K., in collaboration with Van Asperen De Boer, J.R.J. Report Concerning a Preliminary Technical Investigation of Paintings exhibited during the Frans Hals Exhibition, held from May 11 to July 22 1990 in the Frans Hals Museum Haarlem, pp. 45–48. (1991)

Van Dantzig, M.M.: Frans Hals: Echt of Onecht, Naar aanleiding van de Frans Hals-tentoonstelling 1937. Paris, Amsterdam (1937)

Van Dantzig, M.M.: Johannes Vermeer, de Emmausgangers en de Critici. Sijthoff, Leiden (1947)

Van Dantzig, M.M.: Pictology: an Analytical Method for Attribution and Evaluation of Pictures. Brill, Leiden (1973)

Van de Ploeg, P., Runia, E., Van Suchtelen, A., Buvelot, Q., Van der Vinde, L.: Rembrandt, een jongensdroom, exh. cat. Hirmer, Munich (2008)

Van den Brandhof, M.: Een Vroege Vermeer uit 1937: achtergronden van leven en werken van de schilder-vervalser Han Van Meegeren. Het Spectrum, Utrecht/Antwerp (1979)

Van Gelder, R.: Geen laatste woord over Frans Hals (10 May 1990). NRC Handelsblad (1990)

Van Hees, C.A.: Archivalia betreffende Frans Hals en de zijnen. Oud. Holl. **74**, 36–42 (1959)

Van Thiel, P.J.J.: Het portret van Jacobus Hendriksz. Zaffius door Frans Hals. Oud. Holl. **107**(1), 84–96 (1993)

Wheelock, A.K.: The story of two Vermeer forgeries. In: Schneider, C.P., Robinson, W.W., Davies, A.I. (eds.) Shop Talk: Studies in Honor of Seymour Slive: Presented on his Seventy-Fifth Birthday, pp. 271–275. Harvard University Art Museums, Cambridge, MA (1995)

Wolz, S.H., Carbon, C.C.: What's wrong with an art fake? Cognitive and emotional variables influenced by authenticity status of artworks. Leonardo. **47**(5), 467–473 (2014). https://doi.org/10.1162/LEON_a_00869

Zerner, H.: What gave connoisseurship its bad name? In: Strauss, W.L., Felker, T. (eds.) Drawings Defined, pp. 289–290. Abaris Books, New York (1987)

Part II
Supplementing the Eye: The Technical Analysis of Frans Hals's Paintings and Insights from Seventeenth-Century Sources

Chapter 3
Case Study 1: *Portrait of a Young Woman*: Assessing New Technologies

Anna Tummers, Arie Wallert, Katja Kleinert, Babette Hartwieg, Claudia Laurenze-Landsberg, Joris Dik, Roger Groves, Andrei Anisimov, Vassilis Papadakis, and Robert G. Erdmann

Abstract This case study evaluates the respective use of five non-destructive techniques that are at our disposal to research the characteristic style and painting technique of Frans Hals. For this purpose, a typical Frans Hals portrait of around 1635, *Portrait of a Woman* in the Gemäldegalerie in Berlin, was researched using not only observations with the naked eye, x-rays (XR) and infrared reflectography (IRR), but also macro-x-ray fluorescence scans (MA-XRF), hyperspectral imaging also known as reflectance imaging spectroscopy (HI/RIS), and even neutron activated

This article was first published in *The Burlington Magazine*: Tummers, A., Wallert, A., Hartwieg, B., Kleinert, K., Laurenze-Landsberg, C., Dik, J., Groves, R., Anisimov, A., Papadakis, V., Erdmann, R.: Supplementing the eye: the technical analysis of Frans Hals's paintings—I. Burlingt. Mag. 161, 934–941 (2019). The research for this article was funded by the Netherlands Organisation for Scientific Research as part of the projects *Frans Hals or not Frans Hals* (Museumbeurs 2016–2018) and *21st Century Connoisseurship* (NICAS Seed Money Grant 2018–2022); and by the Prins Bernard Cultuurfond Conservatoren Stipendium, 2018. It builds on earlier examinations carried out at the Gemäldegalerie in Berlin (microscope, XR, IR, UV) and at the Helmholtz Zentrum Berlin für Materialien und Energie, the former Hahn-Meitner Institut (NAR). The authors wish to thank their respective institutions for enabling this research project.

A. Tummers (✉)
Ghent University, Ghent, Belgium
e-mail: Anna.Tummers@UGent.be

A. Wallert (✉)
University of Amsterdam, Amsterdam, The Netherlands

Rijksmuseum, Amsterdam, The Netherlands
e-mail: arie.wallert@gmail.com

K. Kleinert (✉) · B. Hartwieg (✉) · C. Laurenze-Landsberg (✉)
Gemäldegalerie, Staatliche Museen, Berlin, Germany
e-mail: K.Kleinert@smb.spk-berlin.de; b.hartwieg@web.de; claudia.landsberg@online.de

J. Dik (✉) · R. Groves (✉) · A. Anisimov (✉)
Delft University of Technology, Delft, The Netherlands
e-mail: J.Dik@tudelft.nl; R.M.Groves@tudelft.nl; A.G.Anisimov@tudelft.nl

A. Tummers, R. G. Erdmann, *Frans Hals or not Frans Hals*, Cultural Heritage Science, https://doi.org/10.1007/978-3-031-59489-2_3

radiographs (NAR). As the portrait is the only painting to date to have been researched with all these techniques, it provided a unique opportunity to compare their respective merits and assess their potential.

Old master paintings are often attributed on the basis of stylistic analysis alone. One such case is *Portrait of a man*, which was published as a work by Frans Hals in *The Burlington Magazine* in 2014 (Buvelot and Ducos 2014). Subsequent technical analysis by Orion Analytical led Sotheby's, which had sold the painting in 2011, to declare in 2016 that it was a forgery. Further tests followed and the work became the subject of lawsuits in France and England (BBC News 2016). Although most of the research into this painting has not yet been made public, the claim that it was a fake led to much debate about the role of traditional connoisseurship in the field of old master painting. According to the British art historian Bendor Grosvenor (2016), 'the system upon which the art market lies for determining authenticity is not working'. In an article discussing not only *Portrait of a man* but also a number of other purported Old Master paintings that have appeared on the market in the past decade and have now become suspect, Sarah Cascone (2016) concluded that 'technical research succeeded where connoisseurship failed'. The present essay seeks to address the role of technical analysis in the attribution of paintings to Hals by publishing findings from the research project *Frans Hals or not Frans Hals*. This was launched in 2016 as a collaboration between specialists from the Frans Hals Museum in Haarlem, the Rijksmuseum in Amsterdam, the Gemäldegalerie in Berlin, the Metropolitan Museum in New York, the University of Amsterdam and the Delft University of Technology to re-examine the criteria that experts have used to attribute works to Hals, and to evaluate and improve them while testing a number of new techniques.

Although Frans Hals (1582/83–1666) was one of the leading painters of the Dutch seventeenth century, comparatively little research has been done into the precise definition of his œuvre, working methods and materials. Since the early twentieth century the attribution of paintings in his manner has been the subject of a number of controversies (see above Chap. 2), including the earliest court case in the Netherlands in which chemical evidence was used to evaluate authenticity, in 1925 (Hofstede de Groot 1925). Both the first overview exhibition of his paintings in 1937 and the one in Washington, London and Haarlem in 1989–1990, were overshadowed by fierce debates about the attribution of pictures on display (Gratama and Van Rijckevorsel 1937; Van Dantzig 1937). Indeed, the two foremost Hals scholars of the twentieth century, Seymour Slive (1920–2014) and Claus Grimm (b.1940), disagreed about roughly a third of Hals's œuvre. In his catalogue raisonné,

V. Papadakis (✉)
Institute of Molecular Biology and Biotechnology (IMBB), FOundation for Research and Technology-Hellas (FORTH), Heraklion, Greece
e-mail: vassilis_papadakis@imbb.forth.gr

R. G. Erdmann (✉)
University of Amsterdam, Amsterdam, The Netherlands
e-mail: r.g.erdmann@uva.nl

published in (1970–1974) and revised in 2014, Slive accepted respectively 222 and 233 paintings as authentic; Grimm's catalogue, published in 1989, is limited to only 145 works, of which 135 were in Slive's selection (Grimm 1990/1989; Slive 2014).[1] No catalogue raisonné has been published since theirs, and the lack of clear criteria for confirming a work's attribution has left Hals vulnerable to further confusion and controversy.

Known for his exceptionally loose yet accurate brushwork, Hals is celebrated for the immediacy of his works, the dynamic poses of the figures and their lively, fleeting facial expressions, captured with great nuance and masterful 'rough' brushwork (Tummers 2013a, b, c). As the seventeenth-century painter and art theorist Samuel van Hoogstraten (1627–1678) explained, a 'rough' ('*ruw*' in Dutch, meaning broad and loose) manner can create a powerful, lifelike effect, since a trained eye will fill in the details: 'it is like recognising a friend in twilight or from a distance'.[2] Understandably, attributions to Hals have rested largely on these stylistic qualities yet, with a few notable exceptions, relatively little is known about precisely how Hals built up his paintings or about his workshop practice and characteristic use of materials. For example, the verve with which Hals applied loose accents, especially in his later works, and the consistency of the paint he used often resulted in raised edges on both sides of these brushstrokes (Hendriks et al. 1991, p. 50).[3] Moreover, his use of indigo blue seems distinctive from his contemporaries, and possibly also his use of black pigments in the top and bottom paint layers when depicting black drapery (Groen and Hendriks 1989–1990, p. 119; Hendriks et al. 1998).

This article analyses five of the many non-destructive techniques that are at our disposal to research Hals's style and technique. Its focal point is the *Portrait of a woman* in the Gemäldegalerie, Berlin, a typical example of a portrait by Hals of around 1635 (Fig. 3.1). Unanimously accepted by Hals scholars, its boasts a solid provenance history (Staatliche Museen zu Berlin, Gemäldegalerie, cat. no. 801). It has been in the Gemäldegalerie since 1841, well before the artist was 'rediscovered' by Théophile Thoré-Bürger (1807–1869) in 1868, before the market value of his paintings increased dramatically and before the many modern copies and imitations of his works were created (Jowell (1989–1990; Von Stockhausen 2000), no. 279, p. 273). What makes the Berlin picture particularly relevant for this study is that it is the only painting by Hals to have been investigated with not only X-radiographs (XR) and infra-red reflectography (IRR) but also with a macro-X-ray fluorescence scanner (MA-XRF), a laser scanner for hyperspectral imaging (HI), also known as reflectance imaging spectroscopy (RIS), and even with neutron activated radiography (NAR). Combined with observations with the naked eye and through the microscope, these techniques yield a number of significant new insights into Hals's characteristic style and working methods, as well as a better understanding of the painting's history and condition.

---

[1] In his forthcoming revised oeuvre catalogue Grimm accepts 199 works as by the master, including also works made in collaboration with his studio, see above Chap. 2 and (Grimm forthcoming).

[2] '*En even gelijk men zijn vriend van verre bespeurende, of by schemerlicht ontmoetende*' (Van Hoogstraten 1678, p. 27).

[3] A copy of this report is available at the Netherlands Institute for Art History (RKD), The Hague.

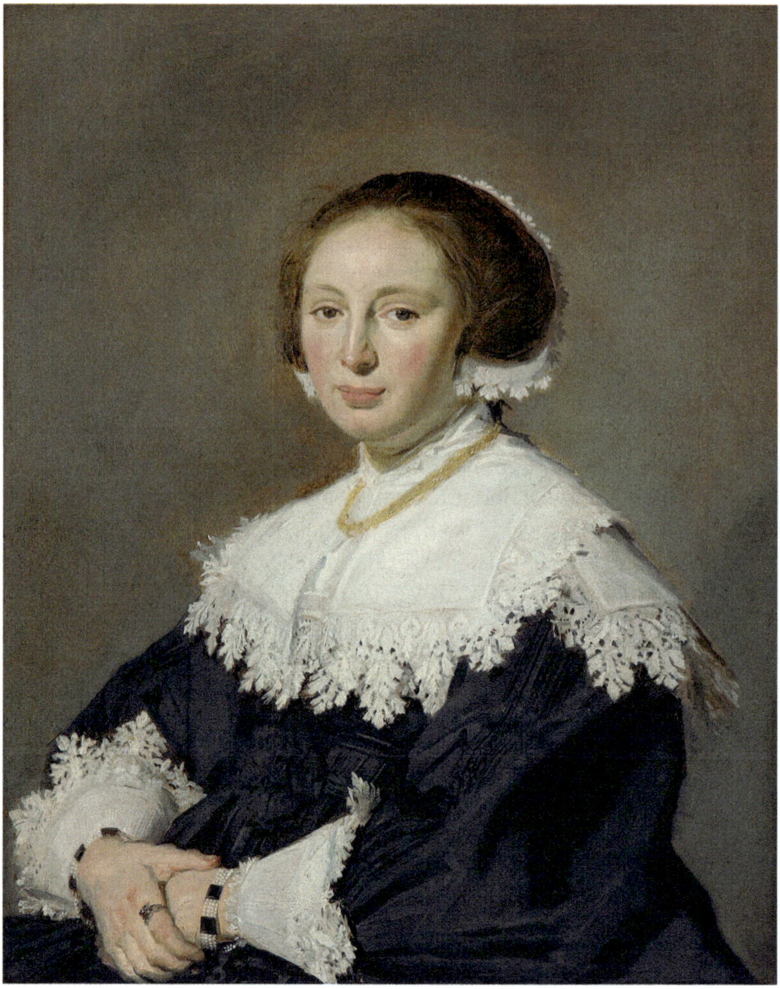

Fig. 3.1 Frans Hals, *Portrait of a woman*, c.1632–1635, oil on canvas, 76 × 61.4 cm. (Staatliche Museen zu Berlin, Gemäldegalerie)

Portrait of a woman now measures 76 by 61.4 cm. A small strip of canvas was added to the lower edge of the painting, which was relined. The paint layer is generally very thin, revealing the sand-coloured ground in some places, and only the white areas show the loose, paste-like paint application typical of Hals. The painting is generally well preserved. There are, however, a few traces of later interventions. In parts of the sitter's golden chain a few old retouchings are discernible and abrasion is visible in the black areas and in the hair. These overpaints as well as the fine retouchings of the last restoration, carried out in 1995, are visible in Ultraviolet (UV) illumination.

In the X-radiograph the characteristic slubs in the canvas weave are clearly visible (Figs. 3.2 and 3.3). That is because the ground layer, which contains lead white, was worked into the interstices of the canvas weave during application with a

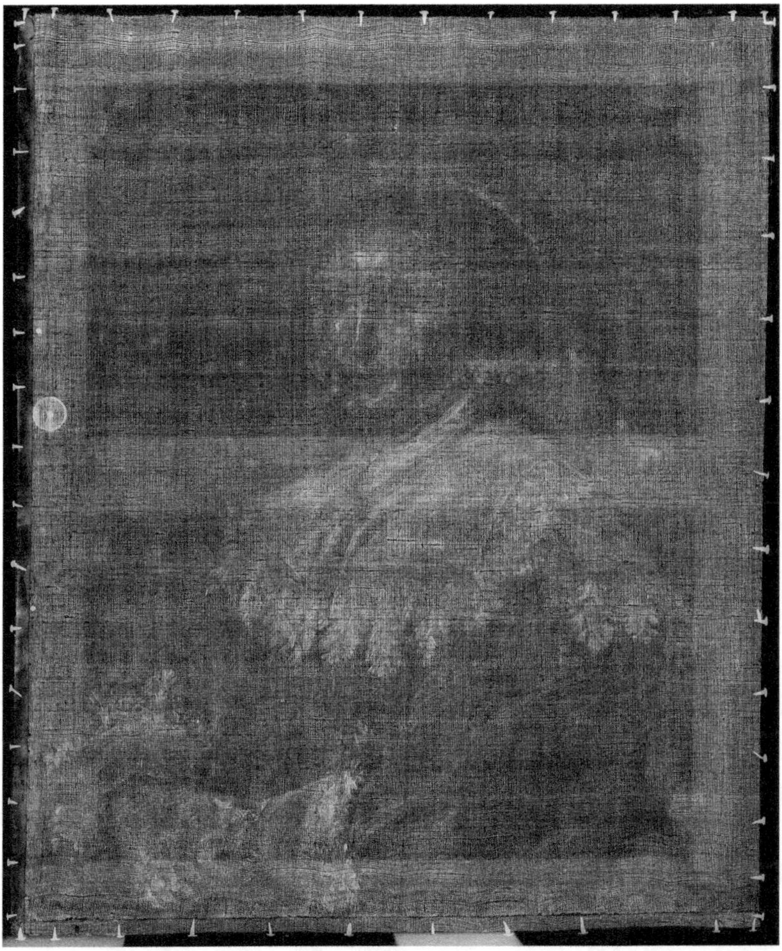

Fig. 3.2 X-radiograph of Fig. 3.1

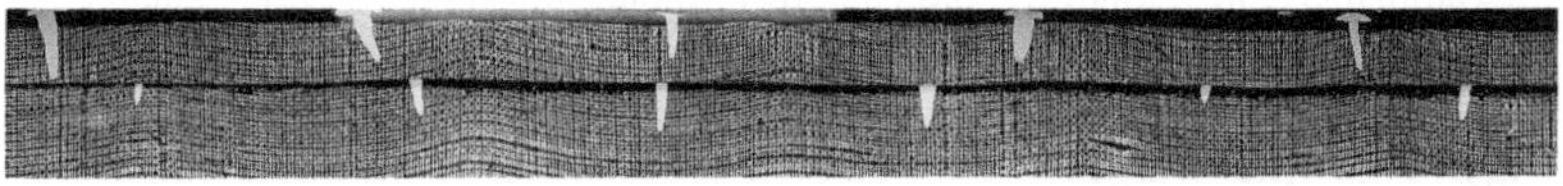

Fig. 3.3 Details of Fig. 3.2 showing bottom strip inverted and aligned with top part

spatula. No corrections or pentimenti were observed. The image shows that there is prominent cusping of the canvas at the upper and lower edges of the painting (Fig. 3.4), especially the latter, minor cusping at the left-hand side and none at the right-hand side, indicating that the painting may have been trimmed especially at the right side (since the cusping is missing entirely there). However, under the microscope a straight line parallel to the painting's proper right edge is visible

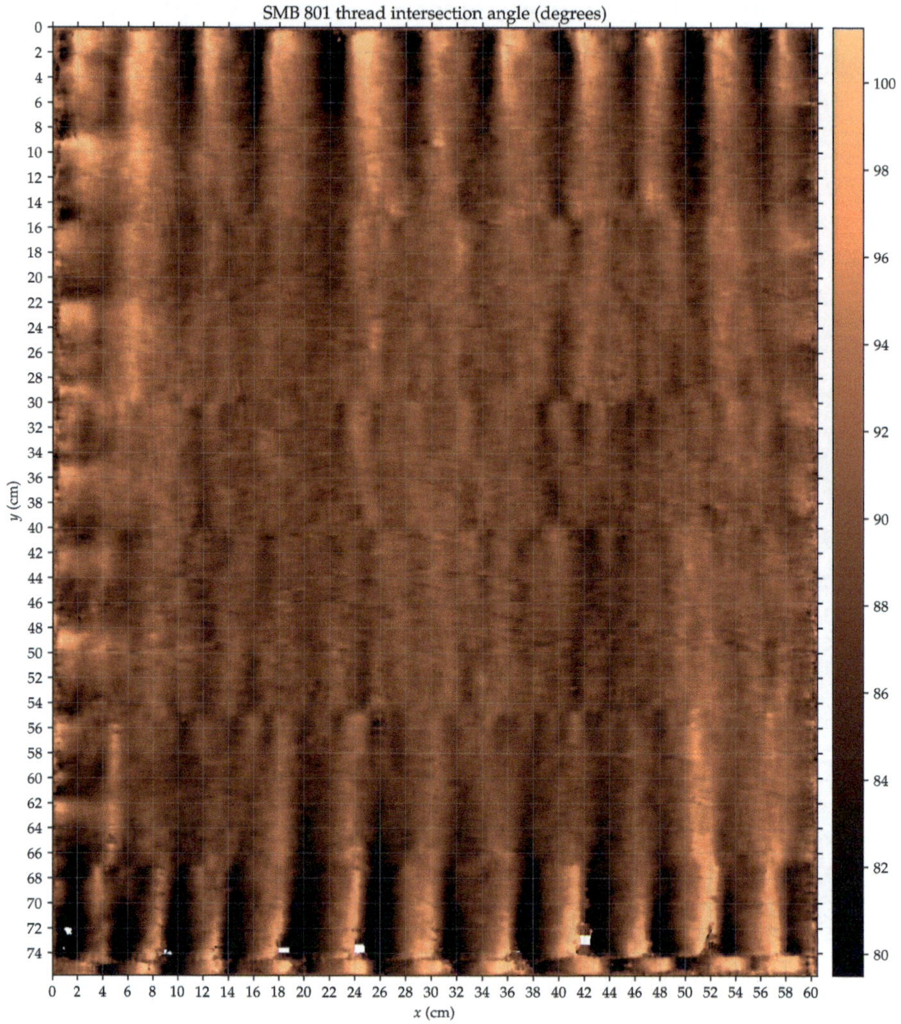

Fig. 3.4 Automated canvas analysis, showing the cusping (thread intersection angle in degrees)

(Fig. 3.5). This line was clearly incised by the artist to indicate the size of the intended image; the paint of the background runs right up to this line, providing an insight into Hals's studio practice that has not before been observed. This proves that Hals used a piece of a larger preprimed canvas that was originally stretched into a much larger frame (as was common practice in his time); the canvas was presumably cut to its specific size in Hals's workshop.

Interestingly, a similar instance of Hals setting out the format of a painting was recently discovered in his *Portrait of Jasper Schade* (National Gallery, Prague) in the form of a black rectangular outline cutting through the clothing at its lower end. (Pokorny 2012; see also Ševčík (ed.) 2012, p. 182; and Tummers (ed.) 2013c, p. 93).

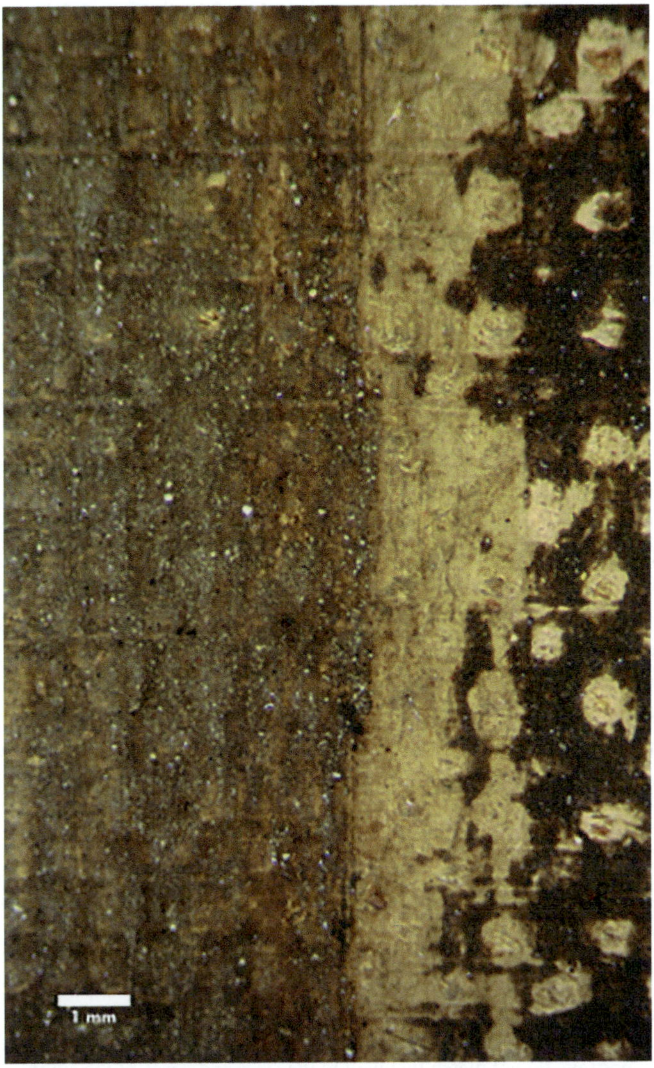

Fig. 3.5 Photomicrograph of Fig. 3.1, showing a fine, light vertical incised line at the right edge, which indicates the size of the intended image

The X-radiograph shows that the canvas added to the bottom edge of the painting is about 1.3 cm. wide. Close examination of the weave pattern of this strip demonstrated that it had been had been cut from the top of the painting and placed at the bottom (Fig. 3.3).[4] This conundrum is explained by interventions in the eighteenth

[4]An identical alteration was recently discovered in the *Singing boy with flute* (c.1627, Staatliche Museen zu Berlin, Gemäldegalerie, cat. no. 801A) (see (Kleinert et al. 2018), pp. 191–194).

century. The painting was probably auctioned in 1767 in Paris.[5] It was sold with a *Portrait of Jan Miense Molenaer (?)* (Gemäldegalerie, Berlin) then believed to be its pendant.[6] At the time of auction, both paintings were recorded as measuring 81.3 by 65 cm. This tallies with the evidence of the cusping in the X-radiograph that the *Portrait of a woman* was originally bigger. Some 20 years later, in 1786, the same pair of paintings was sold again. This time, they were considerably smaller at 75.9 by 59.2 cm. (Fig. 3.6).[7] This is close to the present height of the *Portrait of a woman* but the widths of both paintings were then slightly smaller than they are now. After 1786, but before their acquisition in 1841 by the Gemäldegalerie, the *Portrait of a woman* must have been relined, as a result of which it changed format yet again. During the process of lining, the edges—previously folded around the stretcher—were flattened and integrated into the front surface of the painting. This can be seen in the X-radiograph, where a dark strip on the left-hand side reveals that the painting has been enlarged there by some 1.2 cm.

In spite of these interventions, the painting still provides a good example of Hals's working method. An important aspect of this concerns the function of the sand-coloured ground layer. Under the stereomicroscope, the original ground layer can be observed in areas where it is not, or only partially, covered with paint. This fine-grained buttery mass, which must have been applied with a broad spatula in a horizontal direction, was not covered by an imprimatura. The research group mapped out the areas of the painting where the tan colour of the ground shows through, and compared it to areas of similar colour using hyperspectral imaging (HI), also known as reflectance imaging spectroscopy (RIS). This technique combines the 'normal' two-dimensional visualisation of the painting by optical imaging (such as photography) with optical spectroscopy. It is done in such a way that each

---

[5] Sale, Pierre François Basan, Paris, 17 March 1767, lot no.18 (Lugt Nr.1598; Getty Provenance Index F-A190): '*Deux Portraits, celui d'un homme, celui d'une femme, vus jusqu'aux genoux, peints par Fr. Hals, 30 pouc. de h. sur 24 de larg.*' [c.81.3 by 65 cm.], price 49 livres, buyer La Tour.

[6] Frans Hals, *Portrait of Jan Miense Molenaer (?)*, c.1632–36, Oil on canvas, 76 × 61.4 cm. Staatliche Museen zu Berlin, Gemäldegalerie, Berlin, cat.no.800. Some art historians believe they are pendants, but the canvases are quite different in type, stylistic differences suggest a different dating, and the portraits do not match that well in the sizes and poses of the sitters.

[7] Sale, Jean Baptiste Pierre Lebrun, Paris, 3 May 1786 (and following days), lot no.22 (Lugt Nr. 4025 & 4040): *Deux Tableaux faisant pendans: l'un représente une jeune femme vue à mi-corps, la tête presque de face, coëffée en cheveux, ajustée d'une coëffe de dentelle découpée: elle a le cou & les épaules couverts d'un fichu en forme de rabat bordé d'une dentelle du même genre que la coëffure & les manchettes: elle est vêtue d'une robe noire, ayant les mains croisées l'une sur l'autre. Le pendant offre un homme assis vu à mi-corps: la tête de trois quarts, couverte d'un grand chapeau noir, ayant au cou une fraise blanche & habillé de noir avec manches tailladées, le bras droit appuyé & la main pendante, la main gauche fermée & élevée sur la poitrine. Ces deux Tableaux, d'un faire large & vigoureux, sont de l'effet & de la couleur la plus piquante, & peuvent aller de pair avec les tableaux de Rembrandt & de Vandick. Hauteur 28 pouces, largeur 21 pouces 6 lignes. T'* [c. 75.9 by 59.2 cm]; price 700 francs (760 francs according to the Getty Provenance Index); buyer: Antoine-Charles Dulac. The seller is given as 'M.M***; most of the paintings were probably owned by Lebrun himself.

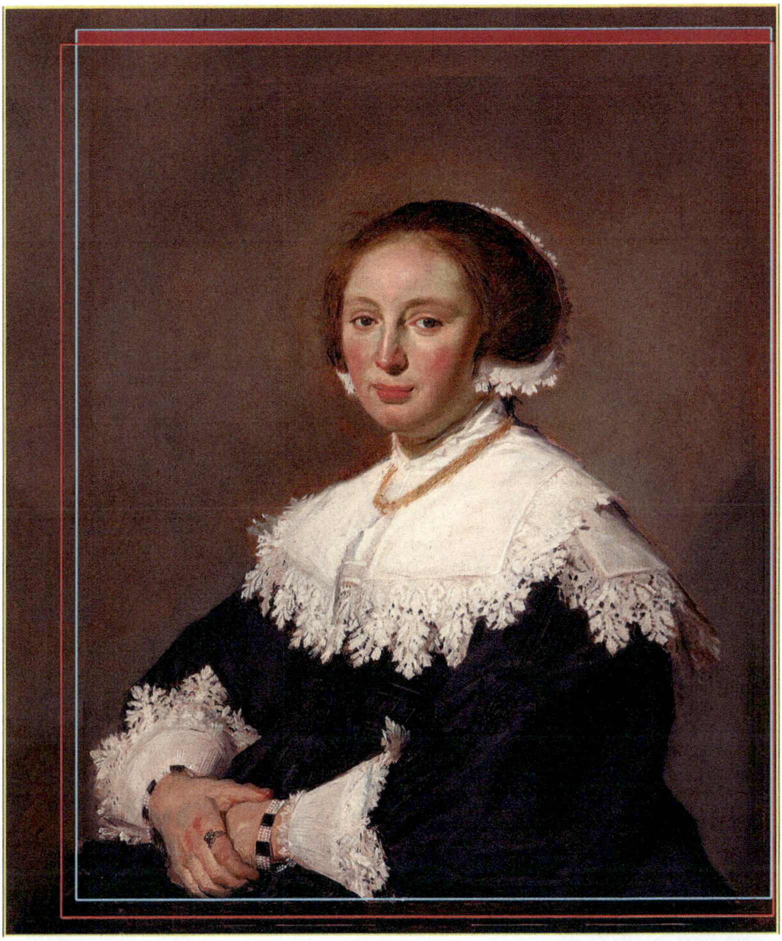

Fig. 3.6 Reconstruction of the changing formats of Fig. 3.1. Yellow margins: format of 1767; probably identical to the original dimensions; blue margins: format of 1786; the image was clearly reduced in size (the cuffs were cut off by the frame at the bottom and slightly at the left edge); red margins: present format; extensions at lower and left edges

pixel of the image is made to represent a complete spectrum of that pixel. Whereas a regular camera records three different wavelengths of the electromagnetic spectrum (corresponding to the colours red, blue and green), modern hyperspectral cameras can nowadays easily differentiate between hundreds of wavelengths, resulting in very precise digital images or 'data cubes'.[8]

This technique allows for selection of the areas where, based on the spectral response, the sand-coloured ground layer remains exposed (Fig. 3.17). A false

[8]The hyperspectral imaging of the *Portrait of a woman* was carried out with the instruments and methods described in (Groves et al. 2018).

colour enhances the legibility of the images, and illustrates how economically and efficiently Hals used the colour of the ground. The hair and many shadowy parts in the face and neck owe much of their colour to the ground, which has been covered with a thin translucent layer. Moreover, a kind of halo effect reveals Hals's fast and very functional painting technique: small areas of exposed ground can be seen around the contours of the heads and the cuffs. Hals must have painted in the clothing and worked up the background, while the head and collar were still drying. Similarly, he painted the hands and cuffs while the blacks and greys of the woman's dress were still wet. By keeping these areas apart he prevented smudging and smearing of the different wet paints. This approach is confirmed in the infra-red image of the painting (Fig. 3.7), where the brushwork of the paint containing black pigment around the head is clearly visible: the strokes at the edges of the painting are large and almost sloppy whereas those closer to her head are smaller and more careful.

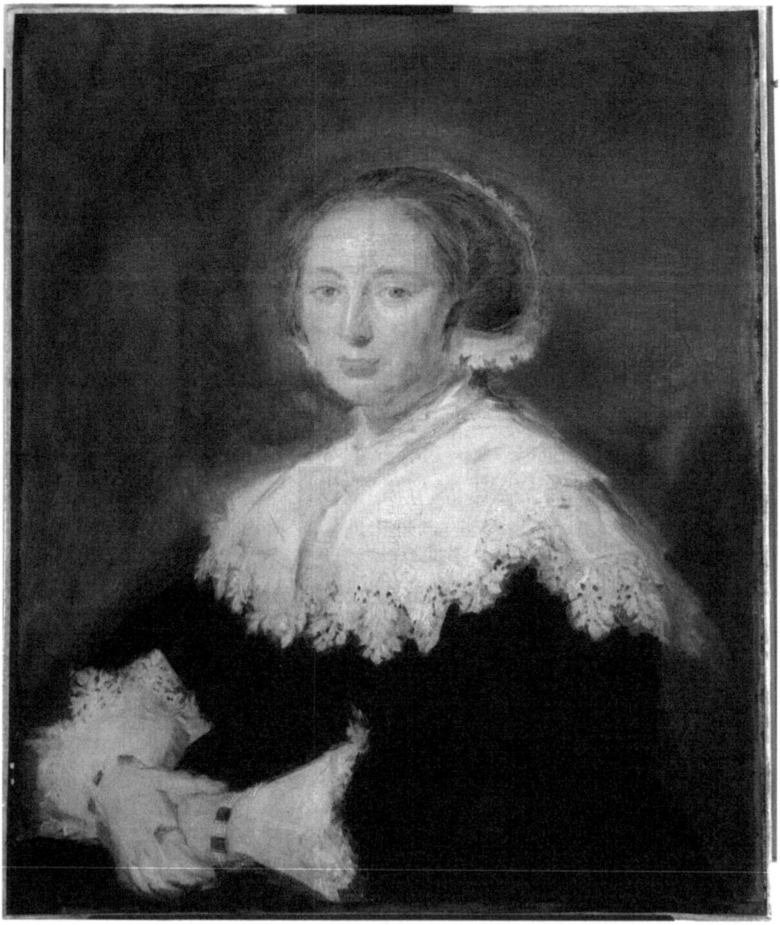

Fig. 3.7 Infra-red reflectogram of Fig. 3.1

Examination with neutron autoradiography (NAR) was very helpful in further understanding the way the portrait was built up.[9] This technique is based on a very moderate, non-destructive radioactivation of the painting with thermal neutrons. In this process a small number of the atoms of the pigments in the paint are temporarily transformed into a radioactive species. About a dozen isotopes, emitting β- (electrons) and γ-radiation, are created. This radioactivity, the β-radiation, is sufficient to blacken a highly sensitive film in direct contact with the painting and reveal the spatial distribution of pigments. As the radioactivity of different elements decays at different rates, a series of films can be made (at different half-life times) that show the different elements, allowing us to represent the distribution of a number of pigments over the painting in separate films.[10] In particular the first, fifth, and second autoradiograph of the series were very informative. The first film confirmed that the sand-coloured ground contains a manganese-based brown (Fig. 3.8).[11] This suggests the use of a composition of iron-based ochres and manganese oxides, commonly known as the pigment umber. In addition, the film showed that the hair and the upper contours of the woman's head were brushed-in with a manganese-based brown, as was the modelling of some of the curls in her hair.

The distribution of ochres and umbers could also be studied with the relatively new technique of macro X-ray fluorescence scanning (MA-XRF) (Alfeld et al. 2011). This technique allows visualisation of the distribution of elements in a flat sample, such as an easel painting, in a non-destructive manner. This is achieved by scanning the surface of the sample with a focused X-ray beam, and analysing the emitted fluorescence radiation. As the X-ray beam scans the whole painting, it produces thousands, sometimes millions, of data points. These can be plotted on elemental distribution maps, which may be interpreted as pigment distribution images.[12] The XRF image for manganese is complementary with the autoradiography for the same element. The somewhat grainy image shows that umber was almost exclusively used to paint the sitter's hair. It also reveals that other areas in the hair, darker touches in her face and the contours of her hands have been touched up with iron-based brown and red ochres (Figs. 3.9 and 3.10).

The fifth NAR film primarily shows the distribution of phosphorus, and therefore of bone black (Fig. 3.11). Here again we can see how the NAR image tallies with the MA-XRF calcium image as some 80–90% of bone black consists of basic calcium phosphate $Ca_5(OH)(PO_4)_3$ (Fig. 3.14). It is reasonable to assume the presence of bone black only where both P and Ca are indicated. These images show how Hals

[9] The Neutronautoradiography was carried out in 1985 with the cooperation of the Gemäldegalerie Berlin by the Helmholtz Zentrum Berlin für Materialien und Energie (HZB, formerly the Hahn-Meitner Institute) in the nuclear research plant in Berlin-Wannsee. The results are published here for the first time.

[10] On the exposure times, methods and setup, see (Denker et al. 2017).

[11] This assumption is based on the non-blackening of the NAR-film, in areas that correspond with attenuation features in the X-radiograph. Many black spots in the film are caused by larger agglomerates of pigment on the painting. Smoother blackening may also relate to retouchings.

[12] The instrument and method used are described in (Alfeld et al. 2013).

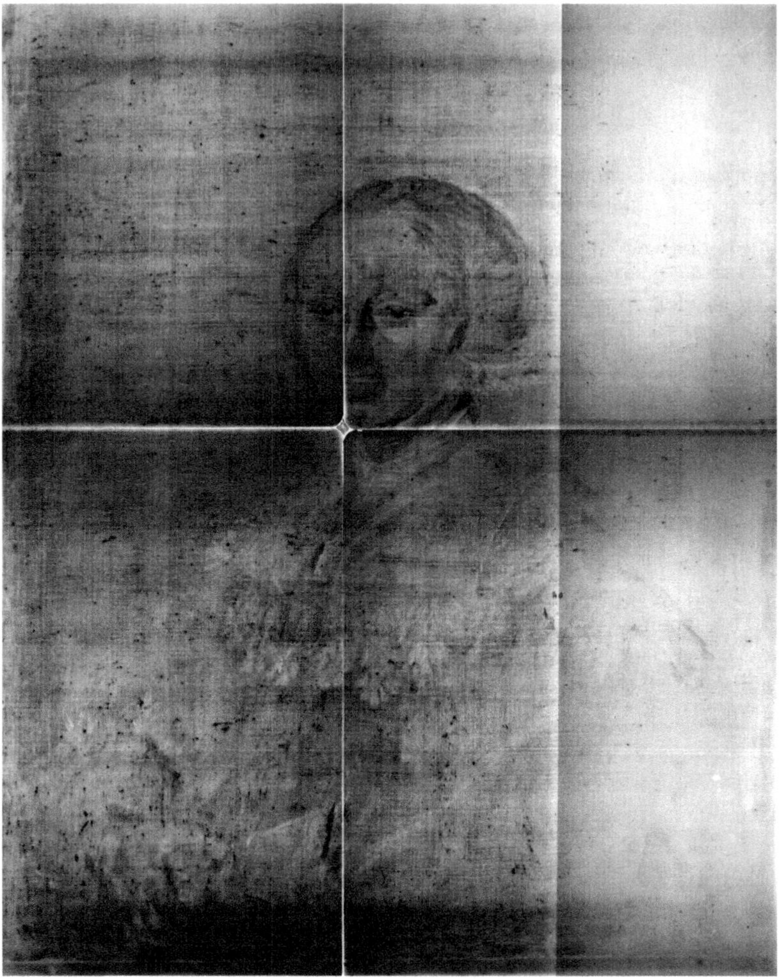

Fig. 3.8 Neutron-Autoradiograph of Fig. 3.1 (first recording)

touched up the dark grey of the sitter's dress, especially around her hands and the white lace of her collar and cuffs, to enhance the contrast with the deepest black possible. From infra-red studies it became apparent that the black for the rest of her dress was done in charcoal black (presumably lamp black). We also see touches of bone black that do not match with the visual impression of the painting in normal light, revealing a quick and economic initial sketch: a beautiful insight into Hals's characteristic 'handwriting' in this deeper layer. The rounded contour of the lace on the woman's left shoulder, the three diagonal longer brushstrokes on the shoulder itself, the split in the front of the white lace collar and some of the folds in the sleeves are all under-painted in bone black. Similarly, the necklace was laid in with a long, fluid brushstroke that—given the intensity of the signal at the borders—must

Fig. 3.9 MA-XRF scan of Fig. 3.1, showing the element manganese (Mn)

have had raised edges on both sides, which, as mentioned above, is one of Hals's hallmark traits. He also briefly indicated some key passages with bone black: the transition from neck to the left shoulder (partially left visible) and the distinction between the sleeve and skirt of the dress. The pupils of the woman's eyes, the shading of her chin and cheek just in front of her ears, well as her hair, have been touched up with bone black in a later stage of the painting.

Comparison with the infra-red image reveals that the bone black pigment was used very locally. It seems to occur only in the deepest blacks and strong accents, to

Fig. 3.10 MA-XRF scan of Fig. 3.1, showing the element iron (Fe)

make the cool white cuffs of the lace stand out against the surrounding blacks. The rest of the woman's dress was executed with a dark brownish paint, consisting primarily of charcoal black, mixed with umber and occasionally some azurite and lead white. This mixture was very dark indeed. There is just a little bit of lightness in her dress to allow for a powerful contrast with the touches of still deeper bone black, indicating the depressions and pleats of the folds of the textile. Interestingly, not all the deepest accents were done with bone black: it was used only for three of the (five) black vertical folds in the dress below the collar at the right and eight (of

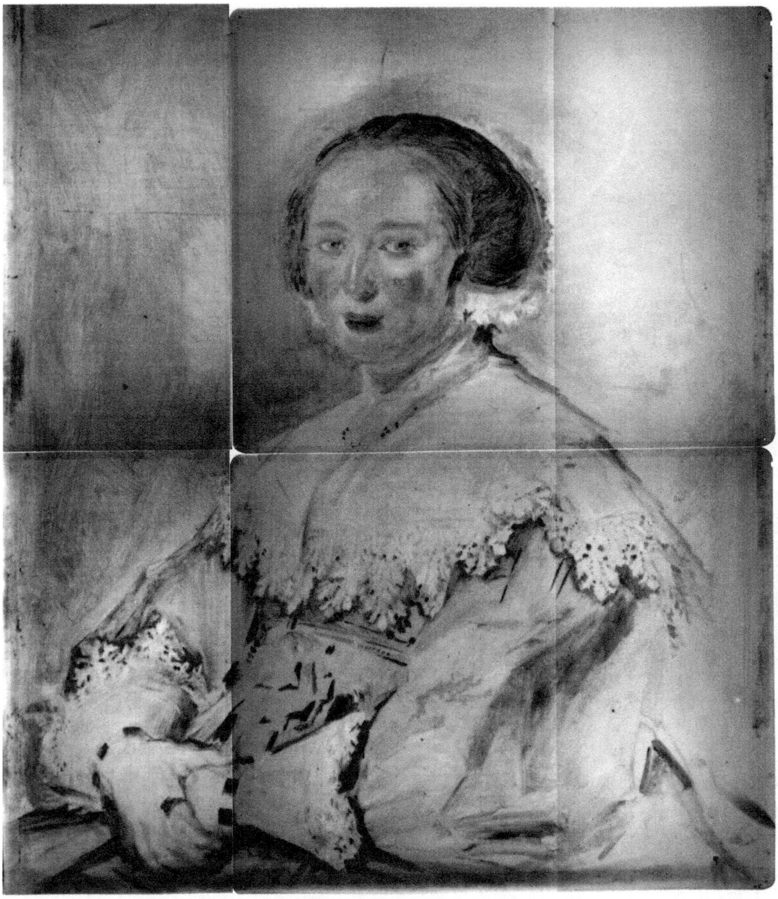

Fig. 3.11 Neutron-Autoradiograph of Fig. 3.1 (fifth recording)

twelve) horizontal dots in the chest decoration; the remainder were added in charcoal black, presumably at a later stage. In the same, fifth, NAR film, we also see the distribution of vermilion, which can be compared to the MA-XRF map of the same element (Figs. 3.11, 3.14, and 3.15). The combination of the two documents reveals that vermilion, a red mercuric sulphide (HgS) was limited to the woman's lips[13] and the flesh tone of her face, where it was added to lead white. This feature is shown here in the hyperspectral false-colour map (Fig. 3.18). There we see that the spectral features of the flesh tone for the face are distinctly different from those of the hands, which are primarily in mixtures of lead white and red ochres. It is therefore likely that the face and hands were executed at different stages of the painting process.

[13]The red of the lips was originally more intense. The Micro-XRF analyses, executed by the Rathgen Forschungslabor of the Staatliche Museen zu Berlin, demonstrated that the dark grey spots on the lips are from vermilion as well, but altered.

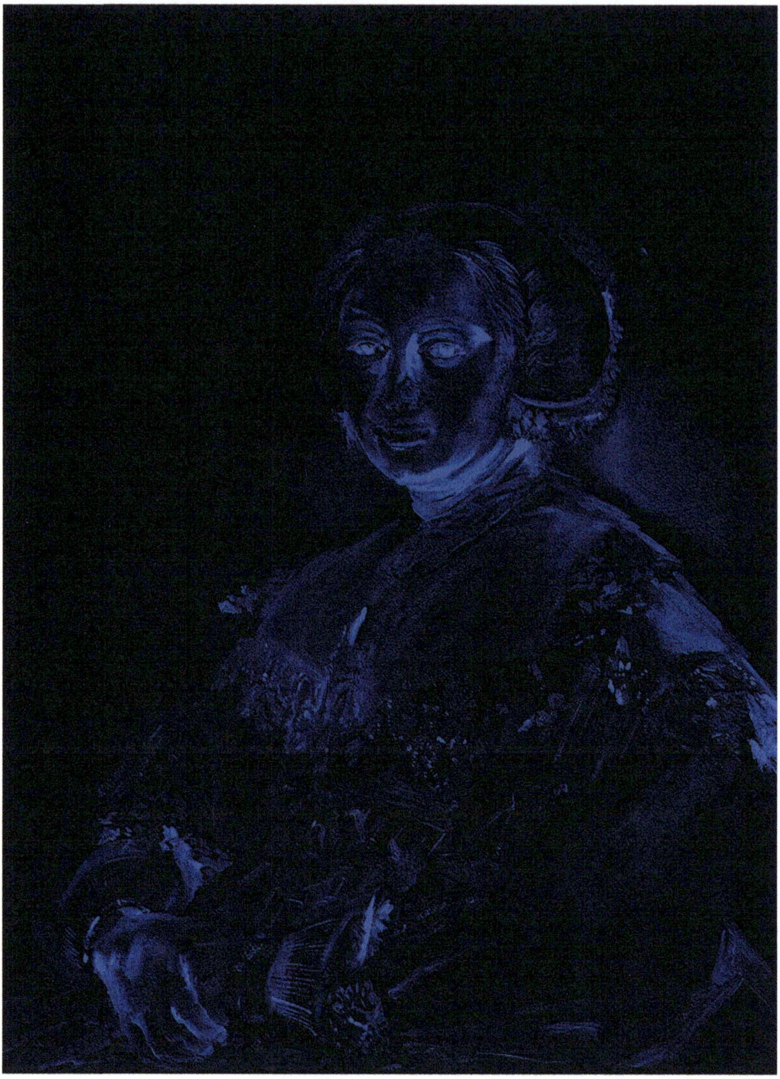

Fig. 3.12 MA-XRF scan of Fig. 3.1, showing the element copper (Cu)

Furthermore, the hyperspectral analysis confirms how little the paint mixtures used for the face and hands have in common.

In the second NAR film we see a blackening of the film in the area of the yellow highlights in the golden chain (Fig. 3.13). This is due to the presence of antimony in Naples yellow, a pigment that appeared on the painter's palette only in the eighteenth century. Examination under the microscope confirmed that these highlights are later retouchings. But the most prominent, and important, feature on this second NAR-film is the blackening in the face, hair and lace cuffs. In the last it was clear

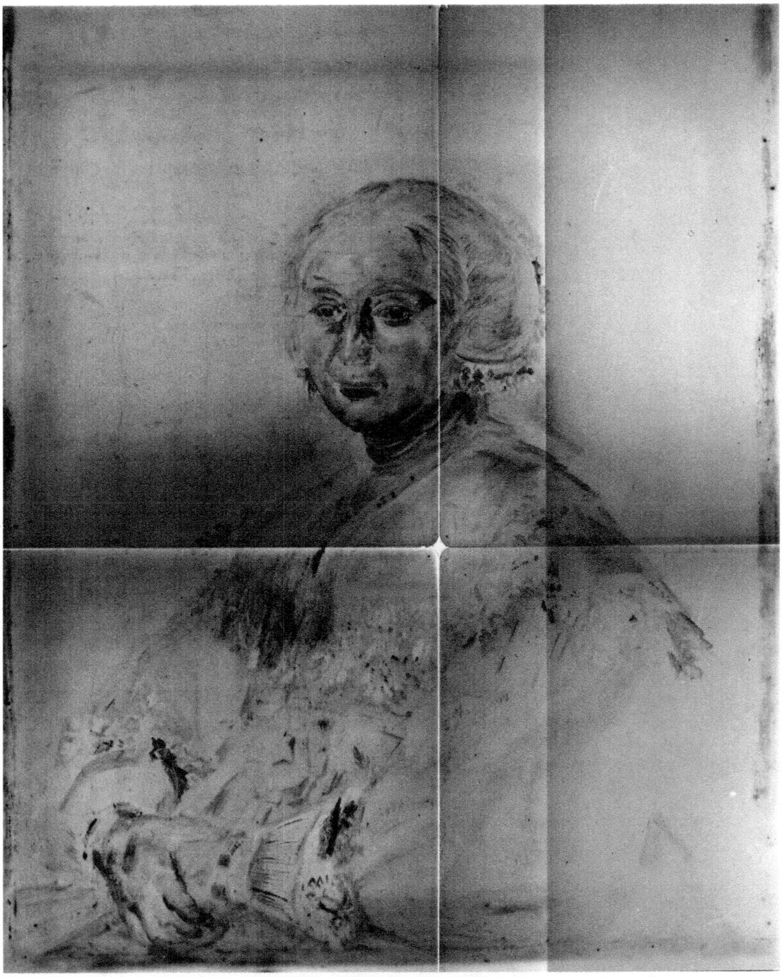

Fig. 3.13 Neutron-autoradiograph of Fig. 3.1 (second recording)

that this signal related to touches of a bright blue pigment. On examination with the microscope, this blue turned out to possess the features of azurite, a common copper carbonate blue pigment. It seems that this same copper-based blue pigment was also responsible for the colour of greyish or blackish areas in the face and hair. Here again, the auto-radiograph complemented the XRF scans (Fig. 3.12). A surprising amount of azurite was found to have been used, not specifically for its colouring properties, but rather to give several spots in the painting a cooler or more greyish tone. Only the touches in the cuffs were undeniably blue; in the black dress, the hair and the shadows in the woman's face, for instance under her mouth and neckline, it served only to create appreciable grey tones. Blue (or green where mixed with yellow) was used to make greys. For example, the shadow at the side of the woman's

Fig. 3.14 MA-XRF scan of Fig. 3.1, showing the element calcium (Ca)

nose and the shadow at the edge of her left eye socket at the right were added with this copper-containing paint. It is tempting to speculate about the use of greens in flesh-tones as an 'anachronistic' *verdaccio*—the popular green underpainting technique used for flesh tones in Renaissance Italy, which often shines through in the shadowy parts.[14] Other areas in the flesh-tone, the shadow around the nose and the

[14] See also below 'Frans Hals & Co: the Civic Guard Portraits and the Attribution of *The Meagre Company*' for a more extensive analysis of the pigments Hals used for skin tones.

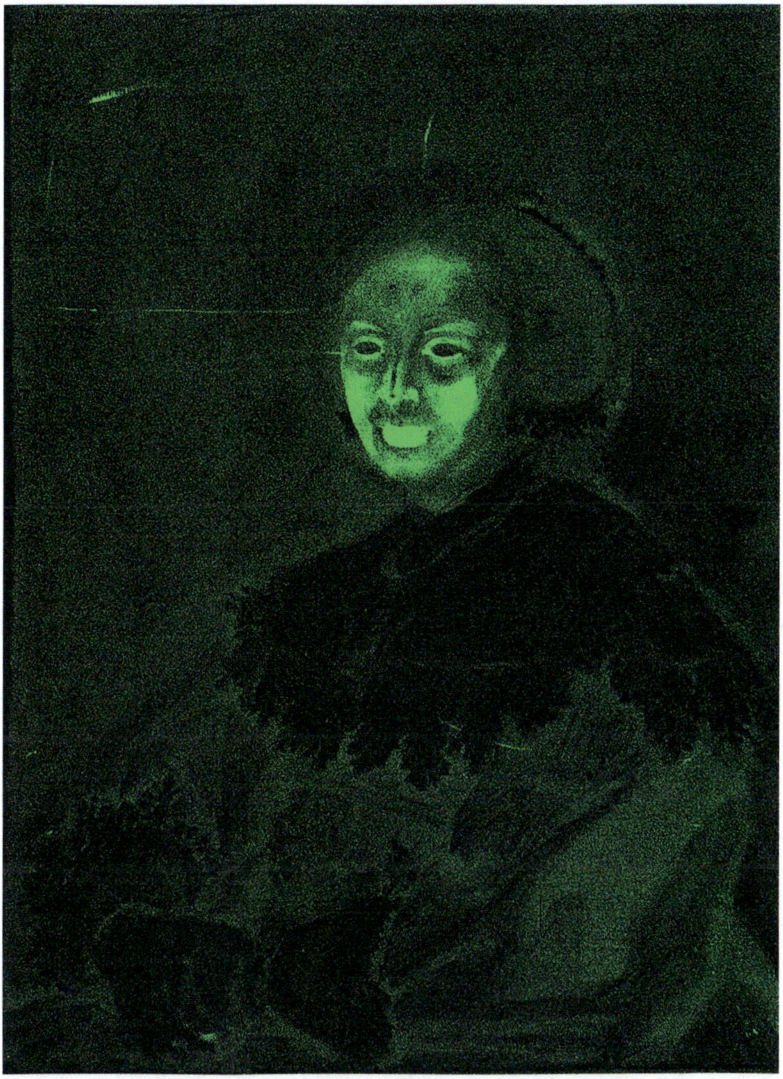

Fig. 3.15 MA-XRF scan of Fig. 3.1, showing the element mercury (Hg)

indication of the eyebrow were blocked in with a grey tone that does not appear to contain copper. The copper-based pigment was also used in the shadow of the woman's hands, as well as in the black at the contour of her elbow, in the blue-greyish shadows of her clothing and in a slight shadow in the background near her neck.

The way in which a painting is built up and blocked in is often just as characteristic of an artist as the final result. Therefore, technical analyses can significantly enhance traditional connoisseurship. The most striking elements revealed by this technical analysis are Hals's very sparse and loose initial sketch, efficient use of the

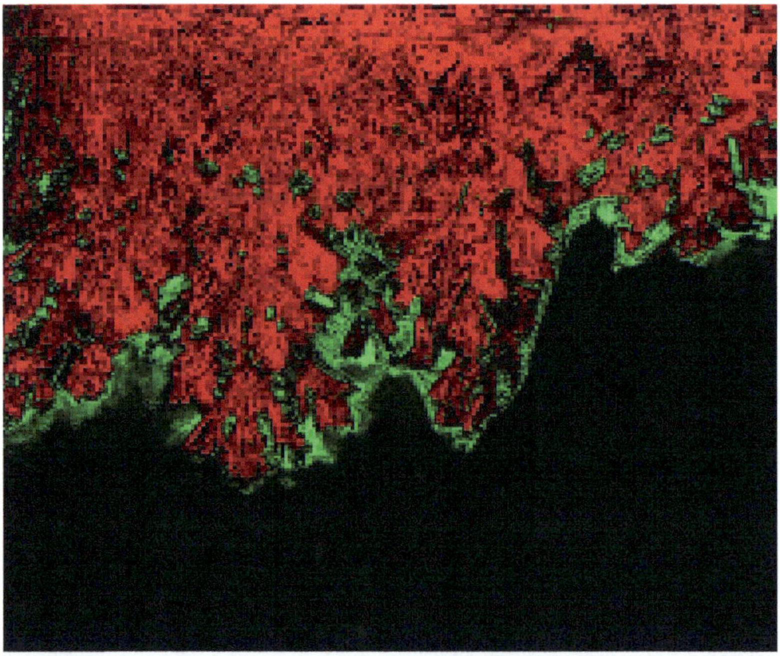

Fig. 3.16 Hyperspectral image of Fig. 3.1 (detail), juxtaposing the ground layer with the adjacent collar

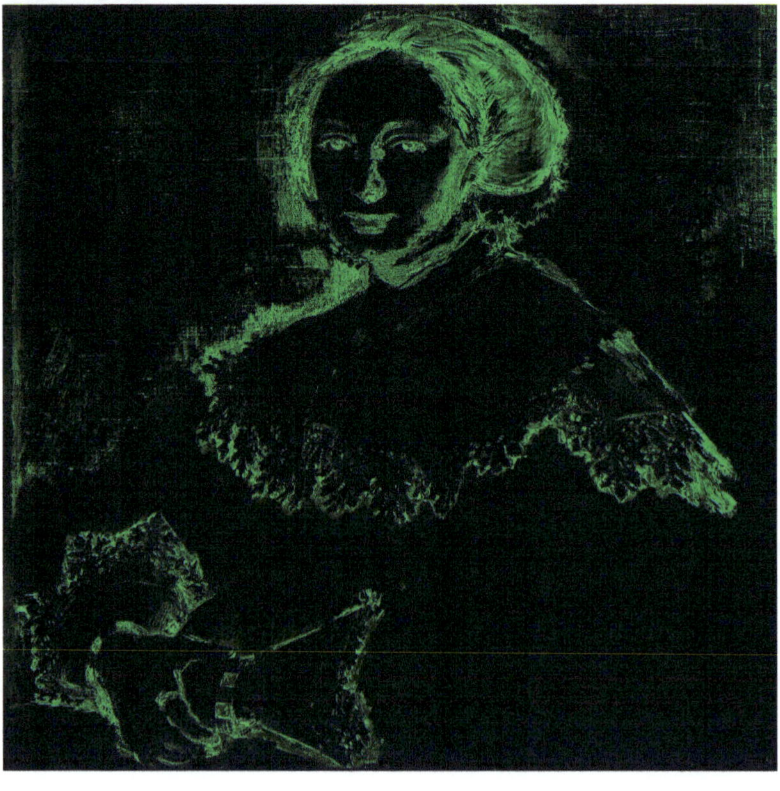

Fig. 3.17 Hyperspectral image of Fig. 3.1, showing the partially exposed ground layer

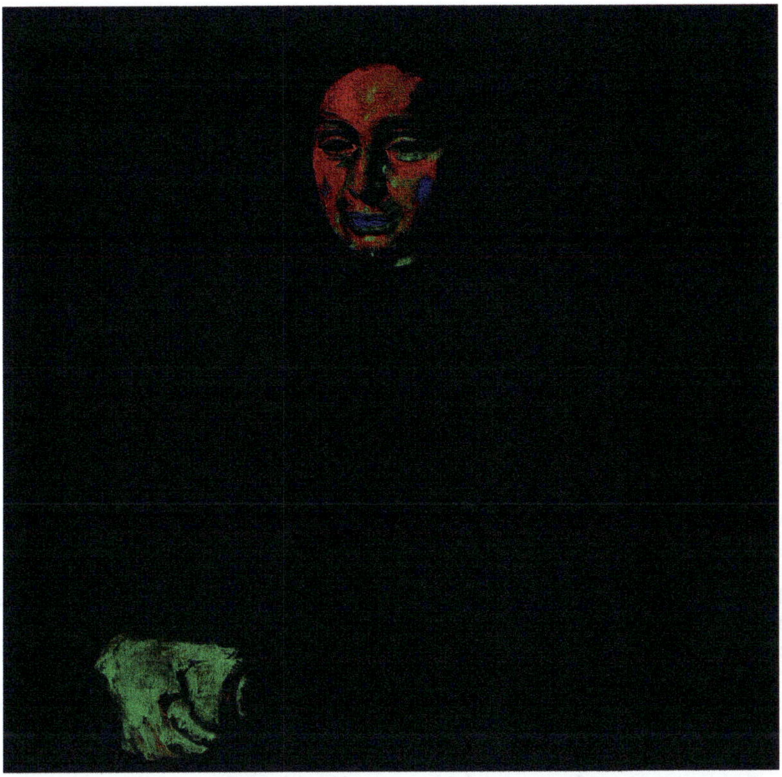

Fig. 3.18 Hyperspectral image of Fig. 3.1, showing the spectral characteristics of flesh tones in the face and hands

ground colour (using translucent glazes over the ground to create shadowy skin tones and brown hair) and speed of application of the paint. To avoid smudging and smearing, he worked up different wet areas as separate 'islands', resulting in a kind of halo effect of exposed ground around the edges of these different areas. A close examination of the pigments used in the painting revealed a very distinctive use of blue azurite to create various types of grey shadows, not only in the flesh tones, but also in the hair, the white lace collars and cuffs and the black dress. Interestingly, several areas that seem to be exactly the same colour were created with different paint mixtures, revealing different stages in the painting process and variation in Hals's working methods: the pink flesh tones in face and hands were created with different pigments (vermilion in the face and red ochre in the hands) and similarly the deepest black accents in the clothing were mostly—but not always—done with bone black. Furthermore, Hals used a pre-primed canvas and defined the format of this picture by incising a border into the ground layer. These characteristics, which had not been evident before, provide a great deal of insight into Hals's characteristic style and technique. In turn, these insights will enable connoisseurs to make more in-depth comparisons with other paintings that might be attributed to him. This will be of use when dealing with imitations or forgeries of his work, but the evidence presented here can also be used to confirm the attribution of paintings that have been doubted, as will be discussed below in the third case study ('A Recent Riddle: The Story of the *Two Fisherboys*' Chap. 5).

References

Alfeld, M., Janssens, K., Dik, J., De Nolf, W., Van Der Snickt, G.: Optimization of mobile scanning macro-XRF systems for the in situ investigation of historical paintings. J. Anal. At. Spectrom. **26**(5), 899–909 (2011). https://doi.org/10.1039/c0ja00257g

Alfeld, M., Pedroso, J.V., Van Eikema Hommes, M., Van Der Snickt, G., Tauber, G., Blaas, J., Haschke, M., Erler, K., Dik, J., Janssens, K.: A mobile instrument for in situ scanning macro-XRF investigation of historical paintings. J. Anal. At. Spectrom. **28**(5), 760–767 (2013). https://doi.org/10.1039/c3ja30341a

Buvelot, Q., Ducos, B.: A rediscovered portrait by Frans Hals. Burlingt. Mag. **156**(1331), 102–103 (2014)

Cascone, S.: Suspected $255 million Old Master forgery scandal continues to rock the art world (11 Oct 2016). Artnet News (2016). https://news.artnet.com/art-world/more-details-old-mastery-forgery-ring-695777. Accessed 10 Oct 2019

Denker, A., Laurenze-Landsberg, C., Kleinert, K., Schröder-Smeibidl, B.: Paintings reveal their secrets: neutron autoradiography allows the visualization of hidden layers. In: Kardjilov, N., Festa, G. (eds.) Neutron Methods for Archaeology and Cultural Heritage, pp. 41–52. Springer, Cham (2017)

Gratama, G.D., Van Rijckevorsel, J.L.A.A.M.: Frans Hals tentoonstelling ter gelegenheid van het 75-jarig bestaan van het Gemeentelijk Museum te Haarlem op 30 juni 1937, exh. cat. De Erven F. Bohn, Haarlem (1937)

Grimm, C.: Frans Hals: The Complete Work (1989). Translated from German by Riehle, J. H.N. Abrams, New York (1990)

Grimm, C.: Frans Hals and his Workshop. RKD, The Hague (forthcoming)

Groen, K., Hendriks, E.: Frans Hals: a technical examination. In: Slive, S. (ed.) Frans Hals, exh. cat, pp. 109–127. National Gallery of Art/Royal Academy of Arts/Frans Hals Museum, Washington/London/Haarlem (1989–1990)

Grosvenor, B.: Fakes, fakes, everywhere? (ctd.) (18 Oct 2016). Art History News (2016). https://www.arthistorynews.com/entries_by_date/2016/October/2. Accessed 10 Oct 2019

Groves, R., Anisimov, A., Papadakis, V., Tummers, A., Wallert, A., Hartwieg, B., and Kleinert, K.: Hyperspectral imaging to guide technical art analysis of Frans Hals paintings. In.: Proceedings of the Lacona (Lasers in the Conservation of Artworks) Conference VII, Paris (2018)

Hendriks, E., Levy-Van Halm, K., in collaboration with Van Asperen De Boer, J.R.J.: Report Concerning a Preliminary Technical Investigation of Paintings exhibited during the Frans Hals Exhibition, held from May 11 to July 22 1990 in the Frans Hals Museum Haarlem. (1991)

Hendriks, E., Van Eikema Hommes, M., Levy-Van Halm, K.: Indigo used in the Haarlem Civic Guard group portraits by Frans Hals. Stud. Conserv. **43**(1), 166–170 (1998). https://doi.org/10.1179/sic.1998.43.Supplement-1.166

Hofstede de Groot, C.: Echt of onecht? Oog of chemie? Beschouwingen naar aanleiding van het mansportret door Frans Hals uit het proces Fred. Muller & Co. contra H.A. de Haas. W.P. Van Stockum, The Hague (1925)

Jowell, F.: The rediscovery of Frans Hals. In: Slive, S. (ed.) Frans Hals, exh. cat., pp. 61–86. National Gallery of Art/Royal Academy of Arts/Frans Hals Museum, Washington/London/Haarlem (1989–1990)

Kleinert, K., Lorenz, B., Reimelt, M., Zielke, M. (catalogue entry in): Gestatten, Suermondt! Sammler, Kenner, Kunstmäzen, exh. cat., pp. 191–194. Belser/Suermondt-Ludwig-Museum, Stuttgart/Aachen (2018),

Pokorny, A.: Restoration Report on 'Portrait of Jasper Schade'. National Gallery Prague, Prague (2012)

Ševčík, A.K. (ed.): National Gallery in Prague: Dutch Paintings of the 17th and 18th Centuries. National Gallery of Prague, Prague (2012)

Slive, S.: Frans Hals, 3 vols. Phaidon, London (1970–1974)

Slive, S.: Frans Hals, revised edn. Phaidon, London (2014)

Sotheby's declares "Frans Hals" work a forgery (6 Oct 2016). BBC News (2016). https://www.bbc.com/news/entertainment-arts-37574411. Accessed 20 Oct 2019

Tummers, A.: Frans Hals: rightly admired by the greatest painters. In: Tummers, A. (ed.) Frans Hals: Eye to Eye with Rembrandt, Rubens and Titian, exh. cat, pp. 13–40. Frans Hals Museum/NAI010 Publishers, Haarlem/Rotterdam (2013a)

Tummers, A.: Frans Hals: Frans Hals: rightly admired by the greatest painters. In: Tummers, A. (ed.) Frans Hals: Eye to Eye with Rembrandt, Rubens and Titian, exh. cat., pp. 13–40. Frans Hals Museum/NAI010 Publishers, Haarlem/Rotterdam (2013b)

Tummers, A. (ed.): Frans Hals: Eye to Eye with Rembrandt, Rubens and Titian, exh. cat Frans Hals Museum/NAI010 Publishers, Haarlem/Rotterdam (2013c)

Van Dantzig, M.M. Frans Hals: Echt of Onecht, Naar aanleiding van de Frans Hals-tentoonstelling 1937. Paris, Amsterdam (1937)

Van Hoogstraten, S.: Inleyding tot de hooge schoole der schilderkonst, anders de zichtbaere werelt: verdeelt in negen leerwinkels, yder bestiert door eene der zanggodinnen. Francois van Hoogstraeten, Rotterdam (1678)

Von Stockhausen, T.: Gemäldegalerie Berlin: die Geschichte ihrer Erwerbungspolitik 1830–1904. Nicolai, Berlin (2000)

Chapter 4
Case Study 2: The New York *Malle Babbe* (*'Mad Barbara'*): Original, Studio Work or Forgery?

Anna Tummers, Arie Wallert, Robert G. Erdmann, Katja Kleinert, Babette Hartwieg, Dorothy Mahon, Silvia Centeno, Roger Groves, Andrei Anisimov, and Joris Dik

Abstract Quite a few of the most innovative and best-known paintings by Frans Hals exist in several variants. Attributing some of these versions or imitations is a notoriously difficult challenge. A case in point is the *Malle Babbe* painting at the Metropolitan Museum in New York, which has a rich attribution history: it

---

The research discussed in this chapter was first presented at the conference *Frans Hals: A Survey of Current Research*, Frans Hals Museum 8–9 Jan 2023 (8 Jan 2023), and first published as an article in the *Journal of Art Crime*, Dec 2023.
The research was funded by the Netherlands Organization for Scientific Research, as part of the projects *Frans Hals or not Frans Hals* (NWO Museumbeurs 2016–2018) and *21st Century Connoisseurship* (NICAS Seed Money Grant 2018–2022). The authors wish to thank the following institutions for enabling this research project: the Frans Hals Museum (Haarlem), the Rijksmuseum (Amsterdam), the Gemäldegalerie, Staatliche Museen zu Berlin, the Metropolitan Museum of Art (New York), the Leiden University Centre for the Arts in Society, the University of Amsterdam, Delft University of Technology and the Vrije Universiteit (Amsterdam).

A. Tummers (✉)
Ghent University, Ghent, Belgium
e-mail: Anna.Tummers@UGent.be

A. Wallert (✉)
University of Amsterdam, Amsterdam, The Netherlands

Rijksmuseum, Amsterdam, The Netherlands
e-mail: arie.wallert@gmail.com

R. G. Erdmann (✉)
University of Amsterdam, Amsterdam, The Netherlands
e-mail: r.g.erdmann@uva.nl

K. Kleinert (✉) · B. Hartwieg (✉)
Gemäldegalerie, Staatliche Museen, Berlin, Germany
e-mail: K.Kleinert@smb.spk-berlin.de; b.hartwieg@web.de

D. Mahon (✉) · S. Centeno (✉)
Metropolitan Museum of Art, New York, NY, USA
e-mail: Dorothy.Mahon@metmuseum.org; Silvia.Centeno@metmuseum.org

A. Tummers, R. G. Erdmann, *Frans Hals or not Frans Hals*, Cultural Heritage Science, https://doi.org/10.1007/978-3-031-59489-2_4

has been called an original Frans Hals, a work by one of his sons, an early imitation with a forged signature and even a modern forgery (Fig. 4.1). This case study sheds new light on its attribution by comparing its style, technique and materials in depth to the well-known original by Frans Hals in Berlin, to the *Malle Babbe* forgery created by Han van Meegeren at the Rijksmuseum in Amsterdam, and by relating it to relevant primary sources and seventeenth-century art theory. New technical research was done on all three paintings specifically for this study, including infrared reflectography (IRR), macro X-ray fluorescence scanning (MA-XRF), hyperspectral imaging or reflectance imaging spectroscopy (HI/RIS) and lead isotope analysis. Advanced digital tools were developed to aid the comparison (sn.pub/3xs1ac; https://images.erdmann.io/Draper/?manifest=/NICAS/Frans_Hals/image_manifest_MB.json).

The storage area of the Metropolitan Museum in New York contains a painting that is rarely on display in the gallery: *Malle Babbe* (or: Mad Barbara) (Fig. 4.1). It clearly bears a relation to Hals's famous painting of the same woman *Malle Babbe* in the Gemäldegalerie in Berlin, one of Hals's most beloved genre pictures, celebrated for both its innovative topic and its virtuoso painting technique (Fig. 4.2). It is the first portrait-like depiction of a woman who was mentally ill in the history of art. Yet the exact nature of the relationship between the two pictures, and thus the attribution of the New York painting, has not been definitively established. It is a familiar challenge for Hals scholars: a lot of variants exist of his most well-known paintings and it can be very challenging to attribute such a picture. Is it a different version created by the master himself, a workshop product or rather a later imitation? Especially his so-called genre paintings, i.e. his depictions of people that did not commission him to paint their portrait such as Malle Babbe, a street musician or laughing children, exist in many variants. These life-size depictions of unassuming and/or marginalized persons in society constitute one of his most important innovations in the art of painting (Tummers (ed.) 2013c, p. 9). No one had depicted these people so prominently and so lifelike before. Hals had a unique ability to capture fleeting moments on his canvases such as a spontaneous laugh, a subtle smile or furtive glance, making his subjects appear as if we encounter them in real life in a very specific moment in time.[1] His paintings of this type must have been well-known in his own time. Quite a few fellow painters cited precisely these innovative character types in their own work (Fig. 4.3), the pictures were often reproduced in

[1] On this effect, often referred to as the 'snapshot'-like quality of his paintings and the relation to seventeenth-century art theory, see also (Tummers 2013a, b, c).

R. Groves (✉) · A. Anisimov (✉) · J. Dik (✉)
Delft University of Technology, Delft, The Netherlands
e-mail: R.M.Groves@tudelft.nl; A.G.Anisimov@tudelft.nl; j.dik@tudelft.nl

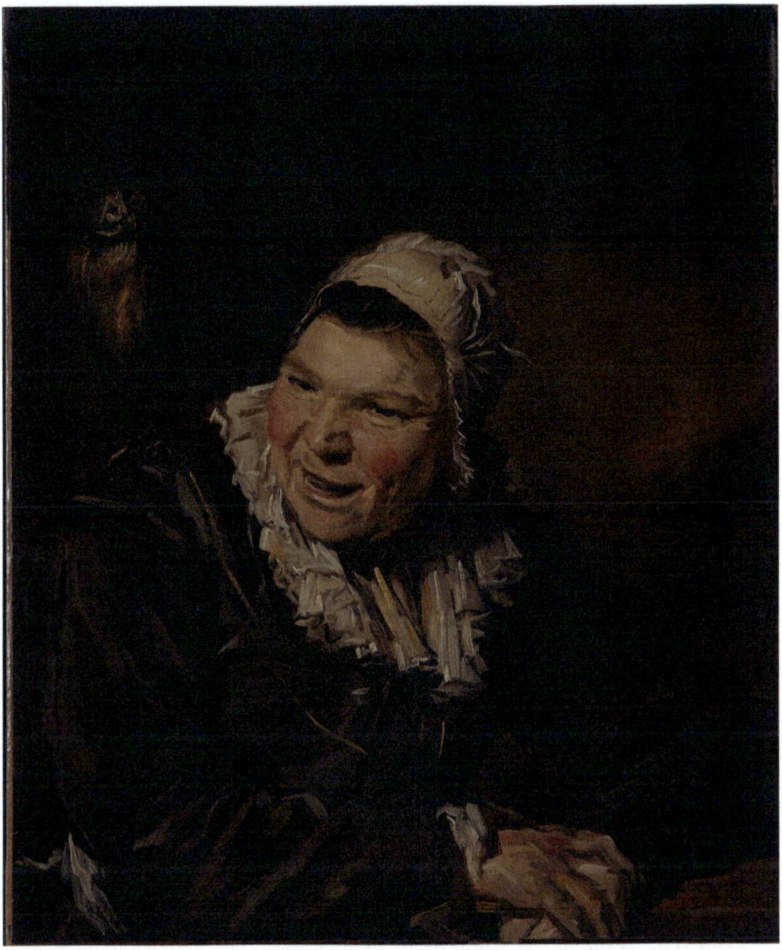

Fig. 4.1 Style of Frans Hals, *Malle Babbe*, c. 1625–1650, oil on canvas, 74.9 × 61 cm. (Metropolitan Museum of Art, New York, acc. no. 71.76)

print and their popularity might also explain the existence of many variants (Slive 2014; Tummers et al. 2017).

4.1 Attribution Debate

The attribution of the New York *Malle Babbe* has long been an issue of debate. It entered the collection of the Metropolitan Museum in 1871 as a valuable original painting by Frans Hals, one of the purchases of which the museum was proudest

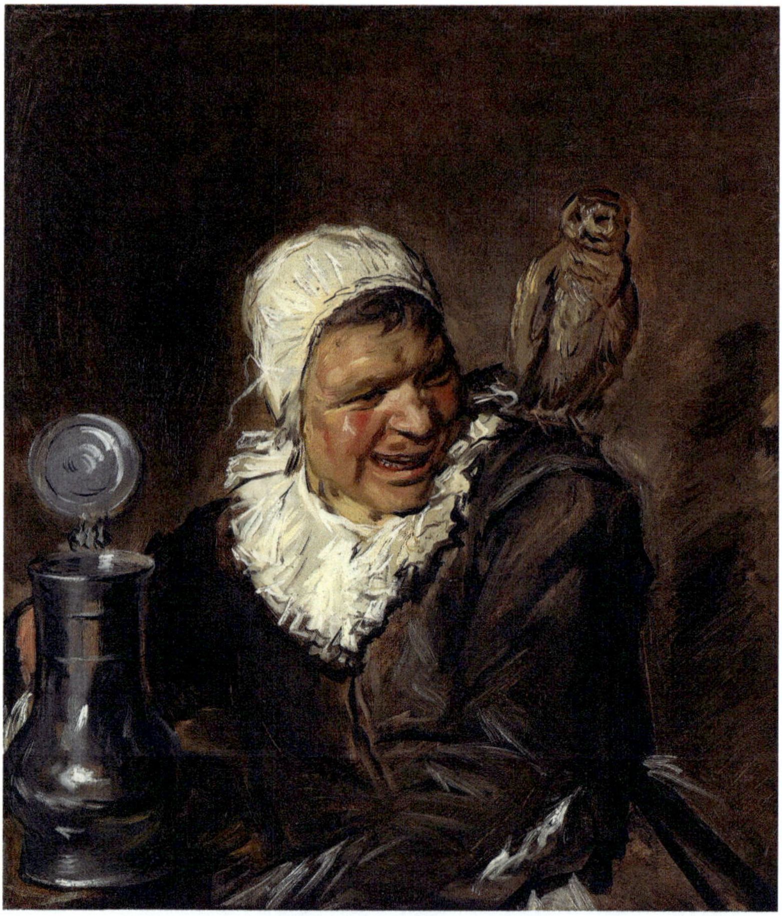

Fig. 4.2 Frans Hals, *Malle Babbe,* before 1646, oil on canvas, 78.5 × 66.2 cm. (Gemäldegalerie, Staatliche Museen zu Berlin, cat. no. 801 C)

that year (Metropolitan Museum of Art 1871, Plate 1, as by Hals). When the museum first opened to the public in 1872, it was even considered –in the absence of a Rembrandt painting– one of its most important works.[2] Already in 1883, however, the German art historian and museum director Wilhelm von Bode classified it as a free repetition by Frans Hals's son Frans Hals the Younger (1618–1669) after the Berlin picture (Von Bode 1883, p. 103). His Dutch colleagues Ernst Wilhelm Moes and Cornelis Hofstede de Groot nevertheless maintained that the painting was an original by the master himself in their monographs on Frans Hals of 1909 and 1910

[2] "this capital chef d'oeuvre of science, color, spirit, life, and boldness would do honor to any museum" (Metropolitan Museum of Art 1872, no. 144 as 'Hille Bobbe Von Haarlem', pp. 54–55; see also Baetjer 2004).

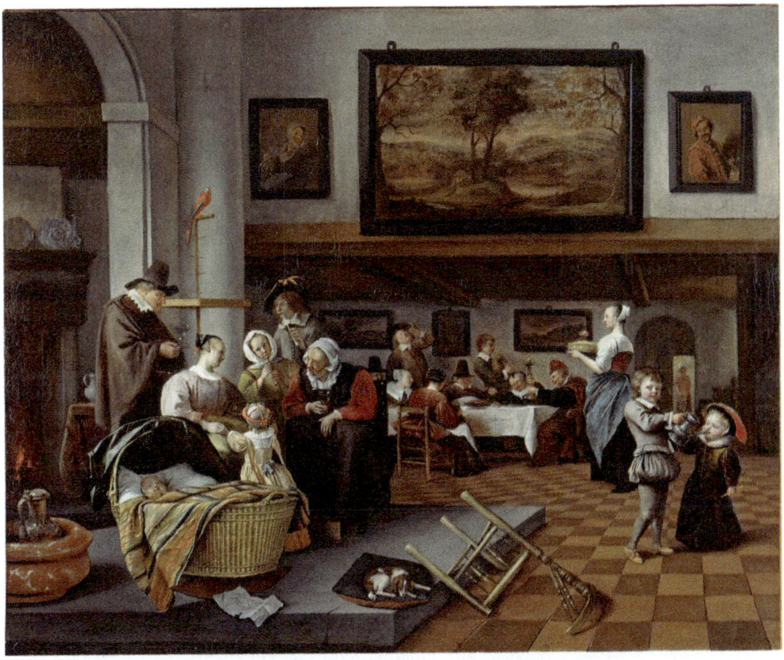

Fig. 4.3 Jan Steen, *As the Old Sing, so the Young Pipe*, c. 1663, oil on canvas, 81.1 × 100.5 cm. (Staatliche Museen zu Berlin, cat. no. 795 D)

(Hofstede de Groot 1907–1927, vol. 3 (1910), no. 109, p. 30; Moes 1909, III, no. 261, pp. 64–65). Moes speculated that Hals must have known Babbe personally and painted her after life, and Hofstede de Groot was the first art historian to state that this painting was the version etched by Louis Bernard Coclers (1741–1817) (Fig. 4.4). Also, the German-American art historian Wilhelm Valentiner attributed the painting to Frans Hals himself in his oeuvre catalogues of (1921, 1923, 1936). Nevertheless, Metropolitan curator Bryson Burroughs claimed in 1931 that "most authorities, including Bode and de Groot, consider the Museum's picture the work of someone close to Hals, probably Frans Hals the Younger" (Burroughs 1931, no. H161-I, p. 152). The first Hals overview exhibition in the Netherlands in 1937 intensified the debate. While the painting was on display as an authentic Hals in the exhibition in Haarlem, many of the attributions in the show were openly questioned and conservator Maurits van Dantzig even dismissed the New York *Malle Babbe* as a downright forgery in the book he wrote in response to the exhibition: *Frans Hals: Echt of Onecht* (*Frans Hals: Real or Fake*) (Van Dantzig 1937, no. 62, p. 103; see also Chap. 2).

The two main Hals scholars of the second half of the twentieth century, Harvard Professor Seymour Slive and his German peer Professor Claus Grimm continued the debate. Slive stated in 1962 that the picture is the closest of the known versions to the painting in Berlin, adding that "whether it is by the master himself or a brilliant

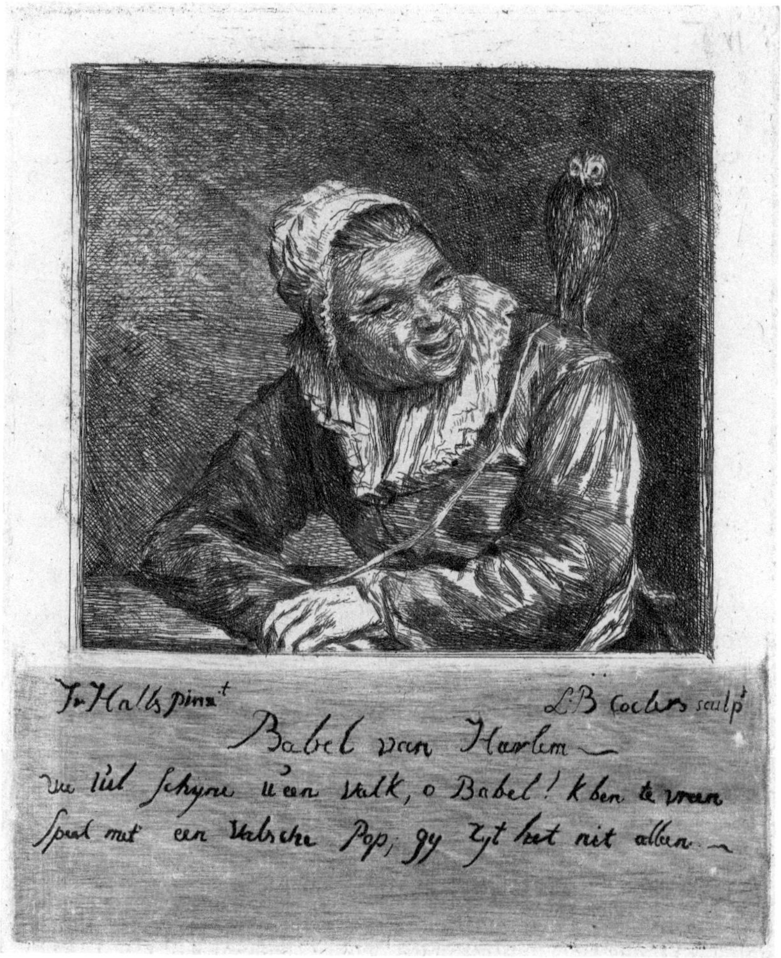

Fig. 4.4 Louis Bernard Coclers after Frans Hals, *Malle Babbe*, 1756–1817, etching, 158 mm × 127 mm. (Rijksmuseum, Amsterdam, RP-P-1883-A-710)

follower is debatable" and rejected the attribution to Frans Hals the Younger (Slive and Van Hees 1962, p. 49). In his extensive oeuvre catalogue of 1970–1974, he subsequently dismissed the attribution to Hals entirely and considered it "the invention of a gifted follower or a copy after a lost original" (Slive 1970–1974, vol. 3, no. D32, p. 140). In 1974, Claus Grimm listed the work among problematic pictures ascribed to Hals in the past, thus confirming its de-attribution (Grimm and Montagni 1974, no. 71, p. 95). Indeed, the picture does not feature in his oeuvre catalogue of 1989, though he recently added it to his latest oeuvre catalogue as a painting created in Hals's workshop (Grimm 1989).[3] Metropolitan curator Walter Liedtke described the picture as "by a contemporary follower" in 1990 and speculated that the signature was forged

[3] Correspondence with the author. Grimm's latest oeuvre catalogue will be published online through the Netherlands Institute for Art History (RKD) in The Hague (Grimm forthcoming).

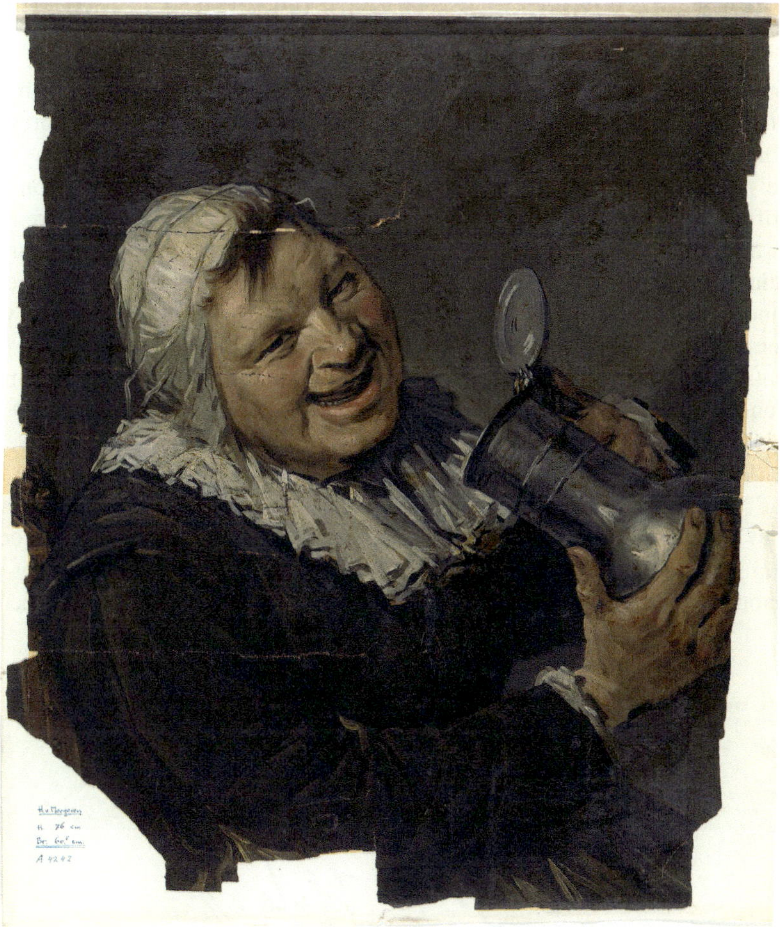

Fig. 4.5 Han van Meegeren, *Malle Babbe*, 1930–1940, oil on canvas 76 × 60 cm. (Rijksmuseum, Amsterdam, SK-A-4242)

(Liedtke 1990–1991, Fig. 19, p. 33). In the museum's collection catalogue of 2007, he listed it as "style of Frans Hals", dated it to the second half of the seventeenth century, and rejected the attribution to Hals's sons Frans, Harmen (1611–1669) and Jan (c. 1620–1654) as well as to other artists in Hals's circle, concluding that –like Slive– he was "unable to offer a plausible attribution" (Liedtke 2007, vol. 1, no. 69, p. 299).

In short, the Metropolitan *Malle Babbe* has been classified as an original by the master, a studio work, a work by a contemporary follower, a copy after a lost original and even as a (partial) forgery. In order to get more insight into its attribution, we have compared its subject, design, style, technique and use of materials closely to, on the one hand, Frans Hals's well-known original at the Gemäldegalerie in Berlin, and, on the other hand, contrasted it to a Malle Babbe forgery by Han van Meegeren at the Rijksmuseum (Fig. 4.5). Furthermore, we have related our observations to relevant primary sources and seventeenth-century art theory.

4.2 An Early Depiction of a Woman Who Was Mentally Ill

As briefly mentioned above, Frans Hals's *Malle Babbe* is exceptional in that it depicts a woman who was mentally ill, life-size, with recognizable features, grinning broadly and painted with very loose, *virtuoso* brushwork. Her name, 'Malle Babbe' ('Mad Barbara'), has come down to us through an old inscription on the back of the stretcher of the Berlin canvas (part of an old stretcher that was reinserted into a more modern one) (Fig. 4.6).[4] She is also mentioned in the only surviving financial document of the Haarlem Workhouse, which was both a house of correction and a charitable institution. In 1653, the Haarlem Burgomasters allowed 65 guilders for the care of 'Malle Babbe'. Hals's mentally impaired son Pieter (†1667) is mentioned in the same document; he was confined in the same institution since 1642 and supported with 35 guilders (Van Thiel-Stroman 1989–1990, Hals Doc. 94, p. 295; 2006, p. 179 and note 38, p. 182).[5] Babbe's full name was discovered in 2013. A document dated 17 February 1646 mentions that "Barbar alias Malle Barbar" was brought to the Workhouse by the captain in charge of the nightwatch in order to prevent "all further instances of disgrace and dishonour that could occur if nothing was done against it" –suggesting that she had behaved indecently or yelled dishonourable things in the streets of Haarlem– and that she was kept there and

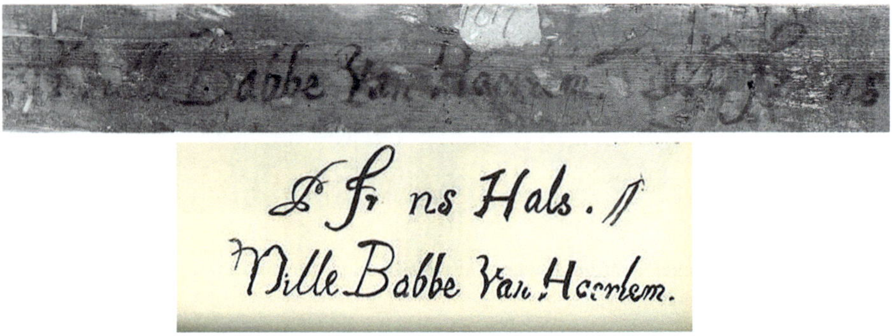

Fig. 4.6 Inscription on the back of the painting *Malle Babbe* by Frans Hals (see Fig. 4.2); Transcription of the inscription, taken from the Berlin collection catalogue of 1883

[4]The full inscription reads: "*Malle Babbe van Haerlem … Fr(a)ns Hals*". The present lining and stretcher date back to before the acquisition for Berlin in 1874; the inscription on the canvas repeats the inscription on the strechter (in the wrong form, as Hille Bobbe...) i.e. it is not conclusive proof that the Doublierung took place in the Netherlands. But since the painting very likely first left the Netherlands when Suermondt bought it and he did not restore it, the Doublierung may well have taken place in the Netherlands.

[5]At the request of her parents in 1642, Hals's oldest daughter Sara was also incarcerated for some time in the Workhouse, on account of fornication (see Van Thiel-Stroman 2006, note 39, p. 183).

would work at the regents' discretion.[6] As of 1646, the Regents of the local hospital, the St. Elisabeth Gasthuis, paid 65 guilders each year for Malle Babbe's care, who is called "Barbara Claes" from 1656 onwards. The last payment dates from 1663 and in the margin her death is mentioned: "*obiit*".

Barbara Claes, or Malle Babbe, is thus a well-documented early example of a person who was mentally ill. She was a very unusual and therefore innovative topic for a life size painting and her picture appears to have been appreciated for exactly that reason already in the seventeenth century—hence the early mentions of her nickname. The provenance history of the painting can in all likelihood be traced back to seventeenth-century Amsterdam. In 1689 the ironmonger Cornelis van Driessche sold 28 paintings to a certain Leendert van Dulcken, one of which is described as "Malle Babbe, by Frans Hals".[7] Possibly, an earlier Amsterdam inventory describes the painting as well. The 1648 inventory of the hat maker Lambers Hermansz. Blaeuw mentions a painting depicting "*een geck*" (a lunatic or fool) by Frans Hals that was appraised by the painter Johannes Collaert at 10 guilders.[8] As the description is very brief, it is hard to determine if it refers to Malle Babbe, a theatrical fool such as Pekelharing or Hals's painting of Verdonck (Fig. 4.7), known through a contemporary print as: '*Verdonk, that outspoken fellow, / whose jawbone attacks one and all, / he paid heed to no one great or small / and so he was consigned to the workhouse*'.[9] Although Verdonk was not explicitly called 'mad' like Barbara, he might have been mentally ill as well. In any case, the print further confirms the interest of Hals and his contemporaries in remarkable local characters, while underscoring Hals's light-hearted, humorous approach.[10] In Malle Babbe's case, the owl on her shoulder emphasizes her folly—owls were common attributes

[6]These documents were discovered by Floris Mulder and presented in a focus exhibition at the Dolhuis Museum, Haarlem in 2013: Alimony register of the St. Elisabeth Gasthuis, 1646–1680, archive EG, no. 37. BR, rood 221, fol. 661: "ale vordere swarichheden van schande en oneere die soude mogen onstaen in gevalle daertegens niet en werde gedaen".

[7]Getty Provenance Index Archival Inventory N-273 Municipal Archive Amsterdam, access 5075, inv. no. 3909 (film 4005), S. 165–167.

[8]"*No. 1 een schilderij, zijnde een geck, met een swart vergulde lijst, geschildert door Frans Halst f 10:—:—*" Montias Database, Municipal Archive Amsterdam N.A. 1914, 1648/05/15. See also Bredius 1927, p. 21; Van Thiel-Stroman 1989–1990, Hals Doc. 126, p. 402). An early eighteenth-century Haarlem inventory also mentions twice "een gek" by Frans Hals, valued at low prices, respectively 1:10:— lc f and —:10:— lc f. Getty Provenance Index N-4993. Alternatively, the inventory could also refer to a depiction of the local character called 'Boontje', see below note 11.

[9]*Verdonk, die stoute gast / wiens kaekebeen eclk een aen tast, / op niemand, groot, noch kleijn, hij past, / dies raeckte hij in 't werkhuis vast.* Translation by Lynne Richards and Philip Clark, taken from (Tummers (ed) 2013c, p. 118).

[10]See also (Tummers et al., 2017) on seventeenth-century humour. Research for the current Frans Hals exhibition in London and Amsterdam (2023–2024) unearthed an eighteenth-century source suggesting that Hals's well-known *Rommelpot-player* depicts a man called 'Boontje' who was 'at the time a well-known fool in Haarlem' (*in dien tijd een bekend gekje te Haarlem*) see (Lammertse 2023–2024, pp. 181–184).

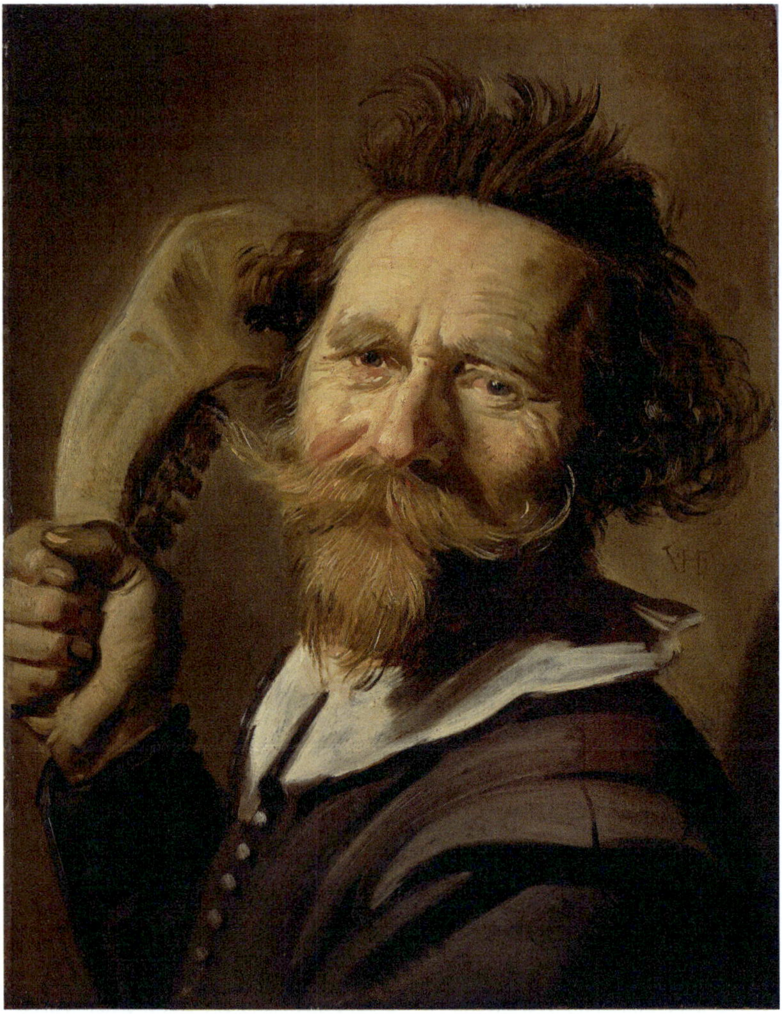

Fig. 4.7 Frans Hals, *Verdonck*, c. 1627, oil on panel, 46.7 × 35.6 cm. (National Gallery of Scotland, inv. no. NG1200)

of fools at the time (see Slive (ed.) 1989–1990, pp. 239–241).[11] Interestingly, Hals must have painted at least one other version of Malle Babbe, a painting showing her smoking, which features as a pendant to Hals's *Pekelharing* (c. 1628–1630) in Jan Steen's Berlin painting *As the Old Sing, so the Young Pipe* of circa 1663 (Fig. 4.3). Frans Hals presumably painted his *Pekelharing* for a popular artists tavern, the *Coninck van Vranckrijck* (the King of France) in the Smedestraat at Haarlem, where

[11] Ironically, owls could also refer to wisdom at the time as they were also the attribute of the Greco-Roman Goddess of Wisdom Athena/Minerva.

it is listed in an inventory dated 1631.[12] His *Malle Babbe smoking* might have been its pendant there. After all, smoking and drinking are an inn's *raison d'être*, and two foolish characters doing just that, would have made a witty pair (see Tummers and Gration 2013, pp. 84–88).

Although it is hard to determine exactly to which version the early inventories refer, the precise spelling of Malle Babbe's name in the 1689 inventory and on the back of the Berlin painting make the latter a likely candidate (see Fig. 4.6 and note 5 above).[13] Since the auction of 1867, the inscription on the reverse of the Berlin painting was, however, repeatedly misread as: "Hille Bobbe van Haerlem f. Frans Hals" (Fig. 4.6).[14] The fact that this inscription was considered important and trustworthy is shown by the fact that Léopold Flameng placed it under the depiction of the Malle Babbe in his etching after the original in 1869 as a kind of poignant quotation (Fig. 4.8).[15] Moreover, after its entry into the Gemäldegalerie, it was included in the Berlin directories as a signature replacement and believed to be "by the painter's own hand" (Meyer and Bode 1883, cat.no. 801C, p. 196). A former owner of the painting, Barthold Suermondt, was also sure that the inscription came from Frans Hals himself.[16] Although the inscription is nowadays no longer attributed to Hals himself, it does appear to be very old. However, as Suermondt noted in a letter to Bode, instead of "Hille Bobbe" it can also be read "Hille Babbe".[17] And, as the transcription of the inscription from 1883 already suggests, the sitter's name must originally have been read as "Malle Babbe".[18] This would also explain why the eighteenth-century mentions of the painting all read "Malle Babbe van Haarlem".[19]

---

[12] *Lijste van verscheijden schilderijen toebehoorende Heijnderick Willemsz. den Abt, die hij meent te verkoopen*, 16 November 1631, Municipal Archive Haarlem, GA 196e. See also (Slive (ed.) 1989–1990, Hals Doc. 58).

[13] The designation was originally on the stenter frame and was preserved in later times by sawing out the corresponding piece of wood and inserting it into the new stenter frame.

[14] Auction catalogue Hoorn 8 Sept 1867 (Lugt no. 29948), no. 69.

[15] Flameng's etching still bears the title "Hille Bobbe" because of its inscription. Léopold Flameng after Frans Hals, *Hille Bobbe*, 1869, etching, 171 mm × 142 mm. Rijksprentenkabinet, Amsterdam.

[16] *Die obige Schrift (mit einer Feder auf dem Holz des Rahmens) scheint von der Hand des Meisters selbst zu stammen, denn die Signatur stimmt vollkommen mit derjenigen überein, die ich auf Gemälden gefunden habe, die er mit dem Pinsel signiert hat, jedenfalls ist sie zeitgleich mit dem Gemälde entstanden."* Letter from B. Suermondt to A. van der Willigen, 22 Jan 1868. (RKD archive. See also von Lützow 1870, p. 78).

[17] Letter from Barthold Suermondt to Wilhelm Bode dated 2/4 Nov 1885, Zentralarchiv der Staatlichen Museen zu Berlin, SMB-ZA, IV/NL Bode 5392. Cf. also the list of restorer Schmidt, who examined and documented Suermondst paintings with regard to their condition in Brussels 1874: No. 79. Frans Hals, Hille Babbe, very well preserved, 17.4.1874, SMB-ZA, I/GG 92.

[18] With thanks to Harmen Snel of the Municipal Archive Amsterdam, who looked at a photograph of the inscription and its transcription again and also came to this conclusion. Email from 21 Dec 2022.

[19] Auction catalogue Cornelis Ploos van den Amstel / Jan Iver, Amsterdam 1 Oct 1778, no. 58 (Lugt no. 2894); Auction catalogue Nijmegen, 10 June 1812 (Lugt no. 8200 / Getty Provenance Index Sale Catalogs N-237), no. 88; Auction catalogue C.- S. Roos Amsterdam 12 May 1834 (Lugt no. 13672), no. 92.

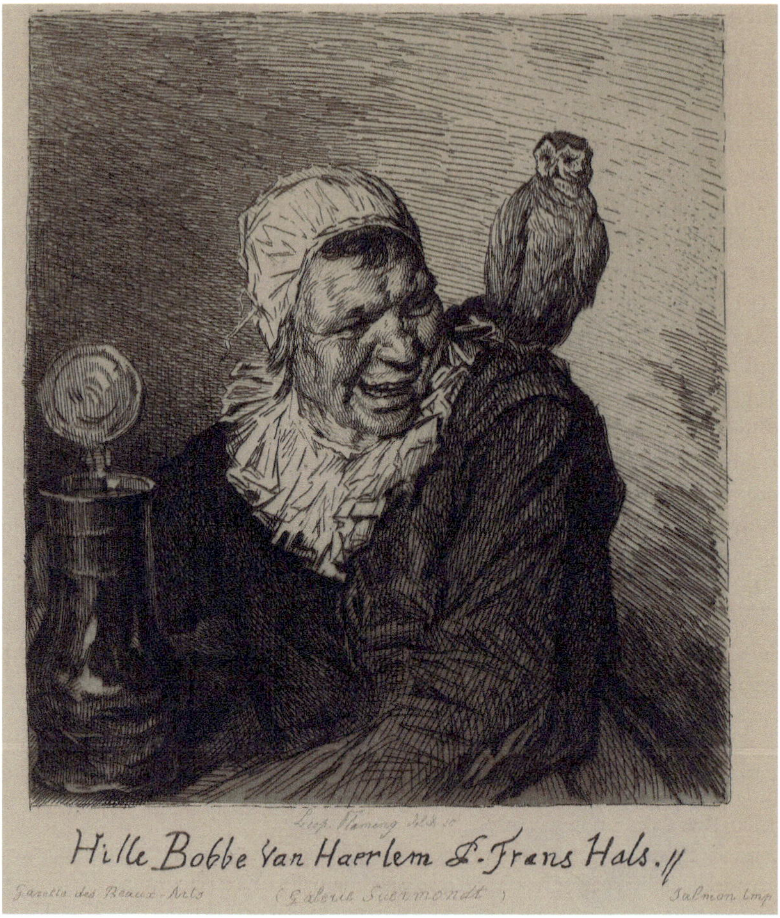

Fig. 4.8 Léopold Flameng after Frans Hals, *Hille Bobbe*, 1869, etching, 171 mm × 142 mm. (Rijksprentenkabinet, Amsterdam, RP-P-1910-3256)

Also in the case of the New York version, Babbe's name appears to have stayed attached to the picture. The earliest evidence of its existence is the etching by the Flemish printmaker Louis Bernard Coclers from the second half of the eighteenth century (Fig. 4.4). Its inscription slightly misspells her name as "Babel", but still remembers her as a foolish Haarlem character painted by Fr(ans) Hals: "*Babel of Haerlem / To you, your owl is a falcon. O Babel! I am glad of it. / Play with an illusion. You are not alone.*"[20] Coclers had not only heard of Babbe's name, but also recognized the pun in the picture: The proverb 'Everyone thinks their owls are falcons' was still a popular one at

[20] *Fr(ans) Hals Pinx(i)t L(ouis) B(ernard) Coclers Sculps(i)t. Babel van Haerlem / Uw uil schijne u een valck, o Babel! Ik ben tevreen / Speel met uw falschen pop, Gij zijt het niet alleen.* Translation from (Slive 2014). See below on the reason we believe this etching depicts the New York painting rather than a lost original.

the time.[21] It adds an interesting layer of meaning to the painting: it does not only portray a specific, foolish individual; it also reminds us of human folly more in general, of our own tendency to exaggerate the importance and qualities of our children and possessions. The mention of both Hals's and Babbe's name is all the more noteworthy since it is generally assumed that Hals was largely forgotten in the eighteenth century, only to be rediscovered in the late nineteenth century (Jowell 1989–1990, p. 84).

As to the dating of Hals's Malle Babbes, he must have created the Berlin picture and the lost version of Malle Babbe smoking before her confinement in the workhouse in 1646. Oil paintings were created in painters' studios in the seventeenth century; mixing paints was quite a complex process (paint tubes were not invented until the second half of the nineteenth century), and as oil paint usually dries rather slowly, pictures had to be protected from dust.[22] It is therefore unlikely that Hals would have painted Barbara Claesz on location at the institution. Stylistically, it is quite complex to date the Berlin painting. Hals varied his painting style depending on the type of picture.[23] In his genre paintings, his brushwork and paint application are more experimental than in his life size commissioned portraits: bolder and looser. Malle Babbe is commonly dated the latest of all his genre paintings on account of its extremely loose, *virtuoso* brushwork—a manner called 'rough'(*ruw*) in the seventeenth century, which was known to be very difficult to master, requiring both considerable talent and experience (Van Mander 1604, Grondt XII, fol. 48v; see also Tummers 2011, pp. 219–221). Another complicating factor is that virtually all Hals's genre pictures appear to date from the 1620s. As noted above, Hals's *Pekelharing* was mentioned in a 1631 inventory listing the paintings in possession of the owner of the artist's tavern *De Koninck van Vrankrijck*, and no genre painting can securely be dated after that year (see above, note 13).[24] In short, there is little comparative material and *Malle Babbe* has variously been dated to circa 1650, circa 1635–1640, circa 1633–1635, circa 1640 and between 1639 and 1646 (Atkins 2012, p. 140 (1630–1633); Baard 1981, p. 118 (c. 1635 or later); Erftemeijer 2014, p. 20 and p. 145 (c. 1633–1635); Grimm 1989, no. 111, p. 280 (c. 1640); Grimm forthcoming, (between 1639 and 1646); Hofstede de Groot 1907–1927, vol. 3 (1910), no. 108, p. 30 (c. 1650); Lützow 1870, p. 80 (1640s); Meyer and Bode 1875, no. 21, p. 29 (c. 1650); Slive 1970–1974, vol. 3, no. D32, p. 140 (1633–1635); Stukenbrock 1993, p. 155 (1645–1655); Thoré 1869, p. 164 (1630–1640); Trivas 1941, no. 33, pp. 35–36 (c. 1628); Unger and Vosmaer 1873, pp. 13–14 (c. 1633); Valentiner 1923, pp. XIII, XXIV–XXV, 142, 316 rmk. 141, 142, 144, 145 (1635–1640); Société néerlandaise de bienfaisance Exposition de tableaux et dessins d'anciens maîtres 1873, no. 17, p. 15 (c. 1650); Von Bode 1871–1872[25] (after 1650)).

---

[21] *Elk meent zijn uil een valk te zijn* (see Slive 2014, p. 189).

[22] Lead-containing paints, however, dried faster than others, and artists used Pb-containing compounds and other driers as well as heat-polymerized oils to speed up the drying process, see (Tumosa and Mecklenburg 2013; White et al. 1998).

[23] Many artists did so in the seventeenth century (see Tummers 2011, ch. 4).

[24] Pekelharing was also reproduced in print in the early 1630s including Hals' signature and with the inscription: 'Frans Hals pinxit'. Jonas Suyderhoef after Frans Hals, *Pekelharing*, c. 1630, engraving, 244 x 221 cm. Teylers Museum, Haarlem. See also (Tummers et al. 2017, pp. 14–15).

[25] Bode in his travel diary 1871–72, entry: Aachen, Galerie Suermondt 22–4- 1872, SMB-ZA, IV/ NL Bode 0042.

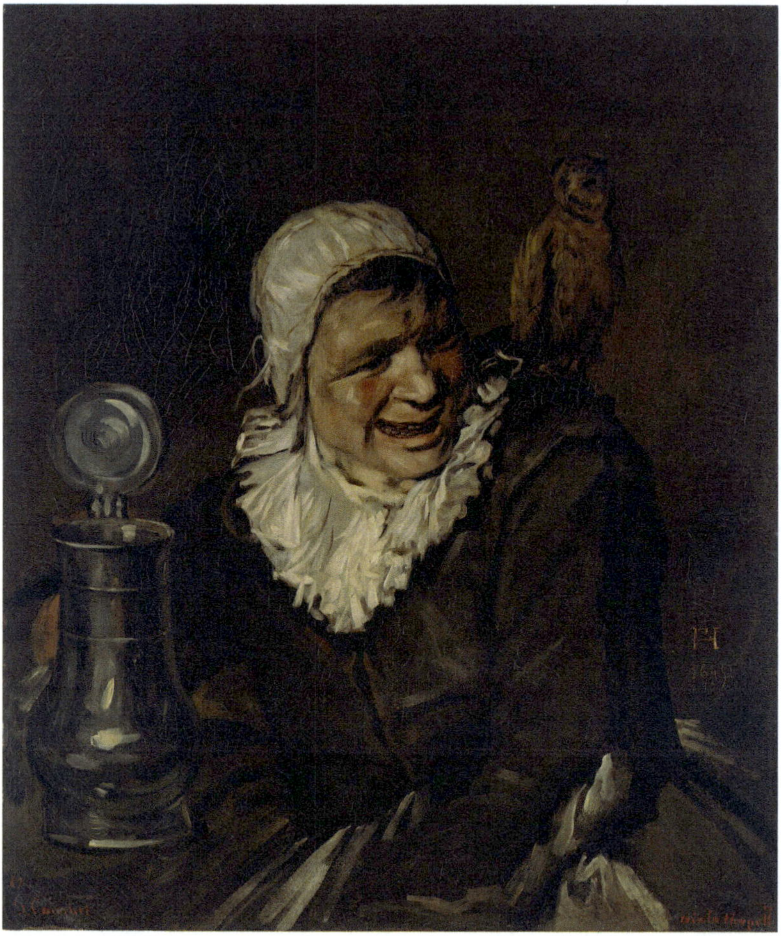

Fig. 4.9 Gustave Courbet, *Malle Babbe, copy after Frans Hals*, 1869, oil on canvas, 85 × 71 cm. (Hamburger Kunsthalle, inv. no. 2262)

Interestingly, a copy of the painting created by the artist Gustave Courbet (1819–1877) in 1869 carries Hals's characteristic monogram and the date 1645—neither of which is visible today (Fig. 4.9). Although the date has often been dismissed as a rather peculiar addition by Courbet (see Jowell 1989–1990, p. 71; Slive (ed.) 1989–1990, cat. no. 37, pp. 236–241, esp. pp. 236 and 238),[26] it is plausible that Malle Babbe gained particular notoriety in Haarlem shortly before her confinement in 1646. It raises the question if Courbet could have seen remnants of a date

[26]On Courbet's signature see, among others, (Krämer in Mensger et al. 2012, cat. no. 65, pp. 241–242; Stukenbrock 1993, p. 154 and note 469, p. 155). Barthold Suermondt, who had acquired the "Malle Babbe" in 1867, also assumed a date of around 1645 for the *Malle Babbe*, see Suermondt; letter of 12 June 1869 to Bode (SMB-ZA, IV/NL Bode 5392): c. 1640–1650; Suermondt letter of 12 March 1871 to Bode (SMB-ZA, IV/NL Bode 5392): 1645.

and possibly a monogram on the picture. If Hals would have added these on top of a first varnish layer, it would have been particularly vulnerable to early cleaning.[27]

4.3 *Alla Prima*: Hals's Virtuoso Painting Technique

A close look at the painting technique of the *Malle Babbe* pictures in Berlin and New York in regular light, through the microscope and with infrared reflectography (IRR) revealed a number of striking similarities as well as some differences. These were further explored with chemical imaging methods (MA-XRF and HI/RIS) and compared to the Amsterdam forgery. Interestingly, both the Berlin and the New York painting are done entirely wet-in-wet in the so-called *alla prima* technique, or -in Dutch- '*ten eerste schier sonder teyckenen schilderen*' ('painting directly without preliminary design).[28] The painter and art theorist Karel van Mander (1548–1606), who is mentioned in three different seventeenth-century sources as Hals's teacher, explained that this technique was only suitable for experienced masters or journeymen ('*werkgesellen*') with a steady hand and an abundance of ideas (Van Mander 1604, fol. 46v-47r).[29] Although early scholars already speculated that Hals mastered and employed this technique, the most extensive research report to date on Frans Hals's painting technique could not subscribe to that conclusion (Hendriks et al. 1991, p. 37).[30] Based on extensive research in the context of the 1989/1990 Frans Hals overview exhibition in Haarlem, the team of researchers concluded that Hals painted in separate stages in all the paintings that were studied in depth and therefore could not be called an *alla-prima* painter.[31] Admittedly, they did not study *Malle Babbe* in depth and in *Pekelharing* they did not distinguish separate stages. In

[27] In cross sections taken from Hals's paintings often layers of varnish have been found in between paint layers, e.g. during the recent restoration of the 1641 regent group portrait at the Frans Hals Museum; Hals must have commonly used varnish while painting, presumably to saturate the colour before adding to the work. On the use of varnishes in the seventeenth century sometimes in combination with pigments, and the possibility that such layers were subsequently cleaned off, see (Taylor 2007, pp. 207–211). However, Suermondt looked very closely at his paintings and had many restored; therefore, one would expect that he would have mentioned a remnant of a date if there was one.

[28] "(...) eenighe wel gheoeffent expeerdich, / En vast in handelinghe cloeck beraden, / (Niet licht'lijck verdolend' in cromme paden, / maer om hun Const zijn Meesters name weerdich, / Gaen toe, en uyt der handt teyckenen veerdich / Op hun penneelen, t'ghene nae behooren / In hun Ide' is gheschildert te vooren. / En vallender aen stracx, sonder veel quellen, / Met pinceel en verw', en sinnen vrymoedich' (Van Mander 1604, Grondt XII, fol. 46v, 04–05). See also (Miedema 2013, p. 25).

[29] Early sources on the relation between Hals and Van Mander cited in (Tummers (ed.) 2013c, p. 16 and note 13, 14 and p. 144), note 15.

[30] A copy of the report by Hendriks, Levy-Van Halm and Van Asperen de Boer is available at the Netherlands Institute for Art History, The Hague (RKD).

[31] "This study corrects the misnomer that Frans Hals was an 'all-prima' painter" (Hendriks et al. 1991, p. 51).

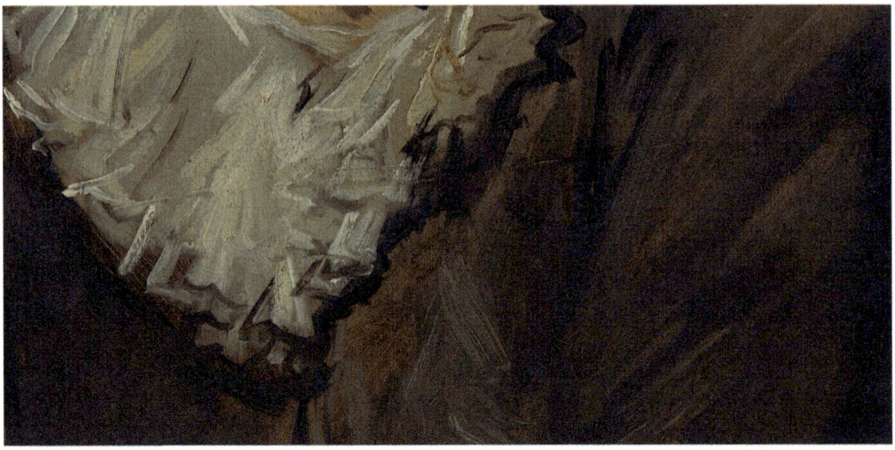

Fig. 4.10 Detail of Fig. 4.2 showing the scratch

fact, hairs scratched through wet flesh paint in *Pekelharing* indicated the absence of underpaint (Hendriks et al. 1991, p. 37). More importantly, the challenge of *alla prima* painting as Van Mander defined it, consisted of being able to design directly on the canvas what one had conceived in mind without needing preliminary sketches or designs. It was about creating the painting directly on the canvas and adding corrections where needed in mid-flight instead of beforehand: "those who have an abundance of ideas, act like the bold, and correct a mistake here or there".[32] A few last corrections or touches thus did not necessarily strip a work of being an *alla prima* painting. The virtuosity was in the direct design and the spontaneous paint application with a steady hand.

In the case of the Berlin *Malle Babbe*, an accidental scratch in the still fresh paint of the collar reveals that the picture was designed directly on the preprimed canvas. At the height of the collar and the black garment two parallel lines were scratched into the paint layer while the picture was still entirely wet, exposing the light brownish ground (Fig. 4.10). Observation through the stereomicroscope showed that the middle part of the white collar was applied over the damage during the painting process and that a small correction at the right beside the black contour line, was added considerably later. A scan of the painting realized with hyperspectral imaging, also referred to with the broader term reflectance imaging spectroscopy (HI/RIS), provides even more clarity.[33] The image in false colours, which highlights the areas in the painting that show chemical similarity to the ground layer, based on the careful observation of reference points through the stereomicroscope, is especially significant (Fig. 4.11). The red colour shows exactly where the light, sand-coloured ground

[32] '(...) die overvloedich / In't inventeren zijn, doen als de stoute, En verbeteren hier en daer een foute' (Van Mander 1604, fol. 46v).

[33] The hyperspectral imaging of the painting was carried out with the instruments and methods described in (Groves et al., 2018). For an explanation of this technique, see above, the first case study or below: Glossary of Techniques.

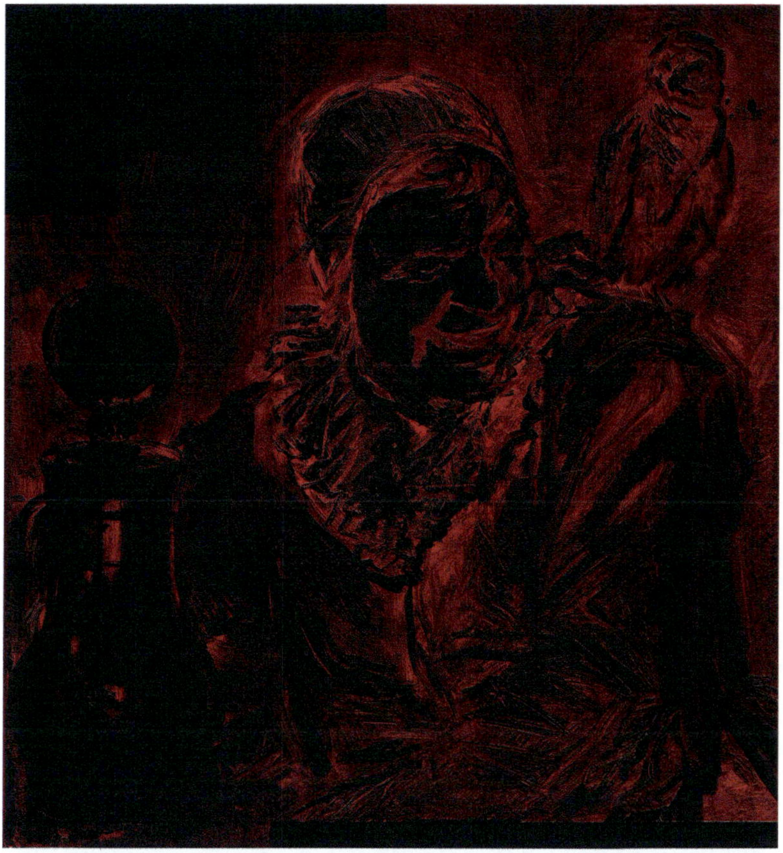

Fig. 4.11 Hyperspectral (HI or RIS) image of Fig. 4.2 showing the ground

is exposed, revealing Hals's fast and efficient painting technique. In several locations he effectively used the sand-coloured ground and let it show through: in the jug, the owl and the greyish black clothing. In the area of the jug and the dark dress he toned down the light ground with a very thin greyish-brown wash which fills the depths of the canvas structure. In Malle Babbe's white cap, the painter allows the colour of the ground to shine through in the darker shadowy area near the contour of her head, while he covered the ground only lightly with a translucent layer in several parts of the face to create a shadow tone, notably to her left of her broad grin, and above her left eye. As we have seen above in the first case study, Hals used a very similar technique for creating facial shadows and the brown hair colour in his *Portrait of an Unknown Woman* of c. 1632–1635 by simply adding a translucent layer directly over the ground (Tummers et al. 2019a, pp. 938–940; and Chap. 3). The image also shows what we called the 'halo effect' in this previous study on Hals's characteristic painting techniques (see Tummers et al. 2019a, p. 938). The ground is left exposed in small areas around the contours of the different shapes: the head, the collar, the owl, the clothing and jug. By keeping these areas apart Hals

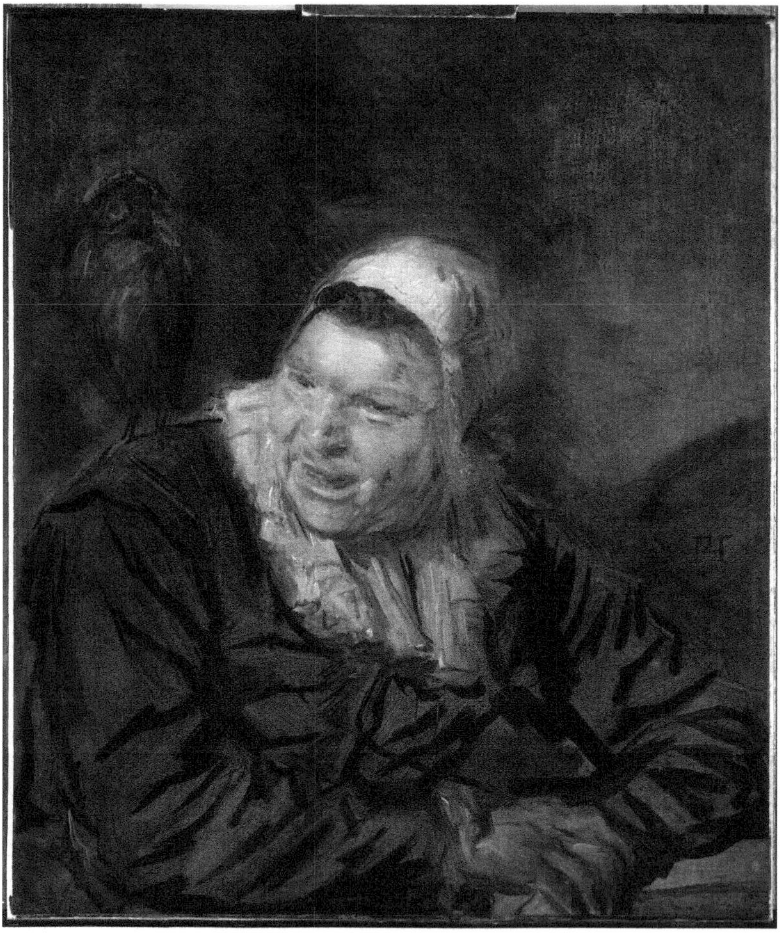

Fig. 4.12 Infrared reflectogram (IRR) of the New York *Malle Babbe* (Fig. 4.1)

prevented smudging and smearing the different wet paints. For the same reason, he laid in the background broadly around Babbe and the owl, and painted more carefully and thinly closer to their contours. Moreover, his firm brushwork with a rather stiff brush left scratches exposing the light ground in many different areas in her clothing, the owl and the background, confirming the *alla prima* execution throughout.

The New York *Malle Babbe* was also executed *alla prima*, directly on the ground layer. Both the high resolution photograph and the IRR are revealing in this respect. The infrared reflectogram (IRR) provides perhaps the clearest evidence (Fig. 4.12). The string attaching the owl's leg to Malle Babbe's hand, was added while the rest of the paint was still wet. The IRR shows the black pigments present in the painting and everywhere the string passed a deep black accent, the pigments were dragged along in the direction in which the stroke was applied: from top left to bottom right.

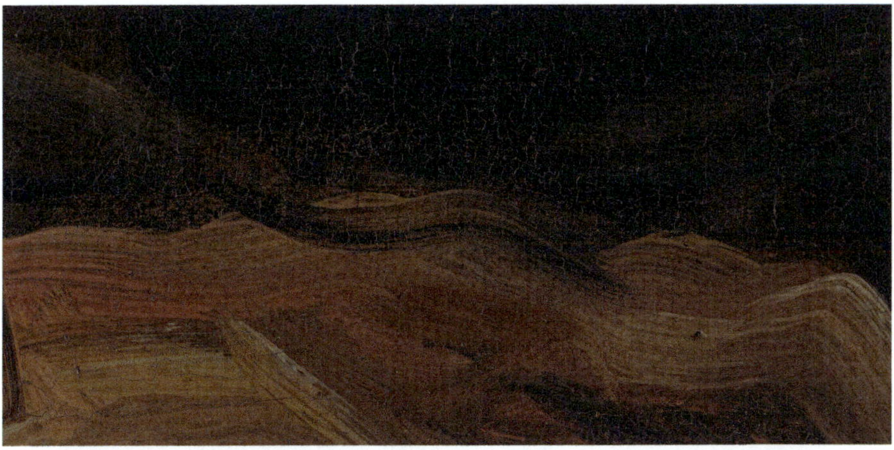

Fig. 4.13 Detail of Fig. 4.1 showing the hand

The high resolution photograph is also very clear at the height of the hand: the stroke clearly mixes the black pigments and flesh tones (Fig. 4.13).

4.4 Further Similarities in Style, Technique, and Use of Materials

Close observation of the IRRs of all three Malle Babbes reveals further similarities in painting technique between the New York and Berlin version, while exposing a strong difference with Han van Meegeren's painting technique (Figs. 4.12, 4.14, and 4.15). At the same time, the comparison also highlights differences in execution between the Berlin *Malle Babbe* and the New York version that relate to noticeable differences in execution visible in regular light. As the IRRs show the black pigments, one can easily compare their application. Both the Berlin and New York version show a rather sparse use of blacks. For example, certain dark accents in the face contain black pigments, but certainly not all darker colours.[34] The IRR of Van Meegeren's *Malle Babbe* on the other hand rather looks like a black and white photograph of the painting: black pigments were used everywhere to create darker colours. Of course, Van Meegeren had never seen an IRR of a seventeenth-century painting and did not realize how sparse and particular seventeenth-century painters were in their use of blacks.

While Van Meegeren's use of blacks is thus uniform and dense, the New York variant is very close to the Berlin version in its technique, yet more hesitant in its execution. Its background is also painted around the main figure and the owl, leaving small areas of ground around the contours exposed. However, it is more opaque and shows two mishaps: directly to the left of Malle Babbe's face the background

[34] As will be discussed more extensively in the next case study, Hals consistently used bone black for certain shadows in fleshtones.

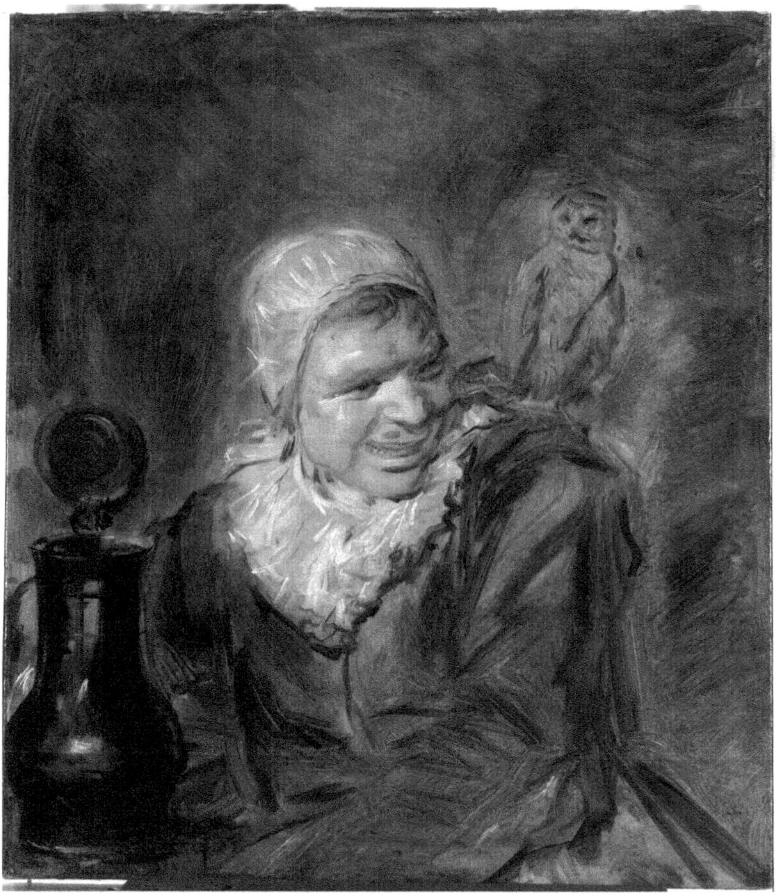

Fig. 4.14 Infrared reflectogram (IRR) of the Berlin *Malle Babbe* (Fig. 4.2)

colour was partially scraped away and at the top right a peculiar dripping pattern is visible. Presumably, the paint contained a surplus of binding medium here. The difference in execution is also very clearly visible in the deepest, darkest accents in Malle Babbe's clothing. While the Berlin version contains just a few efficiently placed accents, the New York variant shows an abundance of accents in the sleeves and body, which do not evoke the three-dimensional shape of the garment as effectively as the Berlin version does. [Interestingly, these black accents contain copper in the New York picture, just like the darkest areas in the black dress of his *Portrait of an Unknown Woman* (Tummers et al. 2019a, pp. 938–939) and above, Chap. 3).]

A similar difference can be seen in visible light in the depiction of the collar. The pleats of the collar in the Berlin version are indicated with rapid, very loose accents in white and black and a greyish middle tone, which are carefully balanced to convincingly suggest the three-dimensional shape of the collar draped around Malle Babbe's neck. The New York version also contains very loose accents in white and black that look very similar from up close but fall short in their overall effect, notably in the

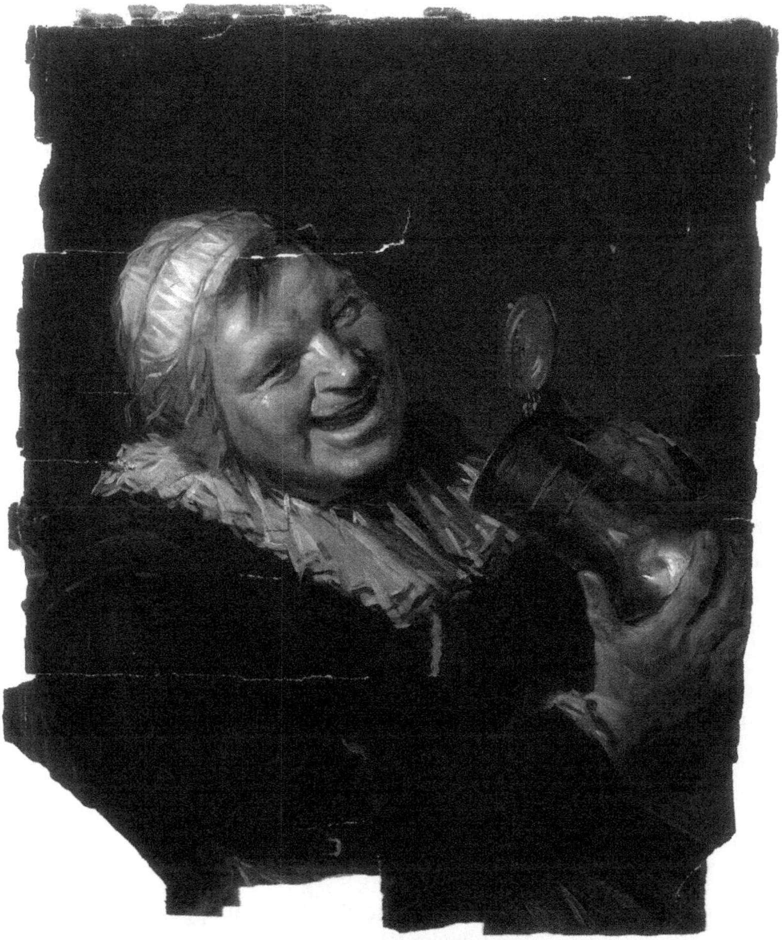

Fig. 4.15 Infrared reflectogram (IRR) of the *Malle Babbe* forgery by Han van Meegeren (Fig. 4.5)

suggestion of three-dimensionality. Moreover, the owl –though similar in colours, pose and use of pigments (including ochre, umber and bone black)– is depicted with shorter, stiffer brushwork. It is precisely for this reason that Slive's theory that Cocler's print could be based on an original of higher quality is not convincing: the depiction of the collar and owl in the print show exactly the same shortcomings (Fig. 4.4). There is thus no reason to assume that the print is based on any other work.

Upon close inspection, another even more striking similarity in technique and use of materials can be seen in the loose, white accents in the Berlin and New York pictures. Hals is known for his so-called ribbon touches: firmly applied loose accents that have raised edges on both sides (See above, Chap. 2; Tummers et al. 2019b). In fact, this feature is so distinctive that the last extensive research report on Hals's technique speculated that they could be unique for Hals (Hendriks et al. 1991, p. 50). Indeed, this type of brushstroke has subsequently been used a lot in attribution issues. However, interestingly, the white accents in the New York picture show precisely this type of raised edges, indicating a similar viscosity of the paint and pressure during the application, while at the same time betraying a certain lack of mastery. While the white accents in Malle Babbe's cap in the Berlin version convincingly suggest a tied ribbon and a few light accents on the fabric, the accents on the cap in the New York version are only superficially similar: loosely applied yet not very suggestive of a concrete knot or shape. Also in the Van Meegeren forgery, some accents mimic Hals's characteristic ribbon touch. However, these lack the raised edges; from up close they rather look like icing on a cake, as if they melted somewhat during the ageing process (Van Meegeren famously baked his forgeries in an oven in order to speed up the drying) (Lopez 2008).[35] In their application, Van Meegeren's brushstrokes are closest to the ones in the New York version. For example, the brushwork in the collar is attractively loose and rhythmical, yet not very effective in creating a convincing illusion of depth.

In both the New York painting and Van Meegeren's *Malle Babbe*, the facial expression is the most successful part of the invention: vivid, convincingly three-dimensional and full of loose accents. It is also closest to the Berlin version, on which both variants appear to be based (though the accents in the Berlin version are more colourful and boldly placed). In the New York variant the lighting is more even as Malle Babbe's face is turned towards the light, while the forgery stays closer to the original, merely lifting Babbe's head backwards. By comparison, the hands in both the New York and Amsterdam variants look less convincing—possibly because a clear example was lacking (the Berlin version contains only a very rudimentary indication of the hand holding the jug).

4.5 The Attribution of the New York *Malle Babbe*

In short, both the New York and the Amsterdam variant appear to be based on the Berlin example, yet do not equal its extraordinary *virtuoso* execution. Moreover, in technique and use of materials, the New York picture is much closer to the Berlin version than the Amsterdam forgery. In-depth analyses of the materials used confirm that the New York variant is consistent with Hals's workshop practice and materials, while Van Meegeren's forgery is of a much later date. Notably, the lead-isotope analysis showed a clear affinity between types of lead-white used in the Berlin and New York version, while the lead white used by Van Meegeren has entirely different characteristics, indicating that the lead ores in the lead white used by Van Meegeren came from a completely different location than the lead ores in Hals's lead white (Fig. 4.16).[36] In the twentieth century lead was often imported into Europe from the United States and Australia, which could explain the difference.

[35] For the court documents, see (Huussen (ed.) 2009).

[36] The samples were analyzed by Gareth Davies and Paolo d'Imporzano of the geochemical Laboratory for Ultra-Low Isotopic Analyses in the Faculty of Sciences, Vrije Universiteit, Amsterdam. See also (Tummers et al. 2019b, p. 999). Vermeer forgeries by Van Meegeren had similar outlier results, as was discussed by Arie Wallert in an unpublished paper (Wallert 2015).

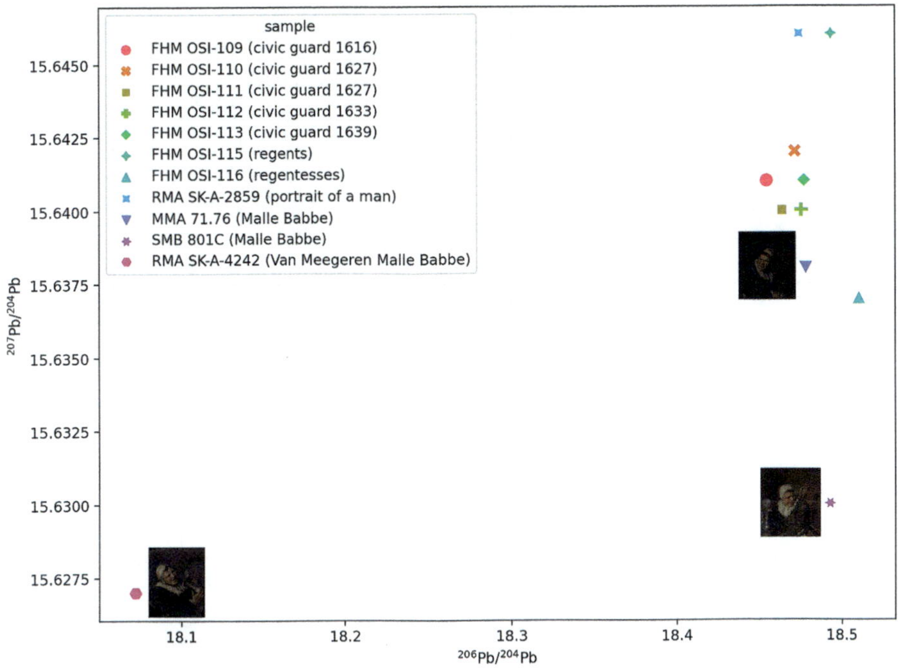

Fig. 4.16 Graph plotting the lead isotope ratios in paint samples taken from nine reference paintings by Frans Hals (including the Berlin *Malle Babbe*) against the New York *Malle Babbe* and Van Meegeren's *Malle Babbe*

A close look at the monogram in the New York variant provides a further clue as to the attribution of this work. Although the monogram was dismissed in the past as a later addition, close observation revealed that it is in fact an integral part of the original paint layer (Fig. 4.17) (Liedtke 2007, cat. no. 69, top matter). A continuous craquelure pattern intersects both the monogram and the paint layer of the background. The picture was thus clearly intended as a 'Frans Hals'. Therefore, the picture was either authenticated by Frans Hals as a work worthy of carrying his name or it is an early forgery, deliberately created to deceive.

Thus far, there is no evidence suggesting that pictures by Frans Hals were forged in his own time, contrary to for example Hans Bol (1534–1593), who reputedly stopped painting because of all the imitations that were sold under his name (Van Mander 1604, fol. 260v; see also Tummers 2011, p. 64). Moreover, the strong similarities in painting technique make the first option by far the most likely. Apart from the challenging *alla prima* technique, the partial exposure of the sand colored ground and the characteristic ribbon touches, the use of pigments is also consistent with Frans Hals's workshop practice. Notably, the use of umber (which contains manganese) is comparable—for some shadows in the face and in the background, as can be seen in

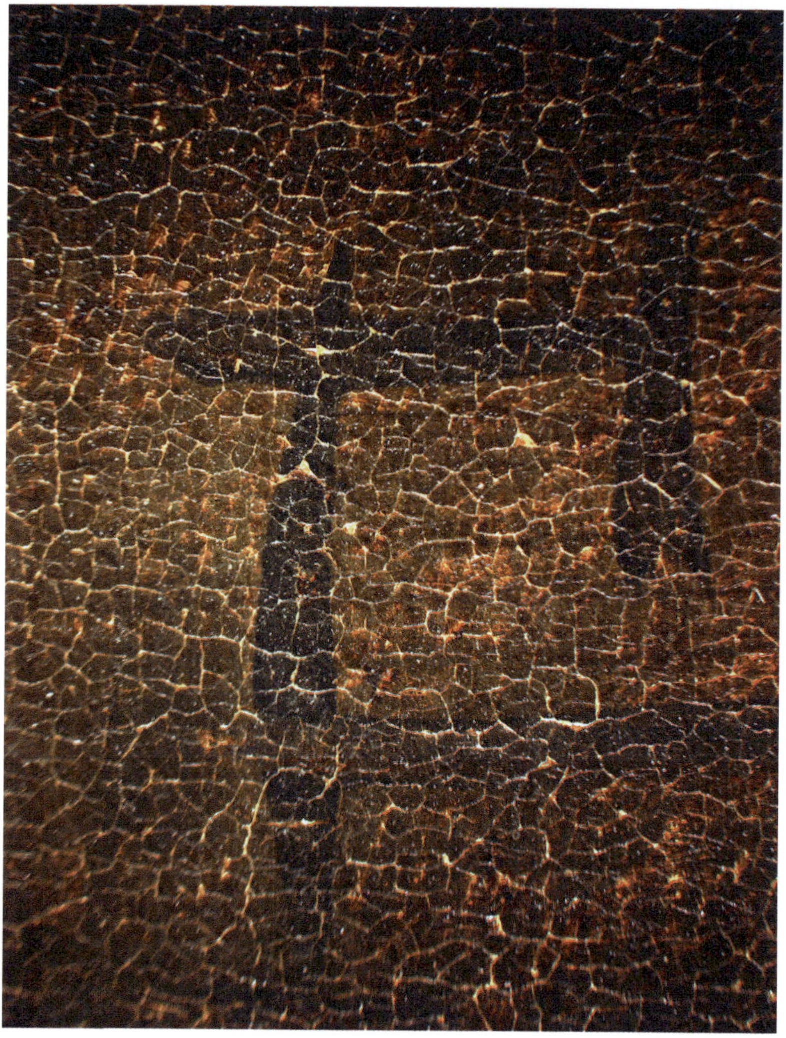

Fig. 4.17 Micrograph of the monogram on the New York *Malle Babbe* (Fig. 4.1)

the MA-XRF maps of the paintings (Fig. 4.18).[37] Similarly, the use of bone black for some facial shadows seems distinctive, as well as the sparse use of vermillion for just a few loose accents in the face (Fig. 4.18). Van Meegeren, on the contrary, used vermillion abundantly for all flesh tones in his forgery. The New York *Malle Babbe* also shows vermillion in the background, which is rare; possibly, the artist simply mixed in some leftover pigment with the background colour so as not to waste materials. Also, in both the Berlin and New York *Malle Babbe*, some of the darkest blacks accents

[37] The instrument and method used are described in (Alfeld et al. 2013). For the Metropolitan picture the spot size was 700 microns, the step size was 1000 microns, and the dwell time 90 msec/pixel.

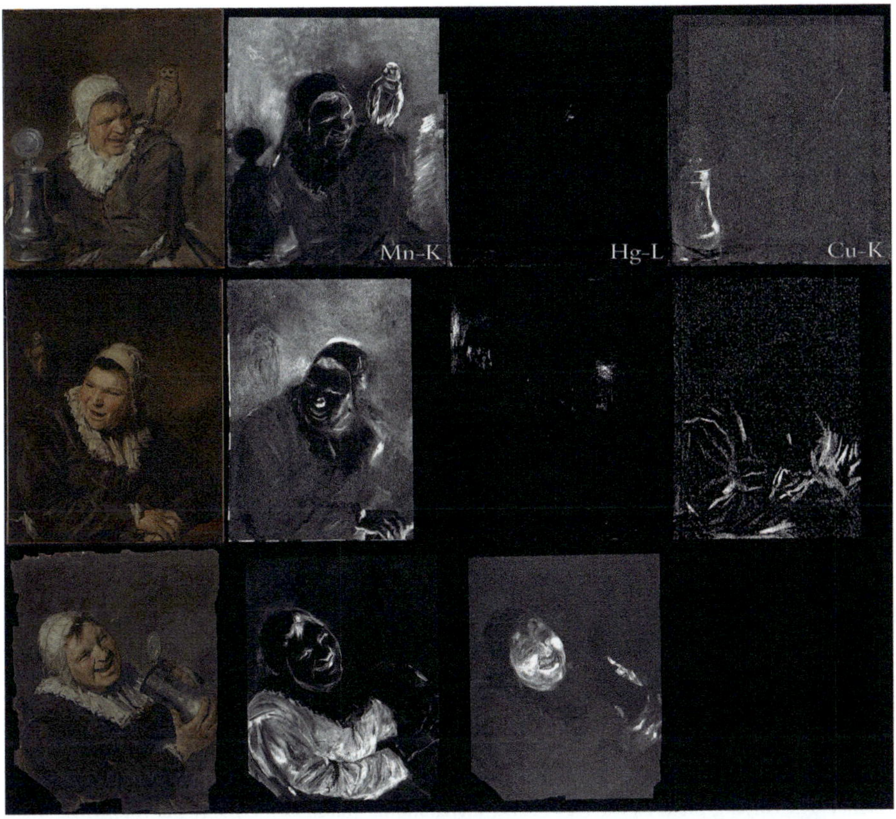

Fig. 4.18 MA-XRF maps of the three Malle Babbe paintings showing the elements manganese (Mn), mercury (Hg) and copper (Cu)

contain copper (possibly used as a drier), though these concern different elements: details of the clothing in the New York version and accents in the tin jug in the Berlin painting (Fig. 4.18). In short, the techniques and use of materials in the New York variant are very similar to the Berlin version but not exactly identical, suggesting that it was not created at exactly the same moment in Hals's studio.

Furthermore, the *alla prima* technique used in both pictures and the documents related to Malle Babbe's confinement give an indication as to the dating of the paintings. The only other paintings by Frans Hals that appear to have been done *alla prima* thus far are his portrait of *Jasper Schade* (dated on the original cartouche: 1645) and a small portrait of a *Traveller* that is dated circa 1650 based on a dendrochronological analysis of the panel on which is it painted; the earliest possible date the latter could have been created is 1649 (Pokorný 2012; Ševčik 2012, p. 182; Tummers (ed.) 2013c, cat. nos. 34 and 47).[38] Since Malle Babbe was confined to the workhouse in 1646, a dating of circa 1640–1646 seems most likely for the Berlin *Malle Babbe*. As the

[38] *The Traveller* measures 35 by 26 cm and is in the Stiftung Heinz Kuckei Collections in Berlin.

New York version is based on the Berlin original rather than painted after life and the use of materials differs slightly, a dating of circa 1640–1650 seems most likely.

The presence of the master's monogram on what appears to be high quality studio work in the New York version is entirely consistent with seventeenth-century workshop practice. If the master deemed the style and quality good enough for their standards he or she was entitled to sign the work and sell it as their own (Tummers 2011, ch. 3). It reminds us that attributing seventeenth-century paintings is somewhat counter-intuitive. While our tendency is to compare paintings in depth and look for telling signs in the brushwork betraying a different hand, it was not the execution by a different hand that necessarily made a difference. Although the New York *Malle Babbe* is a bit lower in quality than the spectacular Berlin version, the Metropolitan Museum of Art was not wrong about its attribution when the museum first opened its doors and displayed it proudly. According to seventeenth-century standards, the picture *is* an original Frans Hals.

References

Alfeld, M., Pedroso, J.V., Van Eikema Hommes, M., Van Der Snickt, G., Tauber, G., Blaas, J., Haschke, M., Erler, K., Dik, J., Janssens, K.: A mobile instrument for in situ scanning macro-XRF investigation of historical paintings. J. Anal. At. Spectrom. **28**(5), 760–767 (2013). https://doi.org/10.1039/c3ja30341a

Atkins, C.D.M.: The Signature Style of Frans Hals: Painting, Subjectivity, and the Market in Early Modernity. Amsterdam University Press, Amsterdam (2012)

Baard, H.P.: Frans Hals. Abrams, New York (1981)

Baetjer, K.: Buying pictures for New York: the founding purchase of 1871. Metrop. Mus. J. **39**, 161–245 (2004). https://doi.org/10.1086/met.39.40034606

Bredius, A.: Hoe een varend gezel schilder werd: Hans Juriaensz van Baden. Oud. Holl. **44**(1), 17–22 (1927). https://doi.org/10.1163/187501727X00039

Burroughs, B.: The Metropolitan Museum of Art: Catalogue of Paintings, 9th edn. Metropolitan Museum of Art, New York (1931)

Catalogue of the Pictures in the Metropolitan Museum of Art. Metropolitan Museum of Art, New York (1872)

Erftemeijer, A.: Het fenomeen Frans Hals. Frans Hals Museum/Nai010 uitgevers, Haarlem/Rotterdam (2014)

Etchings of Pictures in the Metropolitan Museum of New York. Etchings by Jacquemart, J.F. P.and D. Colnaghi, London (1871)

Exposition de tableaux et dessins d'anciens maîtres, organisée par la Société néerlandaise de bienfaisance à Bruxelles. M. Weissenbruch, Brussels (1873)

Grimm, C.: Frans Hals: Das Gesamtwerk. Belser, Stuttgart/Zürich (1989)

Grimm, C.: Frans Hals and his Workshop. RKD, The Hague (forthcoming)

Grimm, C., Montagni, E.C.: L'opera completa di Frans Hals. Rizzoli, Milan (1974)

Groves, R., Anisimov, A., Papadakis, V., Tummers, A., Wallert, A., Hartwieg, B., and Kleinert, K.: Hyperspectral imaging to guide technical art analysis of Frans Hals paintings. In.: Proceedings of the Lacona (Lasers in the Conservation of Artworks) Conference VII, Paris (2018)

Hendriks, E., Levy-Van Halm, K., in collaboration with Van Asperen De Boer, J.R.J.: Report Concerning a Preliminary Technical Investigation of Paintings exhibited during the Frans Hals Exhibition, held from May 11 to July 22 1990 in the Frans Hals Museum Haarlem. (1991)

Hofstede de Groot, C.: A Catalogue Raisonné of the Works of the Most Eminent Painters of the Dutch Seventeenth-Century; Based on the Work of John Smith (1907–1927), 8 vols. Translated and edited by Hawke, E.G. Macmillan and Co., London (1907-1927)

Huussen, H.A. (ed.): Cahiers uit het Noorden XX: Henricus [Han] Antonius van Meegeren (1889-[1945]): documenten betreffende zijn leven en strafproces. Huussen, Zoetermeer (2009)
Jowell, F.: The rediscovery of Frans Hals. In: Slive, S. (ed.) Frans Hals, exh. cat., pp. 61–86. National Gallery of Art/Royal Academy of Arts/Frans Hals Museum, Washington/London/Haarlem (1989–1990)
Lammertse, F.: Lachen. In: Cornelis, B, Lammertse, F., Rinnooy Kan, J., Van der Veen, J. (eds.) Frans Hals, exh. cat. The National Gallery (London) and Rijksmuseum (Amsterdam), pp. 158–187, 197–200. Hannibal, Veurne (2023–2024)
Liedtke, W.: Dutch paintings in America: the collectors and their ideals. In: Broos, B., Buijsen, E., Donahue Kuretsky, S., Liedtke, W. (eds.) Great Dutch Paintings from America, exh. cat., pp. 14–58. Mauritshuis/Waanders, The Hague/Zwolle (1990–1991)
Liedtke, W.: Dutch Paintings in the Metropolitan Museum of Art, 2 vols. Metropolitan Museum of Art/Yale University Press, New York/New Haven (2007)
Lopez, J.: The Man Who Made Vermeers: Unvarnishing the Legend of Master Forger Han Van Meegeren. Harcourt Books, Orlando (2008)
Lützow, C.: Hille-Bobbe van Haarlem, Oelgemälde von Franz Hals in der Galerie Suermondt. Z. bild. Kunst. **5**, 78–80 (1870)
Mensger, A., Eiling, A., Betz, J., Tieze, A., Keazor, H.: Déjà-vu? Die Kunst der Wiederholung von Dürer bis Youtube, exh.cat. Staatliche Kunsthalle, Karlsruhe (2012)
Meyer, J., Bode, W.: Verzeichnis der ausgestellten Gemälde und Handzeichnungen aus den im Jahr 1874 erworbenen Sammlungen des Herrn Barthold Suermondt, 2nd edn. Königliche Museen/Gemälde-Galerie, Berlin (1875)
Meyer, J., Bode, W.: Beschreibendes Verzeichnis der während des Umbaus ausgestellten Gemälde. Königliche Museen/Gemälde-Galerie, Berlin (1883)
Miedema, H.: De ontwikkeling van de kunsttheorie in de Hollandse Gouden Eeuw II. Karel van Mander. Oud. Holl. **126**(1), 17–30 (2013). https://doi.org/10.1163/18750176-90000028
Moes, E.W.: Frans Hals: sa vie et son oeuvre (1909). Translated by De Bosschere, J. Oest, Brussels (1909)
Pokorný, A.: Restoration Report on 'Portrait of Jasper Schade'. National Gallery Prague, Prague (2012)
Ševčík, A.K. (ed.): National Gallery in Prague: Dutch Paintings of the 17th and 18th Centuries. National Gallery of Prague, Prague (2012)
Slive, S.: Frans Hals, 3 vols. Phaidon, London (1970–1974)
Slive, S. (ed.) Frans Hals, exh. cat., National Gallery of Art/Royal Academy of Arts/Frans Hals Museum, Washington/London/Haarlem (1989–1990)
Slive, S.: Frans Hals, revised edn. Phaidon, London (2014)
Slive, S., Van Hees, C.A.: Frans Hals: tentoonstelling ter gelegenheid van het honderdjarig bestaan van het Gemeentemuseum te Haarlem, 1862–1962, exh. cat. Frans Hals Museum, Haarlem (1962)
Stukenbrock, C.: Frans Hals: Fröhliche Kinder, Musikanten und Zecher: Eine Studie zu ausgewählten Motivgruppen und deren Rezeptionsgeschichte. Peter Lang, Frankfurt (1993)
Taylor, P.: Condition: the Ageing of Art. Paul Holberton, London (2007)
Thoré, T. (alias Bürger, W.): Nouvelle études sur la Galerie Suermondt. Gaz. b-arts. **2**, 161–164 (1869)
Trivas, N.S.: The Paintings of Frans Hals. Phaidon, London (1941)
Tummers, A.: The Eye of the Connoisseur: Authenticating Paintings by Rembrandt and his Contemporaries. Amsterdam University Press/Getty Publications, Amsterdam/Los Angeles (2011)
Tummers, A.: Frans Hals: rightly admired by the greatest painters. In: Tummers, A. (ed.) Frans Hals: Eye to Eye with Rembrandt, Rubens and Titian, exh. cat, pp. 13–40. Frans Hals Museum/NAI010 Publishers, Haarlem/Rotterdam (2013a)
Tummers, A.: Frans Hals: Frans Hals: rightly admired by the greatest painters. In: Tummers, A. (ed.) Frans Hals: Eye to Eye with Rembrandt, Rubens and Titian, exh. cat., pp. 13–40. Frans Hals Museum/NAI010 Publishers, Haarlem/Rotterdam (2013b)
Tummers, A. (ed.): Frans Hals: Eye to Eye with Rembrandt, Rubens and Titian, exh. cat Frans Hals Museum/NAI010 Publishers, Haarlem/Rotterdam (2013c)

Tummers, A., Gration, J.: Eye to eye with Frans Hals: the location and reception of some of his masterpieces. In: Tummers, A. (ed.) Frans Hals: Eye to Eye with Rembrandt, Rubens and Titian, exh. cat., pp. 71–93. Frans Hals/NAI010 Publishers, Haarlem/Rotterdam (2013)
Tummers, A., Kolfin, E., Hillegers, J. (eds.): The Art of Laughter: Humour in Dutch Paintings of the Golden Age, exh. cat. Frans Hals Museum/Waanders, Haarlem/Zwolle (2017)
Tummers, A., Wallert, A., De Keyser, N.: Supplementing the eye: the technical analysis of Frans Hals's paintings—II. Burlingt. Mag. **161**, 996–1003 (2019a)
Tummers, A., Wallert, A., Hartwieg, B., Kleinert, K., Laurenze-Landsberg, C., Dik, J., Groves, R., Anisimov, A., Papadakis, V., Erdmann, R.: Supplementing the eye: the technical analysis of Frans Hals's paintings—I. Burlingt. Mag. **161**, 934–941 (2019b)
Tumosa, C.S., Mecklenburg, M.F.: Oil Paints: The Chemistry of Drying Oils and the Potential for Solvent Disruption. In: Mecklenburg, M.F., Charola, A.E., Koestler, R.J. (eds.) New Insights into the Cleaning of Paintings: Proceedings from the Cleaning 2010 International Conference, Universidad Politecnica de Valencia and Museum Conservation Institute, pp. 51–58. Smithsonian Contributions to Museum Conservation. Smithsonian Institution, Washington, DC (2013)
Unger, W., Vosmaer, C.: Etsen naer Frans Hals door William Unger, met eene verhandeling over den schilder door C. Vosmaer. A.W. Sijthof, Leiden (1873)
Valentiner, W.R.: Frans Hals: des Meisters Gemalde in 322 abbildungen. Deutsche Verlags-Anstalt, Stuttgard (1921)
Valentiner, W.R.: Frans Hals: des Meisters Gemalde in 322 abbildungen, 2nd edn. Deutsche Verlags-Anstalt, Stuttgard/Berlin/Leipzig (1923)
Valentiner, W.R.: Frans Hals Paintings in America. F.F. Sherman, Westport (1936)
Van Dantzig, M.M. Frans Hals: Echt of Onecht, Naar aanleiding van de Frans Hals-tentoonstelling 1937. Paris, Amsterdam (1937)
Van Mander, K.: Het schilder-boeck waer in voor eerst de leer-lustighe ieught den grondt der edel vry Schilder-Const in verscheyden deelen wort voorghedraghen. Paschier van Wesbuch boeck-vercooper, Haarlem (1604)
Van Thiel-Stroman, I.: The Frans Hals documents: written and printed sources, 1582–1679. In: Slive, S. (ed.) Frans Hals, exh. cat, pp. 371–415. National Gallery of Art/Royal Academy of Arts/Frans Hals Museum, Washington/London/Haarlem (1989–1990)
Van Thiel-Stroman, I.: Frans Franchoisz Hals. In: Köhler, N., Levy Van Halm, K., Runia, E., Van Thiel, P., Van Thiel-Stroman, I. (eds.) Painting in Haarlem, 1500–1850: the Collection of the Frans Hals Museum, pp. 178–184. Ludion, Gent (2006)
Von Bode, W.: Studien zur Geschichte der Holländischen Malerei. F. Vieweg und Sohn, Braunschweig (1883)
Wallert, A.: Examination of stable isotope ratios: consequences for art history. Paper presented at the Technart Conference, Catania, 2015
White, R., Pilc, J., Kirby, J.: Analyses of Paint Media. National Gallery Technical Bulletin 19, 74–95 (1998). https://www.nationalgallery.org.uk/media/15653/white_pilc_kirby1998.pdf

Chapter 5
Case Study 3: A Recent Riddle: The Story of the *Two Fisherboys*

Anna Tummers, Arie Wallert, and Nouchka De Keyser

Abstract The painting *Two Fisherboys* has long caused confusion among experts: Is it an original Frans Hals? Or rather a forgery? A close comparison of the painting with both a forgery by Han van Meegeren and Frans Hals's *Fisherboy* solved the conundrum while providing valuable insights into the merits and drawback of modern analytical techniques.

When the painting *Two fisherboys* was offered for sale at Christie's in London in July 2017 as a work by Frans Hals (c. 1582/83–1666) its attribution was called into question during the viewing days (Fig. 5.1).[1] One of the causes for concern: A 1965 newspaper article had resurfaced in which Jacques van Meegeren (1912–1977), son of the master forger Han van Meegeren (1889–1947), claimed that the picture was a forgery by his father (Fig. 5.2). As result, the owners of the work and the auction house requested an in-depth technical investigation of the painting. Using a variety of techniques, including macro X-ray fluorescence scanning (MA-XRF), gas chromatography-mass spectrometry (GC-MS) and a lead isotope analysis, as well as observations with the naked eye and a hand-held digital microscope, the authors of this chapter compared the

[1] Sale, Christie's, London, 6 July 2017, Old Masters Evening Sale, lot 28.

A. Tummers (✉)
Ghent University, Ghent, Belgium
e-mail: Anna.Tummers@UGent.be

A. Wallert (✉)
University of Amsterdam, Amsterdam, The Netherlands

Rijksmuseum, Amsterdam, The Netherlands
e-mail: arie.wallert@gmail.com

N. De Keyser (✉)
Rijksmuseum, Amsterdam, The Netherlands
e-mail: N.de.Keyser@rijksmuseum.nl

A. Tummers, R. G. Erdmann, *Frans Hals or not Frans Hals*, Cultural Heritage Science, https://doi.org/10.1007/978-3-031-59489-2_5

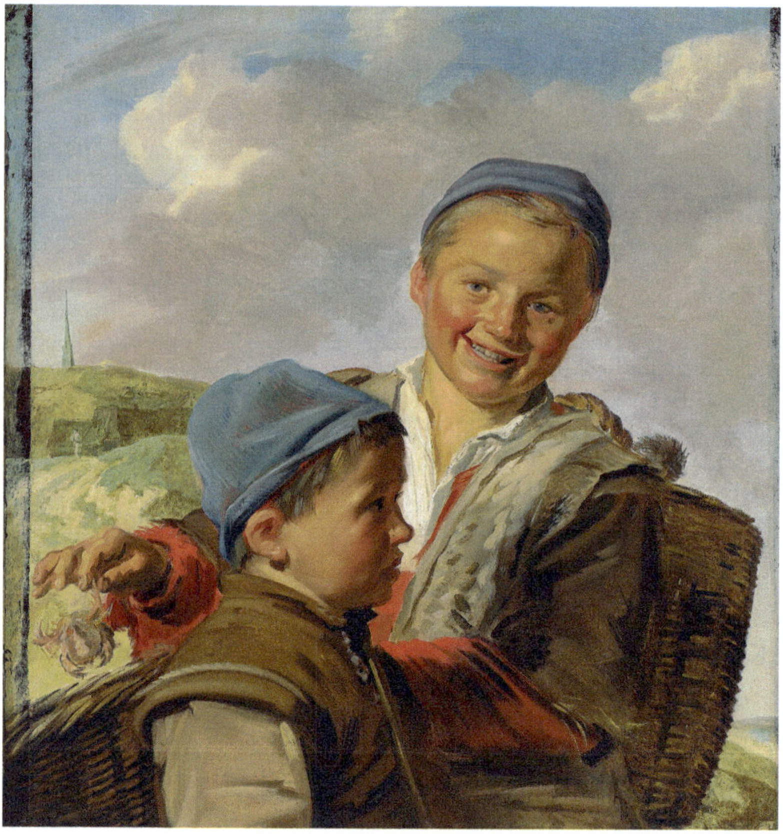

Fig. 5.1 Frans Hals, *Two Fisherboys,* c. 1634–1637, oil on canvas, 74.3 × 65.8 cm. (The Phoebus Foundation)

painting to a forgery by Van Meegeren in the style of Hals and to Hals's *Fisherboy* (Fig. 5.3). The results of the investigation have provided new insights into the status of the disputed painting as well as into the advantages and pitfalls of modern analytical techniques.

Ever since Sotheby's announced in 2016 that a portrait of a man that it had sold privately in 2011 as a work by Hals was a forgery and revealed that it had reimbursed the buyer for the purchase amount of $11.2 million, the art market has been on the alert for potentially fraudulent pictures by Hals (BBC News 2016). The claims by Van Meegeren's son, made to the Dutch newspaper *Trouw* (8 Oct 1965, p. 1) just before the opening of an exhibition of his work at the Café Ruiter in Amsterdam seemed especially disconcerting. 'My father had a great admiration for Frans Hals and owned many reproductions of his work', he stated, adding that he recognised 'without a doubt his father's style' in four works: *Two Fisherboys* (then in an American private collection), *Child with a dog* (Kelvingrove Art Gallery and Museum, Glasgow), *Rommelpot player* (then with the Lilienfeld Galleries,

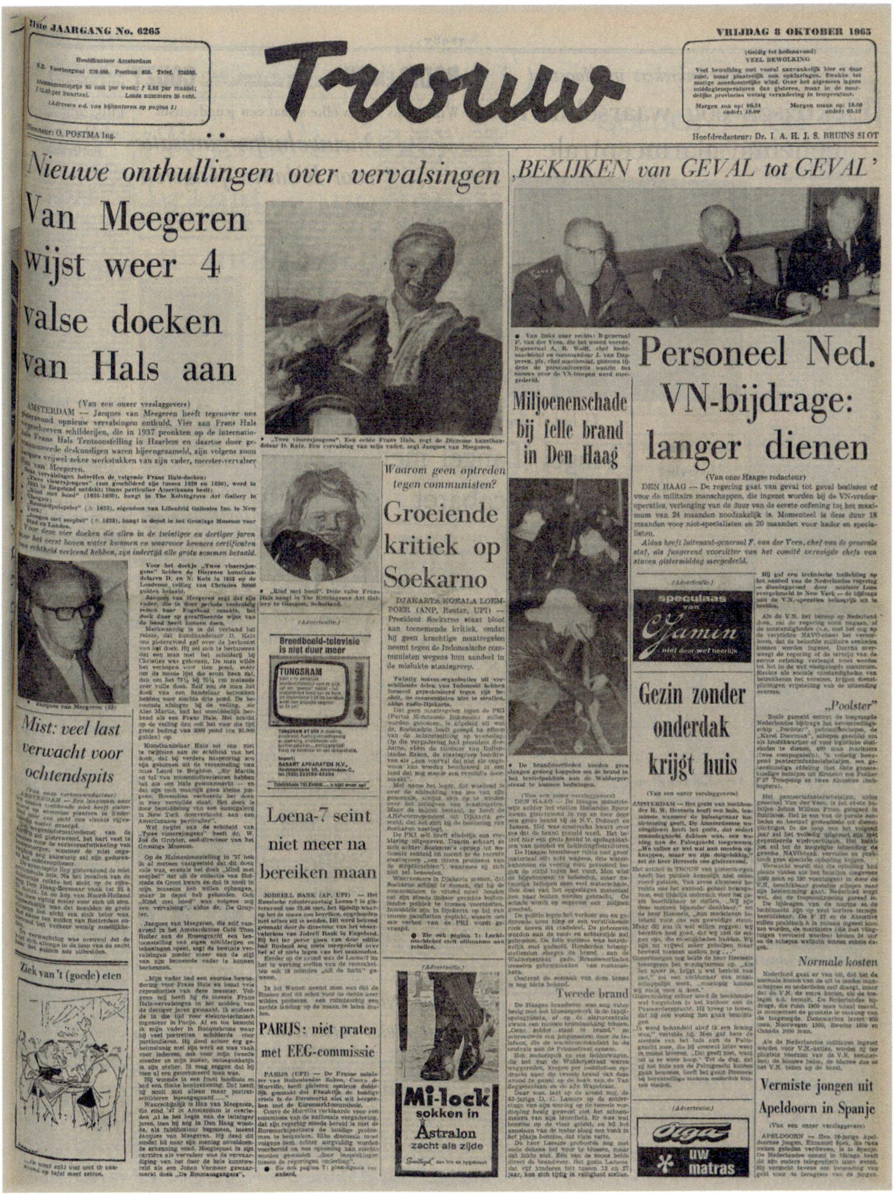

23ste JAARGANG No. 6265

VRIJDAG 8 OKTOBER 1965

Trouw

Directeur: O. POSTMA Ing.

Hoofdredacteur: Dr. I. A. H. J. S. BRUINS SLOT

Nieuwe onthullingen over vervalsingen

Van Meegeren wijst weer 4 valse doeken van Hals aan

(Van een onzer verslaggevers)

AMSTERDAM — Jacques van Meegeren heeft tegenover ons opnieuw vervalsingen onthuld. Vier aan Frans Hals toegeschreven schilderijen, die in 1937 prijkten op de internationale Frans Hals Tentoonstelling in Haarlem en daartoe door gerenommeerde deskundigen waren bijeengezameld, zijn volgens zoon vrijwel zeker werkstukken van zijn vader, meester-vervalser Han van Meegeren.

„Twee visserjongens". Een echte Frans Hals, ...

Jacques van Meegeren (53)

Mist: veel last verwacht voor ochtendspits

Ziek van 't (goede) eten

Waarom geen optreden tegen communisten?

Groeiende kritiek op Soekarno

DJAKARTA-KOEALA LOEMPOER (ANP, Reuter, UPI) — President Soekarno staat bloot aan toenemende kritiek, omdat hij geen krachtige maatregelen neemt tegen de Indonesische communisten wegens hun aandeel in de mislukte staatsgreep.

Breedbeeld-televisie is niet duur meer

TUNGSRAM

Loena-7 seint niet meer na bereiken maan

PARIJS: niet praten met EEG-commissie

Mi-lock sokken in Astralon zacht als zijde

,BEKIJKEN van GEVAL tot GEVAL'

Miljoenenschade bij felle brand in Den Haag

Personeel Ned. VN-bijdrage: langer dienen

(Van onze Haagse redacteur)

DEN HAAG — De regering gaat van geval tot geval beslissen of voor de militaire manschappen, die ingezet worden bij de VN-vredesoperaties, verlenging van de duur van de eerste oefening tot het maximum van 24 maanden noodzakelijk is. Momenteel is deze duur 18 maanden voor niet-specialisten en 20 maanden voor kader en specialisten.

speculaas van Jamin niet duur wel heerlijk

Gezin zonder onderdak krijgt huis

Tweede brand

„Poolster"

Normale kosten

Vermiste jongen uit Apeldoorn in Spanje

Auping uw matras

Fig. 5.2 Front page of the Dutch newspaper *Trouw*, 8 Oct 1965

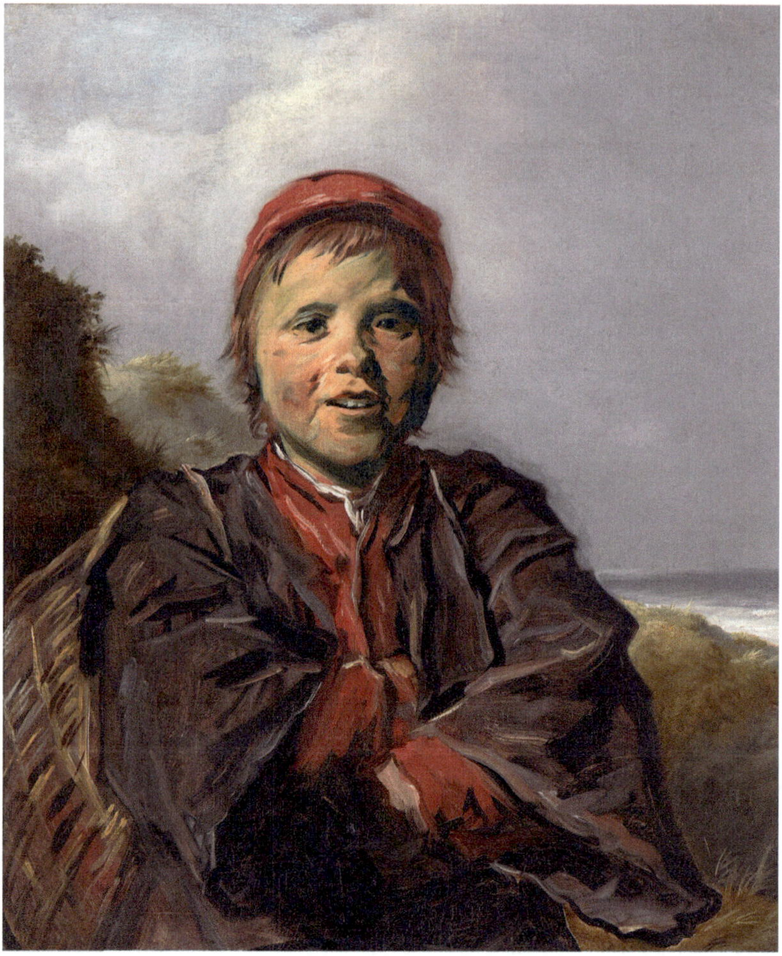

Fig. 5.3 Frans Hals, *Fisherboy*, c. 1630–1632, oil on canvas, 74 × 61 cm. (Royal Museum of Fine Arts, Antwerp)

New York) and *Boy with a soap bubble* (then at the Groninger Museum, Groningen).[2] Jacques van Meegeren suspected that his father had created the paintings in the mid-1930s while working in Roquebrune, France; he remembered that Han did not let anyone enter the back room of his studio there, not even his second wife or his daughter. This story was retold by three other newspapers, who also interviewed various experts about the works (Leidsche Courant 8 Oct 1965, p. 7; Utrechts Nieuwsblad 9 Oct 1965, p. 15; Provinciale Zeeuwsche Courant 9 Oct 1965, p. 9).

[2] '*Mijn vader had een enorme bewondering voor Frans Hals en bezat vele reproducties van deze meester.*' '*Jacques van Meegeren* [. . .] *zegt de bewuste werken zonder meer aan de stijl van zijn vader te kunnen herkennen*' (Trouw 8 Oct 1965, p. 1).

This was not the first time that *Two fisherboys* had been questioned. In 1937 the Dutch restorer Maurits van Dantzig (1903–1960) wrote a book *Frans Hals: Echt of Onecht* (*Frans Hals: Real or Fake*) on the attribution of the paintings shown that year in Haarlem in the first Hals monographic exhibition. He classified *Two fisherboys* as a forgery by an unidentified late nineteenth-century English artist (Van Dantzig 1937, cat. no. 88, p. 100). In his opinion, no fewer than thirteen paintings - eleven of which depicted fisher children - were all by this hand. These pictures lacked the forty-four qualities he deemed illustrative of an authentic work by Hals and possessed fifteen characteristics that in his view betrayed a different hand (Van Dantzig 1937, pp. 5–37 and 97–100). For example, he believed that the facial expressions were revealing. Perceiving solemnity, grace and spirit (‘*statigheid, zwier en geest*’) in the expressions of all Hals's protagonists - which he attributed more to the painter's character than to that of his sitters - Van Dantzig felt that the group of paintings of fisher children fell short in this respect. Their expressions were ‘either too surly or too dull’ (*‘òf norsch òf bête’*) and their smiles resembled harsh grins. Moreover, the skin colours were unusually dark, and many details in the clothing, hands and attributes were too roughly executed and therefore unclear, while the backgrounds contained a surplus of meaningless details.

Other art historians have also dismissed *Two fisherboys*. When Seymour Slive compiled his Hals catalogue raisonée in 1970–1974, he deemed the work, which he had assessed on the basis of photographs, ‘a nineteenth-century painting done in Hals's style’ (Slive 1970–1974, III, no. D16, p. 133). In an article published in 2007 Eddy de Jongh referred to the painting as an example of a Hals forgery (De Jongh 2007). Moreover, the provenance of the painting cannot be traced back further than 1935, when it was reputedly bought at a sale on the South Coast of England for £3 and subsequently sold for 2800 guineas at Christie's (The Illustrated London News 1936, p. 8).

Before studying the painting in relation to Hals's œuvre, the authors of this chapter first focused on establishing whether or not *Two fisherboys* was a seventeenth-century painting or a later imitation. To assess its technical characteristics, the authors compared it to Van Meegeren's *Malle Babbe*, a forgery in the style of Hals[3] (Figs. 5.4 and 5.5), which was confiscated by the police when they searched Van Meegeren's workshop in France in 1945, and to Hals's *Fisherboy*. Although the precise attribution of the latter painting—to Hals alone or with participation by assistants—is the topic of some debate,[4] its origin in his workshop is not. The picture bears Hals's characteristic monogram (‘FH’ in ligature), which close study with a dino-lite microscope has confirmed was an integral part of the original paint layer (Fig. 5.4). Moreover, the boy in this picture was cited (a similar figure in pose and costume) in a painting signed by Jan Miense Molenaer, *Beach scene with fisher folk*

[3]Van Meegeren based the forgery on Frans Hals's famous *Malle Babbe* at the Gemäldegalerie in Berlin as well as on another prototype, then believed to be an original Frans Hals too (Metropolitan Museum of Art, acc. no. 71.76), with thanks to Rosa Hoogenboom for this observation. See also Chap. 4.

[4]The painting was dismissed as not by Hals in (Trivas 1941) and (Grimm 1972, p. 214).

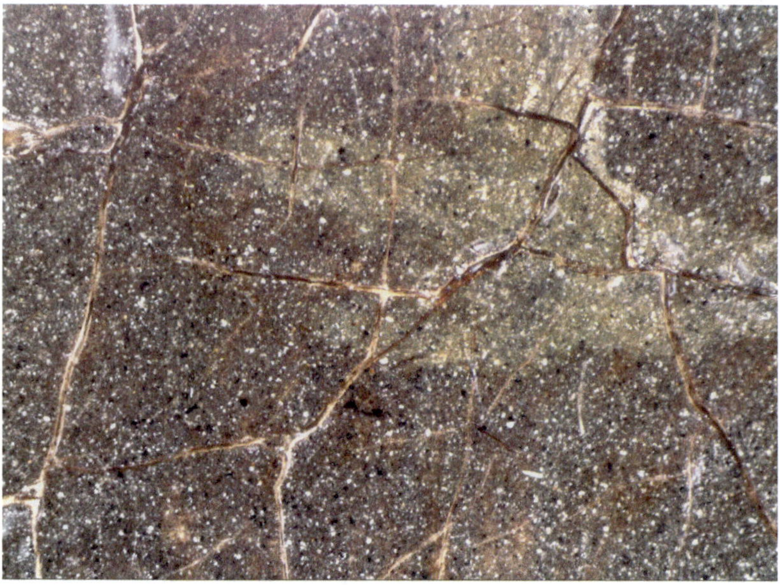

Fig. 5.4 Detail of Fig. 5.1 taken with a dino-lite microscope (X65) showing craquelure over Hals's monogram and surrounding areas

(private collection).[5] Both the monogram and this early allusion strongly indicate that, whatever the level of workshop involvement, Hals deemed the picture worthy to carry his name.[6]

Van Meegeren is known to have painted his forgeries on top of seventeenth-century originals, making it impossible to distinguish their support and ground from those of an authentic old work. Furthermore, he made an effort to select pigments that were consistent with the period (Wallert 2011). However, a curious characteristic of his technique concerns the binding medium in the layer he added on top. To create a convincing pattern of craquelure in this layer Van Meegeren used a binding medium consisting of oils mixed with a resin containing phenol formaldehyde, a twentieth-century material better known as 'bakelite'. After he had baked his forgeries in an oven, the top layer hardened as if the painting had genuinely aged, enabling him to create intricate patterns of cracks by bending the canvases.

The identification of phenol formaldehyde played a key role in the court case against Van Meegeren in 1945–1946 (Tummers 2011, p. 25 and note 8, p. 253), and has also been crucial in later studies of potential forgeries by him, such as the *Procuress* in the Courtauld Institute of Art, London.[7] The most common way to

[5] Sale, Sotheby's London, 24 Oct 1973, lot 129. See also (Slive 1970–1974, Fig. 34d, p. 228).

[6] On *Beach scene with fisher folk* see (Slive (ed.) 1989–1990, Fig. 34d, p. 228). On seventeenth-century views on authenticity see (Tummers 2011, esp. pp. 81–112).

[7] This painting was identified as a work by Van Meegeren in an episode of the BBC television series *Fake or Fortune?*, first transmitted on 3 July 2011.

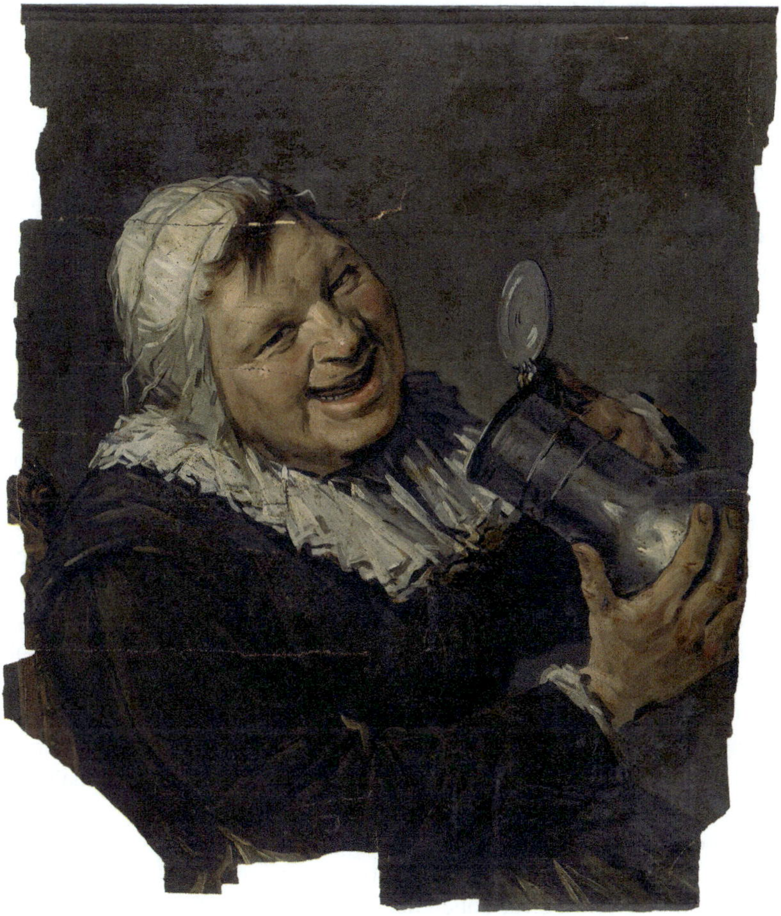

Fig. 5.5 Hans van Meegeren *Malle Babbe*, 1930–1940, oil on canvas, 76 × 60 cm. (Rijksmuseum, Amsterdam)

identify phenol formaldehyde now is by GC-MS, which provides a method of separating and identifying complex mixtures of organic molecules (Colombini 2018). The analysis of *Two fisherboys* was carried out by Henk van Keulen and Saskia Smulders at the Cultural Heritage Agency of the Netherlands, who used a piece of Van Meegeren's resin as a reference.[8] Samples were taken from both *Two fisherboys* and *Malle Babbe*. Initially the GC-MS analysis seemed to provide a clear and positive result. Analysis of the sample taken from *Two fisherboys* showed nothing but drying oil and natural resins. The oil had not been prepolymerised (which would have indicated artificial ageing). The results indicated that the binding medium used

[8] This resin was analysed by Wiebo Froentjes in the context of the 1945–1946 court case, see report cited at note 10 below. The sample was kindly provided by Jaap Mosk.

in the painting is linseed oil, which is consistent with seventeenth-century practice, and that the (recently applied) varnish on top contains pine resin and damar resin. No indication of the presence of a phenol formaldehyde resin was found, nor was there a match with the components in the reference sample.

However, when a sample from the *Malle Babbe* forgery was analysed with the same instrument using the same method, there was—surprisingly—also no match with the components in the reference sample. In other words, the absence of phenol formaldehyde in the GC-MS analysis of *Two fisherboys* did not prove that the painting could not be a forgery by Van Meegeren. In fact, the analysis of *Malle Babbe* had produced a false negative. In the 1945–1946 court case *Malle Babbe* had tested positive for phenol-formaldehyde in tests done with both a sulphuric acid solution and with an ammonia solution, which caused yellow and blue discolorations.[9] Both natural and artificial materials age and the amount that can be detected with GC-MS diminishes over time, which could explain the false negative.

Subsequent in-depth analysis of the lead-white pigments used in the paintings did, however, provide conclusive evidence. The chemical element lead (Pb) can be identified with remarkable precision. Radioactive decay causes slight variations in the quantities of three of the four isotopes it contains (207, 206, 208), while the other (204) remains constant. Lead ores from different mines and ages have distinctive ratios of these isotopes, which function as a kind of signature. For the lead isotope analysis, the authors took samples from *Two Fisherboys* and the *Fisherboy*, and compared them to samples analysed for the *Frans Hals/not Frans Hals* project (funded by the Netherlands Organisation for Scientific Research), which included nine paintings by Hals and Van Meegeren's *Malle Babbe*. The samples were analysed by Gareth Davies and Paolo d'Imporzano of the Geochemical Laboratory for Ultra-Low Isotopic Analyses in the Faculty of Science, Vrije Universiteit, Amsterdam.

As can been in Fig. 5.6, the samples taken from paintings by Hals cluster: the lead isotope ratios found in nine well-known masterpieces dating from between 1616 and 1663/64 show a clear coherence. Samples taken from *Two fisherboys* and *Fisherboy* fall well within the Hals cluster, whereas the sample from *Malle Babbe* shows entirely different isotope ratios (206/204; 207/204; 208/204), indicating that the lead ores in the lead white used by Van Meegeren came from a completely different location than Hals's lead white. These lead isotopes ratios are not consistent with materials used by seventeenth-century painters (Fortunato et al. 2005). In the twentieth century lead was often imported into Europe from Australia and the United States, which could explain this difference. In Van Meegeren's time it was not yet known that lead ores would be identified with such great precision in the future. The great difference between the isotope ratios of the lead white used in *Two fisherboys* and the isotope ratios of Van Meegeren's lead white rules out the possibility that *Two fisherboys* was painted by Van Meegeren.[10] The great similarity with

[9] For the official report by Coremans, P., Froentjes, W., Plenderleith, H.J., Rawlins F.I.G. and De Wild, A.M., see (Huussen (ed.) 2009, pp. 92–100 esp. p. 99). On the then current chemical analyses see (Wallert and Van de Laar 2019).

[10] Vermeer forgeries by Van Meegeren had similar outlier results, as was discussed by A. Wallert (2015).

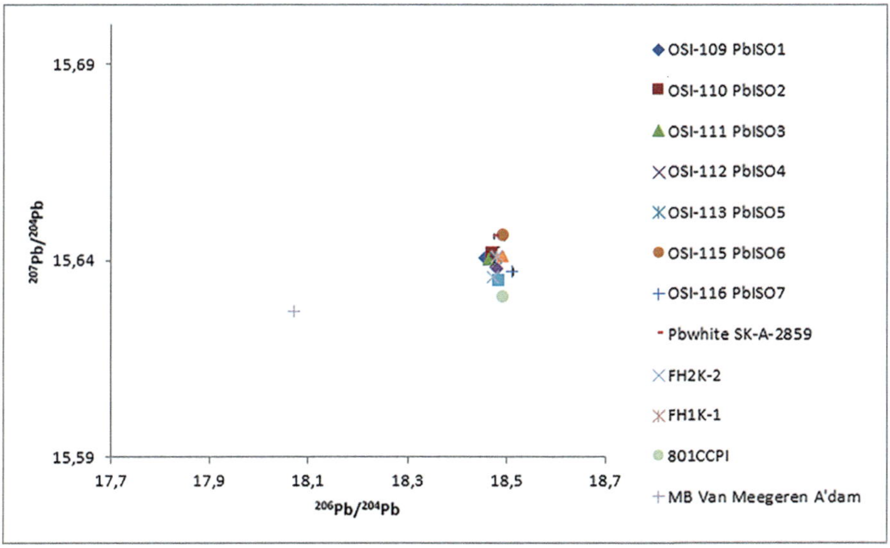

Fig. 5.6 Graph plotting the distribution of the samples for the $^{206}Pb/^{204}Pb$ ratio against $^{207}Pb/^{204}Pb$ ratio in nine paintings by Frans Hals, the *Two fisherboys* (sample FH2K-2), the *Fisherboy* (sample FH1K-1) and Van Meegeren's *Malle Babbe* (sample MB Van Meegeren A'dam)

lead white used by Frans Hals and his workshop is also telling: the lead white in both *Two fisherboys* and the *Fisherboy* is entirely consistent with the materials used by Hals and his workshop.

Jacques van Meegeren had, therefore, clearly not been truthful. Although his family connection potentially makes him a well-informed witness, he was a notorious liar. Like his father, he was unable to make a living creating original art, and in the 1960s he resorted to fraud to get by. Profiting from his father's reputation, he created fake forgeries in Han's style, which he used as payments (Kreuger 2007, pp. 193, 202, 203, 208 and 211). Presumably, Jacques based his 'discovery' that the works were forgeries on Van Dantzig's book, and spread the fake news to attract attention to the exhibition of his own paintings. His claims were questioned at the time. George Buchanan, curator at the Kelvingrove Art Gallery, pointed out that *Boy with a dog* was bought at a London auction in 1894 and thus could not possibly be by Han van Meegeren, who was only 5 years old at the time (see Trouw 9 Oct 1965, p. 5). Furthermore, the associate director of the Groninger Museum stated that *Boy with a soap bubble* could not have been made in the 1930s as it was documented in the 1920s (see Provinciale Zeeuwsche Courant 9 Oct 1965, p. 9).

The attribution of the group of paintings of fisher children in Hals's style has long been the subject of debate for reasons other than the issue of forgery. Their brushwork seems somewhat less confident than in Hals's other genre works and there are dissimilarities within the group. Several Hals scholars have therefore rejected the attribution of the entire group (see note 4 above). It seems more likely, however, that the

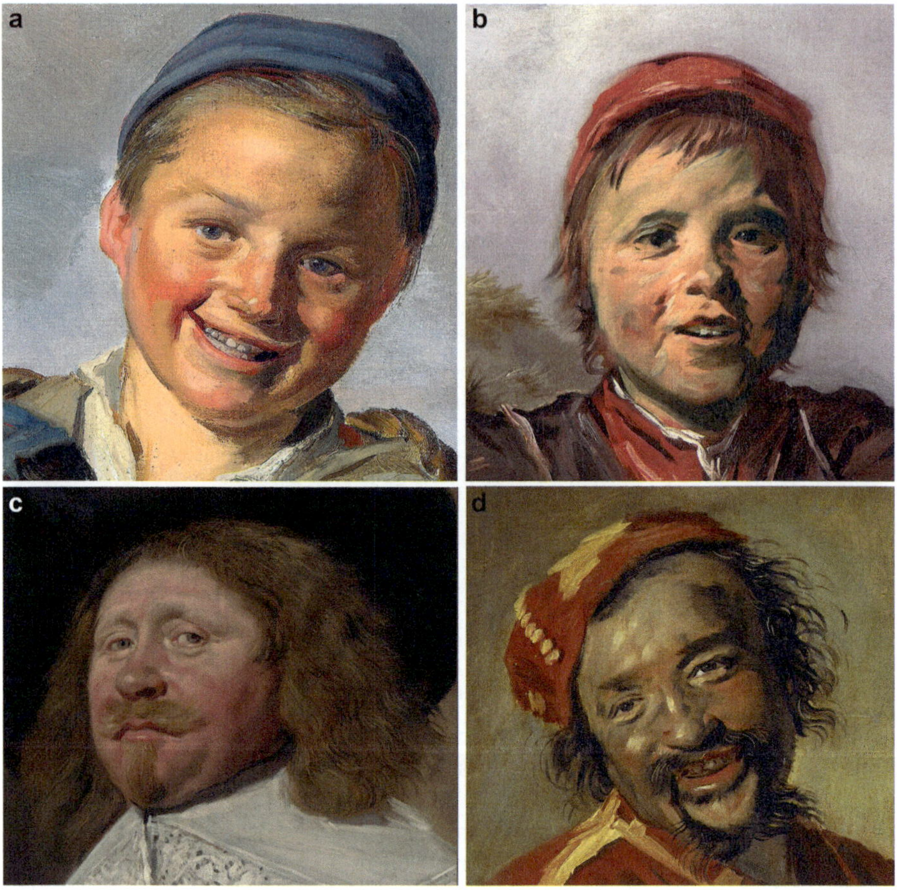

Fig. 5.7 (**a**) Detail of Fig. 5.1; (**b**) Detail of Fig. 5.3; (**c**) Detail of (Frans Hals, *Portrait of a man, possibly Nicolaes Pietersz Duyst van Voorhout*, c. 1636–1638, oil on canvas, 80.6 × 66 cm. (Metropolitan Museum of Art, New York) showing highlights on the face of the sitter); (**d**) Detail of Frans Hals, *Pekelharing*, c. 1628–1630, oil on canvas, 75 × 61.5 cm. (Gemäldegalerie, Kassel)

variations in quality within the group are due at least in part to workshop assistance.[11] It is evident, for example that the landscape in both *Two fisherboys* and the *Fisherboy* is by a different hand than the figures. Such workshop participation is also evident in Hals's commissioned portraits, especially in the late 1630s.

Two fisherboys is different in colour and tonality from the *Fisherboy*, especially in the strikingly tanned skin colour of the boys (Figs. 5.7a, b). Also, for a genre painting the brushwork is relatively smooth, as for example in the diffused highlight on the boy's forehead, painted in a way that resembles a commissioned portrait (for example, Fig. 5.7c). Hals's genre paintings generally possess more pronounced brushwork, for example in the distinctive loose accents in his *Pekelharing* (Fig. 5.7d).

[11] For this reason, Claus Grimm will use new categories in his upcoming revised oeuvre catalogue, indicating the level of involvement of the master, see above Chap. 2 and (Grimm forthcoming).

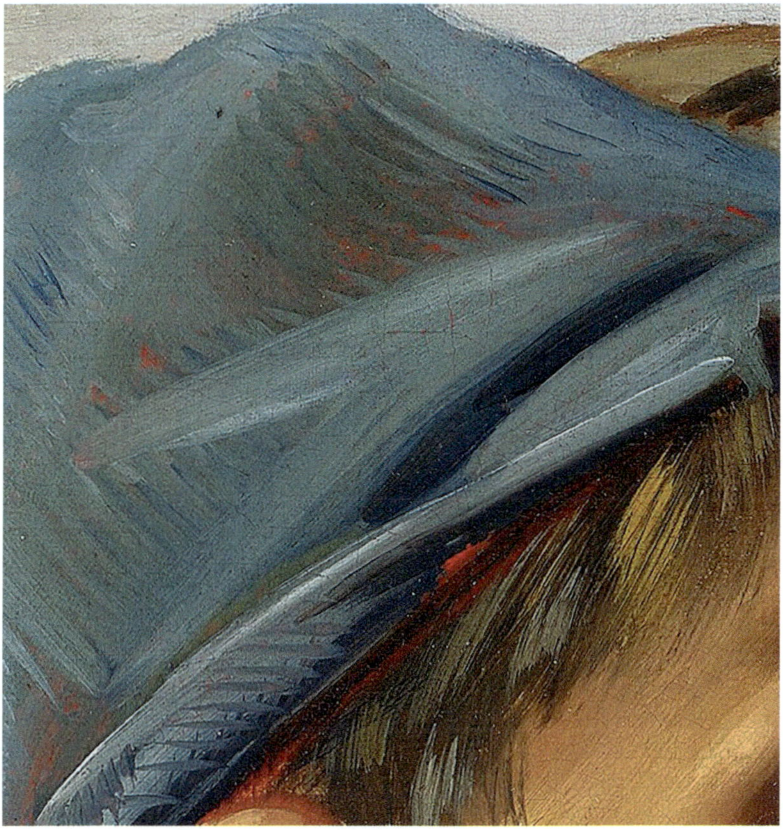

Fig. 5.8 Detail of Fig. 5.1, showing the hat of the boy on the left

Although the Antwerp *Fisherboy* has similar pronounced, loose accents, its execution is more hesitant. The faces in *Two fisherboys*, on the other hand, although more smoothly executed, are painted with great ease and accuracy. Moreover, the paint application in *Two fisherboys* displays several hallmark traits of Hals's style, such as the hatching-like wet-in-wet painting technique used to indicate, for example, the eyebrows and the hat, almost as if the artist were drawing with a brush (Figs. 5.8 and 5.9), and the so-called 'ribbon touch'—a brushstroke with raised edges on each side—in the thick white and red accents (Figs. 5.10a, b). Similar 'ribbon' touches can also be found in the *Fisherboy* (Fig. 5.10c).

Furthermore, *Two fisherboys* and the *Fisherboy* show a similar use of materials, which is consistent with Hals's workshop and with seventeenth-century practice in general, although there are also distinctive differences from the latter. To study the paintings' material characteristics, the authors used the relatively new technique of MA-XRF (see Alfeld et al. 2011). This allows visualisation of the distribution of elements in a flat sample, such as an easel painting, in a non-destructive manner. This is achieved by scanning the surface of the sample with a focused X-ray beam and analysing the emitted fluorescence radiation. As the x-ray beam scans the

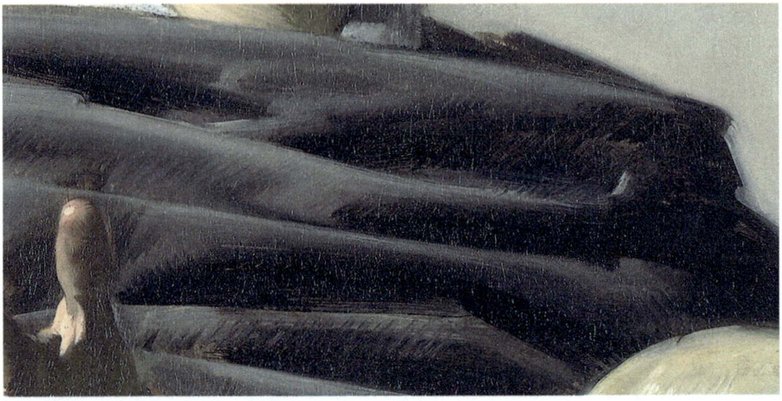

Fig. 5.9 Detail of Frans Hals, *Young man with a skull*, c. 1626–1628, oil on canvas, 92.2 × 80.8 cm. (National Gallery, London)

painting, it produces thousands and sometimes millions of data points. These can be plotted as elemental distribution maps, which may be interpreted as pigment distribution images.[12]

Both paintings contain bone black (Ca & P), umber (Mn and Fe), ochres (Fe), azurite (Cu), smalt (Co, Ni and K) and lead white (Pb). Their use is consistent with Hals's workshop practice, as for example in the use of bone black for shadows in the flesh tones (Figs. 5.11a, b). Copper is found only in the landscape backgrounds (in the form of azurite),[13] not in shadows in the face or clothing, which is helpful in dating the pictures, since so far as is known azurite occurs in facial shadows only in Hals's paintings created between circa 1627 and the late 1630s (see Chaps. 3 and 6). On stylistic grounds, differing dates were proposed for *Two Fisherboys* in the 2017 auction catalogue: c. 1627 (Pieter Biesboer), 1627–1630 (Norbert Middelkoop) and 1634–1637 (Claus Grimm).[14] From a material technical point of view, the last date is the most likely.

Distinctive differences between the works concern the use of lead tin yellow (Sn) and mercury or vermilion (Hg). Lead tin yellow occurs only in the *Fisherboy*, in the light green part of the background landscape at the right, confirming its seventeenth-century date, as the pigment went out of use around 1750 and was rediscovered only in 1941 (Jacobi 1941). Vermilion is used quite generously in *Two fisherboys*, appearing both in the faces and in the red clothing (Figs. 5.11c, d), whereas the *Fisherboy*

[12] We used the instrument and method described by (Alfeld et al. 2013) and processed the data using the method described by (Alfeld and Janssens 2015). Both paintings were mapped with a step size of 700 micron and a dwell time of 70 ms/step. PyMCA and Datamuncher software packages were used to process the collected XRF data cubes.

[13] Paint sample analyses by Jaap Boon and Nicholas Eastaugh confirmed that the background landscape of *Two fisherboys* contains azurite (see Boon 2017, esp. Appendix III with cross section (BLA3C); and Christie's 2017).

[14] Sale, Christie's, London *Old Masters Evening Sale*, 6 July 2017, lot 28.

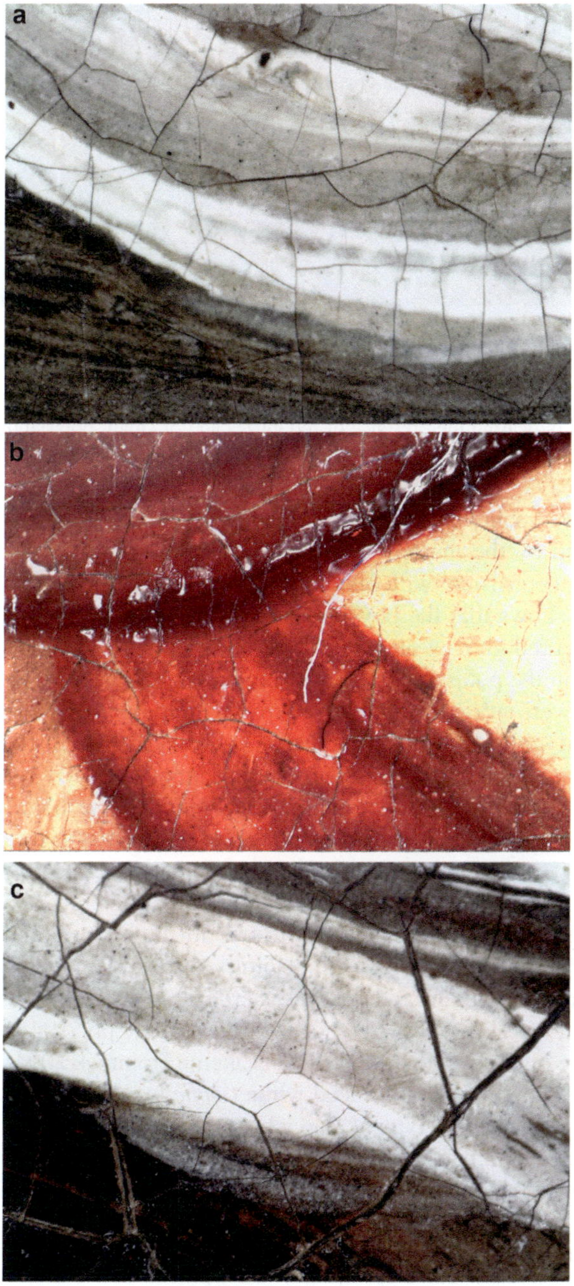

Fig. 5.10 (**a**) Detail of Fig. 5.1, taken with a dino-lite microscope (x65), showing the so-called 'ribbon touches'; (**b**) Detail of Fig. 5.1, taken with a dino-lite microscope (x65), showing a brush-stroke with raised edges; (**c**) Detail of Fig. 5.3, taken with a dino-lite microscope (x65), showing cvidcncc of 'ribbon touch'

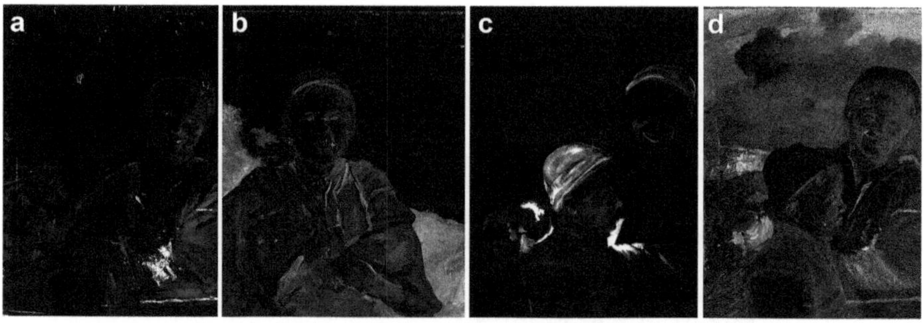

Fig. 5.11 (**a**) Calcium map of Fig. 5.1, showing bone black in shadows; (**b**) Calcium map of Fig. 5.3, showing bone black in shadows; (**c**) MA-XRF map of Fig. 5.1, showing the element mercury (hg), indicating the presence of vermilion; (**d**) MA-XRF map of Fig. 5.1, showing the element potassium (k) and indicating the presence of red glazes and of smalt; potassium is also present in earth pigments

is painted with red ochres, apart from one red brushstroke using vermilion, presumably a last touch by the master. Hals's use of vermilion varies, and both types of application seem consistent with his workshop practices. The vermilion map of *Two fisherboys* also shows that the boys' hats were originally painted in red, mostly in vermilion with a red glazing on top, visible around the lower edge of the hat of the boy on the left in the potassium (K) map (probably alun, used as a substrate for red lake). The hats were subsequently painted over with indigo blue, a pigment that cannot be identified with MA-XRF analyses; its presence was confirmed in two earlier analyses of paint samples using Raman spectroscopy (Boon 2017; Christie's 2017). Both the hatching-like wet-in-wet painting technique in the blue hats, mentioned above, and the rather open manner of painting, exposing parts of the red layer underneath, especially around the contours, are characteristic of Hals's style.[15]

The use of indigo is also significant. In the Northern Netherlands, Hals was the first to use indigo in the uppermost paint layer in important commissions around 1627. In the first half of the seventeenth century its use seems to have been largely limited to Haarlem painters and four Amsterdam contemporaries, two of whom were influenced by Hals (Van Eikema Hommes 2004, pp. 104–109). Moreover, Hals's use of the pigment is distinctive (Van Eikema Hommes 2004, p. 155 and note 80, p. 159). He mixed it with pure lead white ('*schelpwit*'), whereas his contemporaries mostly used a cheaper type of lead white mixed with chalk for this. Moreover, some of the lead white particles in Hals's mixture are unusually large (32–70 μm) since he used a coarse type of lead white that his fellow Haarlem painters Johannes Cornelisz Verspronck and Hendrik Gerritsz Pot reserved for the lower layers in their pictures (Hendriks et al. 1998, p. 167). The indigo in the fisherboys' hats is entirely

[15] On this open manner of painting and the 'halo' effect around the contours (see Tummers et al. 2019).

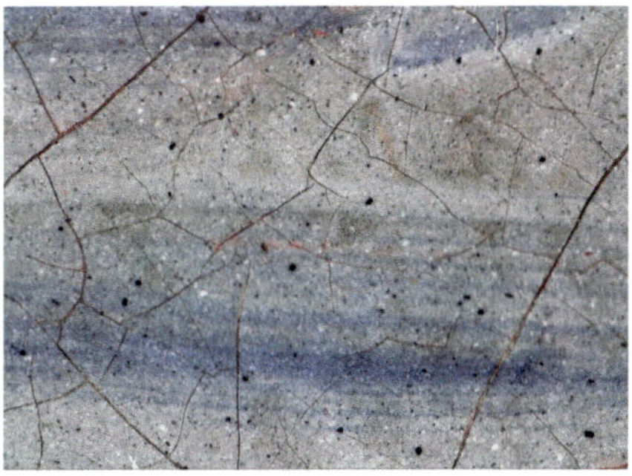

Fig. 5.12 Photograph of Fig. 5.1 taken with a dino-lite microscope (X65), showing indigo blue on top of the red hat

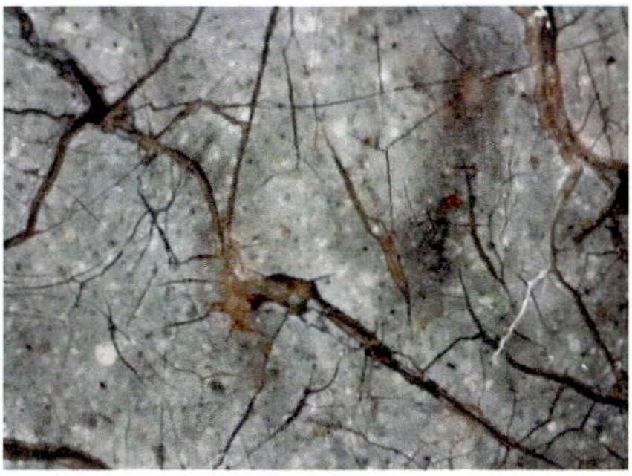

Fig. 5.13 Detail of Frans Hals, *Officers and Sergeants of the St George Civic Guard*, 1639, oil on canvas, 218 × 421 cm. (Frans Hals Museum, Haarlem). Micrograph taken with a dino-lite microscope (X65), showing detail of the blue sash of the third figure from the right

consistent with Hals's practice (Figs. 5.12 and 5.13),[16] which not only confirms the genesis of the work in Hals's studio, but also disproves Van Dantzig's theory that this picture is the work of a nineteenth-century British forger. Natural indigo was no longer in use as a pigment in the nineteenth century and its use in the seventeenth century by such painters as Hals was rediscovered only in 1928 by A.M. de Wild—a finding that was not widely known for a decade (De Wild 1928, pp. 49–50). In 1937 Van Dantzig speculated that Hals would have used Prussian blue and ultramarine to paint blue colours (Van Dantzig 1937, p. 25).

To conclude, despite its seemingly modern appearance, *Two Fisherboys* is not a forgery by Van Meegeren or a nineteenth-century artist. A close look at its style and technique confirms that the painting was produced by Hals and his studio. Hals must have deemed it worthy of bearing his name.

References

Alfeld, M., Janssens, K.: Strategies for processing mega-pixel X-ray fluorescence hyperspectral data: a case study on a version of Caravaggio's painting supper at Emmaus. J. Anal. At. Spectrom. **30**(3), 777–789 (2015). https://doi.org/10.1039/C4JA00387J

Alfeld, M., Janssens, K., Dik, J., De Nolf, W., Van Der Snickt, G.: Optimization of mobile scanning macro-XRF systems for the in situ investigation of historical paintings. J. Anal. At. Spectrom. **26**(5), 899–909 (2011). https://doi.org/10.1039/c0ja00257g

[16]The presence of indigo in *Two fisherboys* was confirmed in two technical studies in 2017, see note 14.

Alfeld, M., Pedroso, J.V., Van Eikema Hommes, M., Van Der Snickt, G., Tauber, G., Blaas, J., Haschke, M., Erler, K., Dik, J., Janssens, K.: A mobile instrument for in situ scanning macro-XRF investigation of historical paintings. J. Anal. At. Spectrom. **28**(5), 760–767 (2013). https://doi.org/10.1039/c3ja30341a

Art Analysis and Research: Report AAR0828.A (17 July 2017). Christie's (2017)

Boon, J.J.: A Material Study of the Painting "Vissersjongens" (4 July 2017). JAAP Enterprises for Art Scientific Studies, Amsterdam (2017)

Bought for £3 and sold for 2940: the romance of a Frans Hals (4 Jan 1936). The Illustrated London News (1936)

Colombini, M.P.: Gas chromatography coupled with Mass spectrometry. In: De Jongh, I., Den Leeuw, M., Mass, J., Pinna, D., Shindell, L., Spapens, O. (eds.) Technical Art History: a Handbook of Scientific Techniques for the Examination of Works of Art, pp. 91–92. Authentication in Art Foundation, The Hague (2018)

De Jongh, E.: Halsen van latere makelij. Kunstschift. **2**, 32–39 (2007)

De Wild, A.M.: Het natuurwetenschappelijk onderzoek van schilderijen. Levisson, The Hague (1928)

Fortunato, G., Ritter, A., Fabian, D.: Old masters' lead white pigments: investigations of paintings from the 16th to the 17th century using high precision lead isotope abundance ratios. Analyst. **130**(6), 898–906 (2005). https://doi.org/10.1039/b418105k

Grimm, C.: Frans Hals: Entwicklung, Werkanalyse, Gesamtkatalog. Gebr. Mann, Berlin (1972)

Grimm, C.: Frans Hals and his Workshop. RKD, The Hague (forthcoming)

Hendriks, E., Van Eikema Hommes, M., Levy-Van Halm, K.: Indigo used in the Haarlem Civic Guard group portraits by Frans Hals. Stud. Conserv. **43**(1), 166–170 (1998). https://doi.org/10.1179/sic.1998.43.Supplement-1.166

Huussen, H.A. (ed.): Cahiers uit het Noorden XX: Henricus [Han] Antonius van Meegeren (1889-[1945]): documenten betreffende zijn leven en strafproces. Huussen, Zoetermeer (2009)

Jacobi, R.: Über den in der Malerei verwendeten gelben Farbstoff der Alten Meister. Z. Angew. Chem. **54**, 28–29 (1941)

Jacques van Meegeren noemt opnieuw aantal valse "Frans Halsen" (9 Oct 1965). Provinciale Zeeuwsche Courant (1965)

Kreuger, F.H.: A New Vermeer: Life and Work of Han van Meegeren. Quantes, Rijswijk (2007)

Niet van Hals, maar van Van Meegeren sr. (9 Oct 1965). Utrechts Nieuwsblad (1965)

Nieuwe onthulling over vervalsingen: "Van Meegeren wijst weer 4 valse doeken van Hals aan" (8 Oct 1965). Trouw (1965)

Slive, S.: Frans Hals, 3 vols. Phaidon, London (1970–1974)

Slive, S. (ed.) Frans Hals, exh. cat., National Gallery of Art/Royal Academy of Arts/Frans Hals Museum, Washington/London/Haarlem (1989–1990)

Sotheby's declares "Frans Hals" work a forgery (6 Oct 2016). BBC News (2016). https://www.bbc.com/news/entertainment-arts-37574411. Accessed 20 Oct 2019

Trivas, N.S.: The Paintings of Frans Hals. Phaidon, London (1941)

Tummers, A.: The Eye of the Connoisseur: Authenticating Paintings by Rembrandt and his Contemporaries. Amsterdam University Press/Getty Publications, Amsterdam/Los Angeles (2011)

Tummers, A., Wallert, A., De Keyser, N.: Supplementing the eye: the technical analysis of Frans Hals's paintings—II. Burlingt. Mag. **161**, 996–1003 (2019)

Van Dantzig, M.M. Frans Hals: Echt of Onecht, Naar aanleiding van de Frans Hals-tentoonstelling 1937. Paris, Amsterdam (1937)

'Van Meegeren's zoon noemt vier valse doeken' (8 Oct 1965). Leidsche Courant (1965)

Van Eikema Hommes, M.: Changing Pictures: Discoloration in 15th–17th-Century Oil Paintings. Archetype Publications, London (2004)

Wallert, A.: Werkwijzen, trucs, materialen en technieken: vervalsingen op het hoogste niveau. In: Lammertse, F., Garthoff, N., Van Laar, M., Wallert, A. (eds.) De Vermeers van Van Meegeren: kennerschap en de techniek van het vervalsen, pp. 66–99. Museum Boijmans Van Beuningen, Rotterdam (2011)

Wallert, A.: Examination of stable isotope ratios: consequences for art history. Paper presented at the Technart Conference, Catania, 2015

Wallert, A., Van de Laar, M.: Expertise in the Van Meegeren case: the Contributions by Coremans, Froentjes and De Wild. In: Deneffe, D. and Van Wijnsberghe, D. (eds.) A Man of Vision: Paul Coremans and the Preservation of Cultural Heritage Worldwide, Proceedings of the International Symposium Paul Coremans Held in Brussels, 15–17 June 2015. Scientia Artis, vol. 15, pp. 202–214. Koninklijk Instituut voor het Kunstpatrimonium (KIK), Brussels (2019)

Chapter 6
Case Study 4: Frans Hals & Co: the Civic Guard Portraits and the Attribution of *The Meagre Company*

Anna Tummers, Arie Wallert, Robert G. Erdmann, Joris Dik, Nouchka De Keyser, Annelies van Loon, and Erma Hermens

Abstract This study concerns Hals's civic guard portraits, his largest and most prestigious commissions. All five Haarlem civic guard portraits, which were completed under Hals's supervision, are studied in depth (using super-high-resolution photography, MA-XRF scans and infrared reflectography), as well as a contested Amsterdam civic guard portrait, *The Meagre Company*. That last work was begun by Hals and – because of a conflict with his patrons - eventually finished by another painter, Pieter Codde (1599–1678). Hals experts have long disagreed about the exact extent of Hals's and Codde's contribution to the painting. This case study offers deeper insights into Hals's technique, style and workshop practice as well as the differences with Codde, and subsequently sheds new light on the attribution of *The Meagre Company*.

The research for this article was funded by the Netherlands Organisation for Scientific Research (NWO) as part of the projects *Frans Hals or not Frans Hals* (Museumbeurs 2016–18) and *21st Century Connoisseurship* (NICAS Seed Money Grant 2018–2022); by the Prins Bernard Cultuurfonds Conservatoren Stipendium, 2018, by a private benefactor and by a grant from the Leiden University Centre for the Arts in Society (LUCAS). The authors would like to thank Lidewij de Koekkoek, Ann Demeester, Liesbeth Abraham, Mireille de Marvelde, Jessica Roeders, Marrigje Rikken, Norbert Middelkoop, Ulrich Roldanus, Herman Bessels, Wilfred Zwart and Judith Harren at the Frans Hals Museum; Taco Dibbits, Robert van Langh, Pieter Roelofs, Gregor Weber, Friso Lammertse, Petria Noble, Anna Krekeler, Susan Smelt, Gwen Tauber, Gerrit Albertson, Nienke Woltman, Laura Raven, Jolanda van Iperen, Carola van Wijk, Rik Klein Gotink, Suzanne Plaum, Ton Nuijen, Ivo Rijkers, Gijs Smit, Tessa Hogeboom, Frederique van Reij and Pauline Kintz at the Rijksmuseum for their assistance with the research. Furthermore, Maarten van Bommel, Ella Hendriks, Eric Jan Sluijter, Carol Pottasch, Rosa Hoogenboom, Jasper Hillegers and Lotte Kokkedee -as well as the experts mentioned above- provided valuable feedback during five expert meetings organised in the context of the research projects, and Eric Jan Sluijter also gave thoughtful comments on an earlier draft for this chapter.

A. Tummers (✉)
Ghent University, Ghent, Belgium
e-mail: Anna.Tummers@UGent.be

A. Tummers, R. G. Erdmann, *Frans Hals or not Frans Hals*, Cultural Heritage Science, https://doi.org/10.1007/978-3-031-59489-2_6

The third and largest case study concerns Hals's civic guard portraits, his largest and most prestigious commissions. These works depict the officers of the city guards, the men who both protected the city from –occasional– threats from the outside and – more commonly– maintained the order within, including by patrolling the streets each night before the city gates were closed (the so-called 'night watch').[1] As we have seen in the previous chapter, Malle Babbe was arrested on one of these rounds and brought to the Workhouse by a captain of the Haarlem civic guard. After three years of service, the city of Haarlem invited the civic guard officers to a banquet to mark the end of their term of office. Hals depicted this occasion in three of his civic guard group portraits. He shows us elegantly dressed men with sashes and banners gathered around a sumptuous banquet. Sources reveal that the gentlemen did themselves proud at such banquets, feasting for days and nights on end and consuming large amounts of food and wine (Knevel 1994). An early seventeenth-century account from the St George Civic Guard in Haarlem shows that wine to the value of 298 pounds and 11 stivers was drunk during an officers' farewell banquet—a sum equivalent to the annual salary of a skilled craftman.[2] In 1633, the Haarlem burgomasters even issued a new rule because the banquets often lasted for seven full days and nights and cost the city exorbitant amounts of money: from then on, the feasts could only last four days and the guardsmen were no longer allowed to bring their wives and children (see Knevel 1994, pp. 298–299; Tummers 2011a, cat. no. 43, pp. 152–153).

[1] On the civic guards in the Netherlands see (Carasso-Kok et al. 1988), especially the essay by Paul Knevel, 'De kracht en de zenuwen van de Republiek: de schutterijen in Holland 1580–1650'. The title refers to (Schrevelius 1648, pp. 320–321), who explained that the civic guards were founded 'to protect the cities'(*tot bewaringhe vande Steden*) and that the guardsmen served as the 'power and nerves of the Republic' (*kracht en zenuwen van de Republijck*).

[2] In Holland most councils still used the pound as a monetary unit; the value was the same as a guilder (Kurtz 1979, p. 17; Rietveld in Tummers (ed.) 2013b, cat. no. 26). Comparative salary information provided by Marten Jan Bok from the University of Amsterdam.

A. Wallert (✉)
University of Amsterdam, Amsterdam, The Netherlands

Rijksmuseum, Amsterdam, The Netherlands
e-mail: arie.wallert@gmail.com

R. G. Erdmann (✉)
University of Amsterdam, Amsterdam, The Netherlands
e-mail: r.g.erdmann@uva.nl

J. Dik (✉)
Delft University of Technology, Delft, The Netherlands
e-mail: j.dik@tudelft.nl

N. De Keyser (✉) · A. van Loon (✉)
Rijksmuseum, Amsterdam, The Netherlands
e-mail: N.de.Keyser@rijksmuseum.nl; A.van.Loon@rijksmuseum.nl

E. Hermens (✉)
Hamilton-Kerr Institute and Conservation and Science Division, Fitzwilliam Museum, University of Cambridge, Cambridge, UK
e-mail: eh707@cam.ac.uk

For this research project, all five Haarlem civic guard portraits that were finished under Hals's supervision were studied in depth, as well as the Amsterdam civic guard portrait *The Meagre Company* (Figs. 6.1, 6.2, 6.3, 6.4, 6.5 and 6.6). The latter was begun by Hals and –because of a conflict with his patrons– eventually completed by a local painter, Pieter Codde (1599–1678). Hals experts have long disagreed about the exact extent of Hals's and Codde's contributions. The aim of this case study is to provide deeper insights into Hals's characteristic style, technique, use of materials and workshop practice, to explore the key differences with Codde, and subsequently to shed new light on the attribution of *The Meagre Company*. Specifically for this study, all six civic guard portraits were examined in depth using visible light, a hand-held

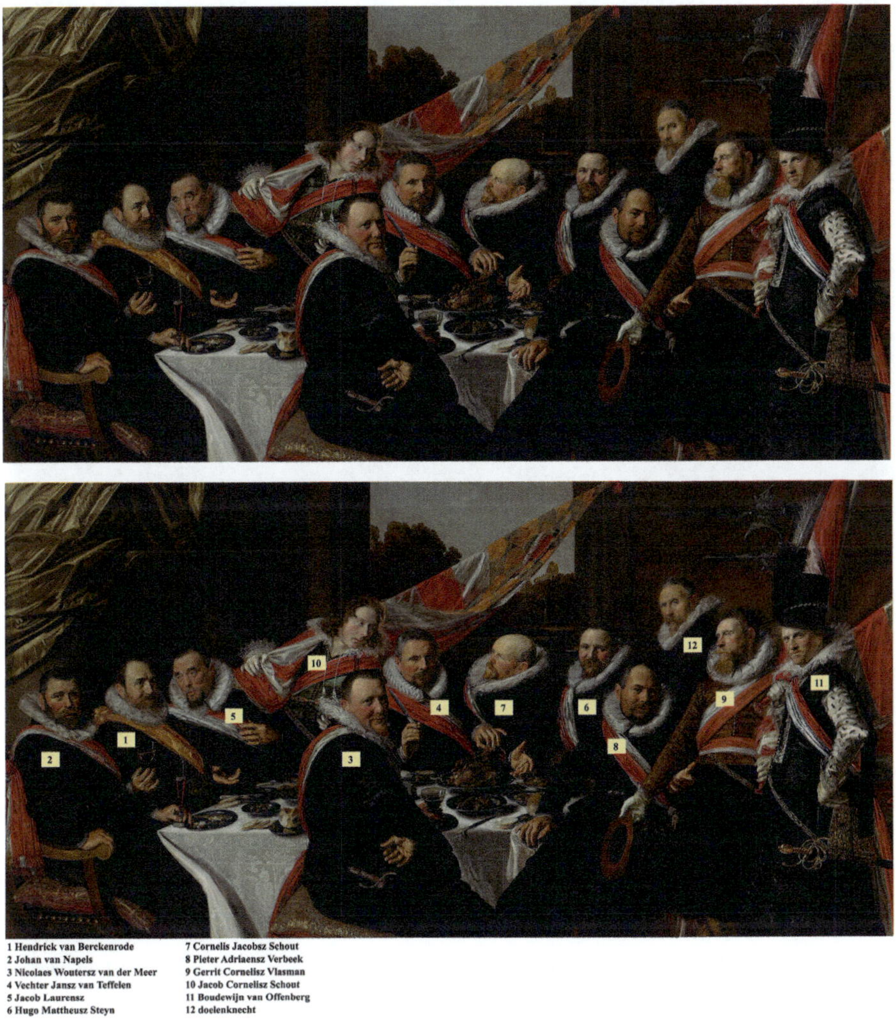

Fig. 6.1 Frans Hals, *Banquet of the Officers of the St. George Civic Guard*, 1616, oil on canvas, 175 × 324 cm. (Frans Hals Museum, Haarlem) (sn.pub/rxn6bc)

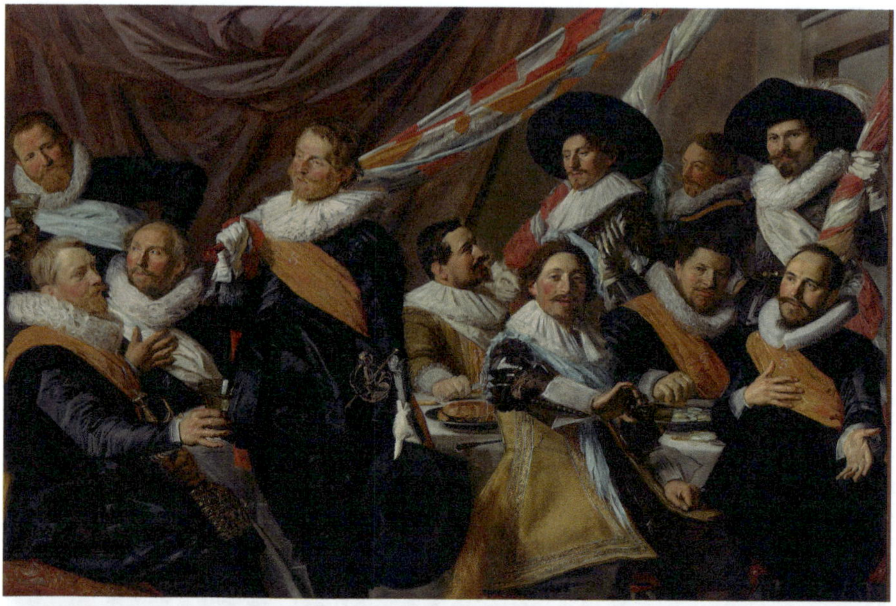

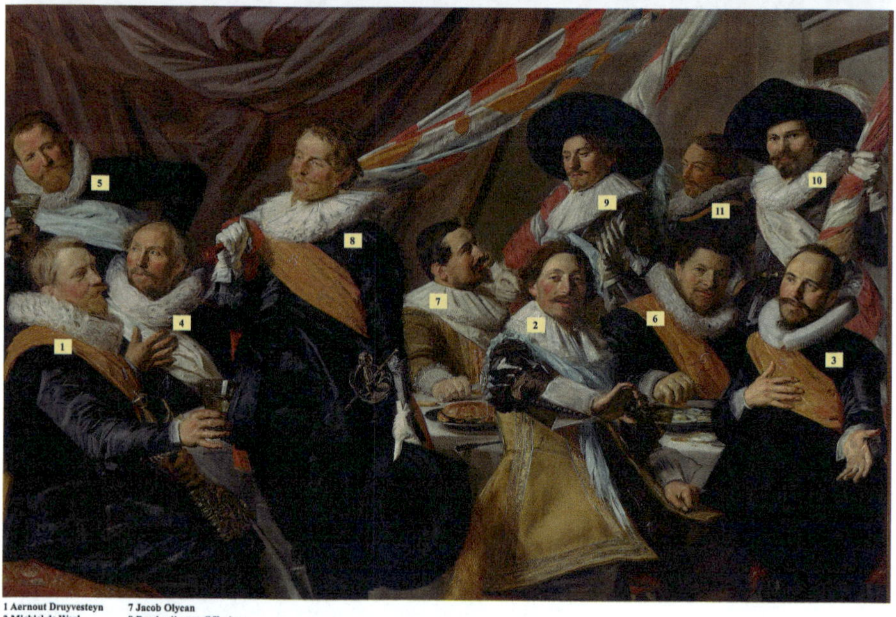

Fig. 6.2 Frans Hals, *Banquet of the Officers of the St. George Civic Guard*, 1627, oil on canvas, 179 × 257.5 cm. (Frans Hals Museum, Haarlem) (sn.pub/9v813n)

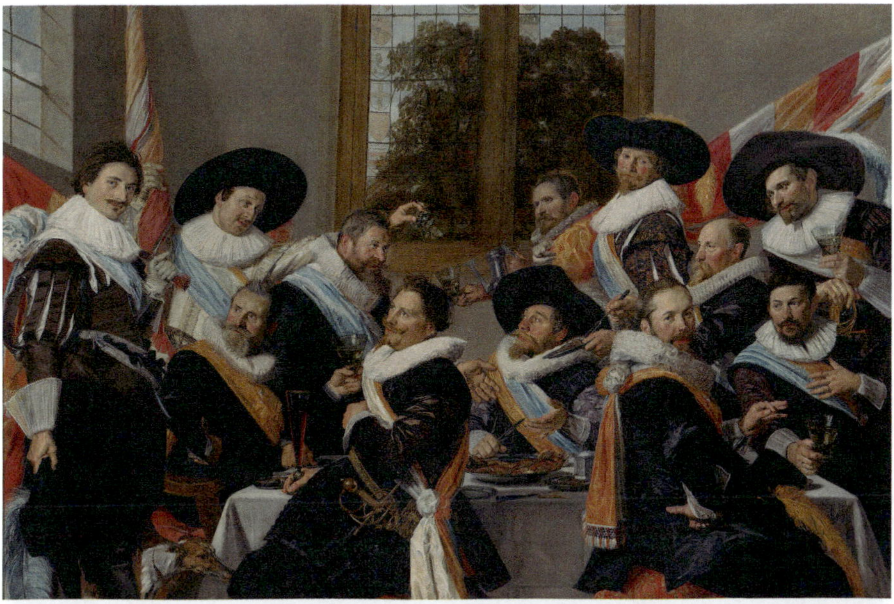

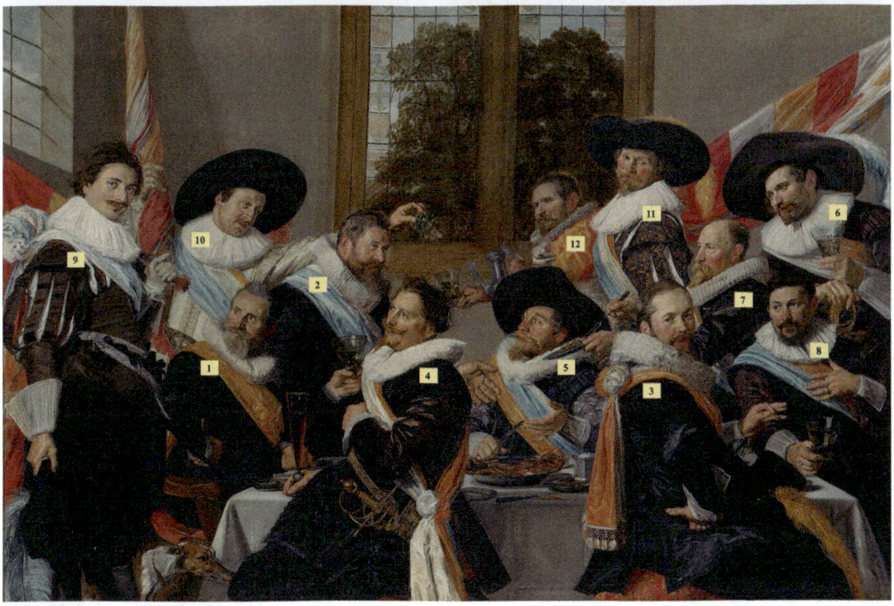

1 Willem Claesz Vooght
2 Johan Damius
3 Willem Warmont
4 Johan Schatter
5 Gilles de Wildt
6 Nicolaes van Napels
7 Outgert Akersloot
8 Matthijs Haeswindius
9 Adriaen Matham
10 Loth Schout
11 Pieter Ramp
12 Willem Ruychaver

Fig. 6.3 Frans Hals, *Banquet of the Officers of the Calivermen Civic Guard*, 1627, oil on canvas, 183 × 266.5 cm. (Frans Hals Museum, Haarlem) (sn.pub/2ma71r)

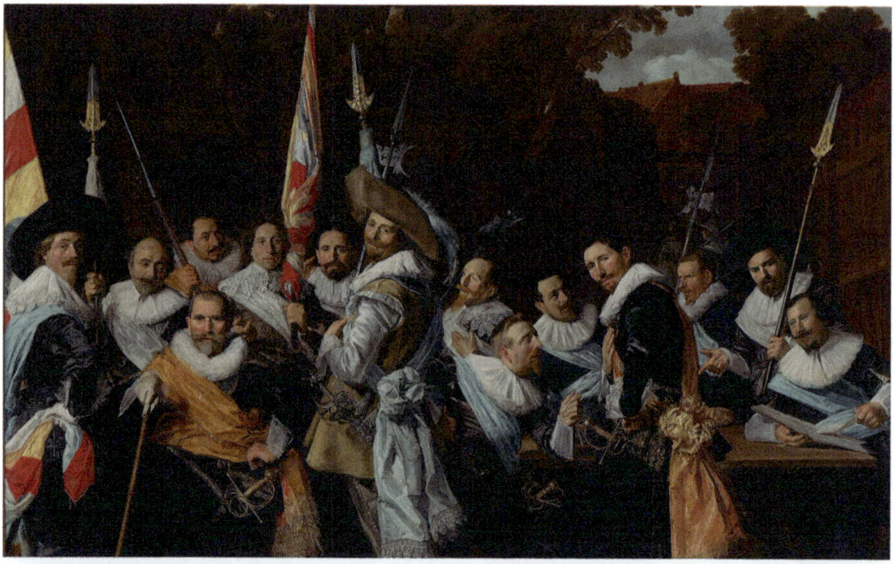

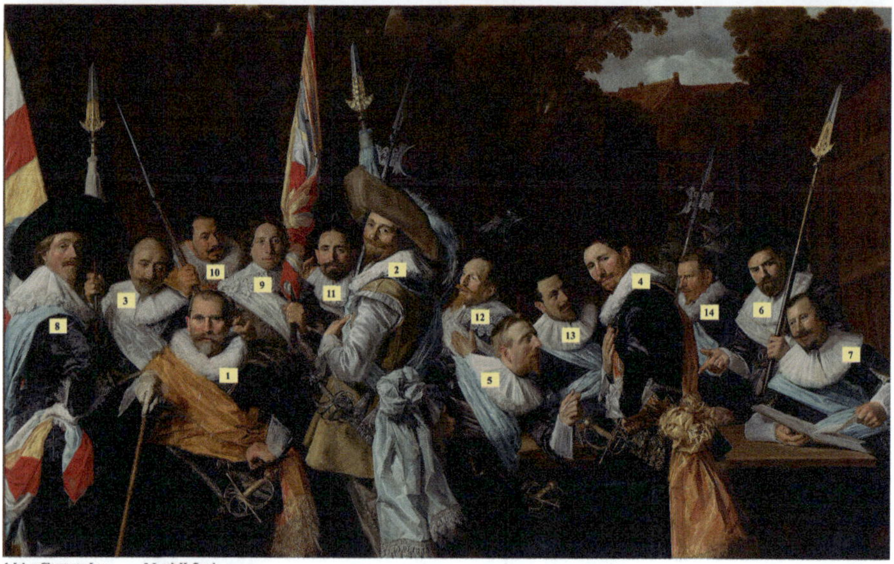

Fig. 6.4 Frans Hals, *Officers and Subalterns of the Calivermen Civic Guard*, 1633, oil on canvas, 207 × 337 cm. (Frans Hals Museum, Haarlem) (sn.pub/pmyayh)

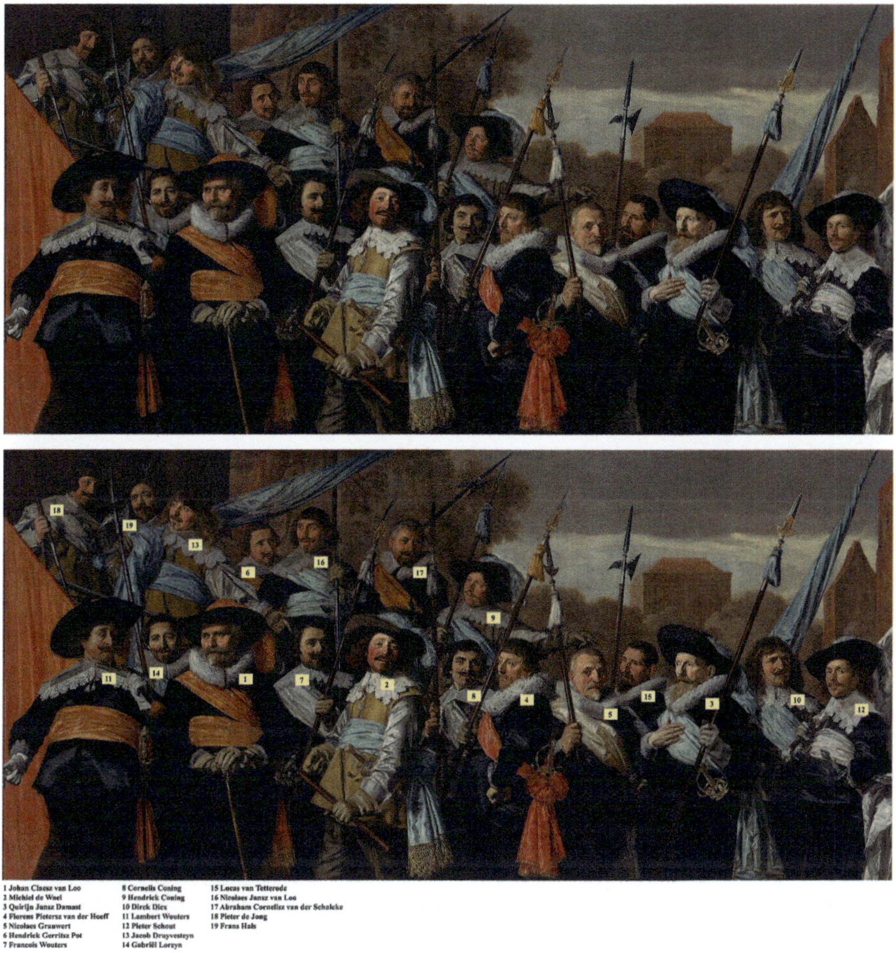

Fig. 6.5 Frans Hals, *Officers and Subalterns of the St. George Civic Guard*, 1639, oil on canvas, 218 × 421 cm. Frans Hals Museum, Haarlem) (sn.pub/9cja2b)

digital microscope and high-resolution photography, as well as various analytical techniques including infrared reflectography (IRR) and macro X-ray fluorescence spectroscopy imaging (MA-XRF).[3] Furthermore, existing documentation was re-

[3]The 20 micrometer pixel high resolution photography was done by Carola van Wijk and Rik Klein Gotink; the latter also made the infrared reflectograms (IRR). At the Rijksmuseum, the MA-XRF scanning of *The Meagre Company* was done by a large team consisting of: Annelies van Loon, Nouchka De Keyser, Anna Krekeler, Susan Smelt, Gwen Tauber, Erma Hermens, Gerrit Albertson, Nienke Woltman, Laura Raven and Petria Noble. The MA-XRF scanning of the five civic guard paintings at the Frans Hals museum was led by Nouchka De Keyser with assistance from Jennie Allred and Rosa Hoogenboom. All paintings were mapped with a step size of 700 micrometer and a dwell time of 70 ms/step, using a Bruker M6 Jetstream, and the method described by (Alfeld et al. 2013).

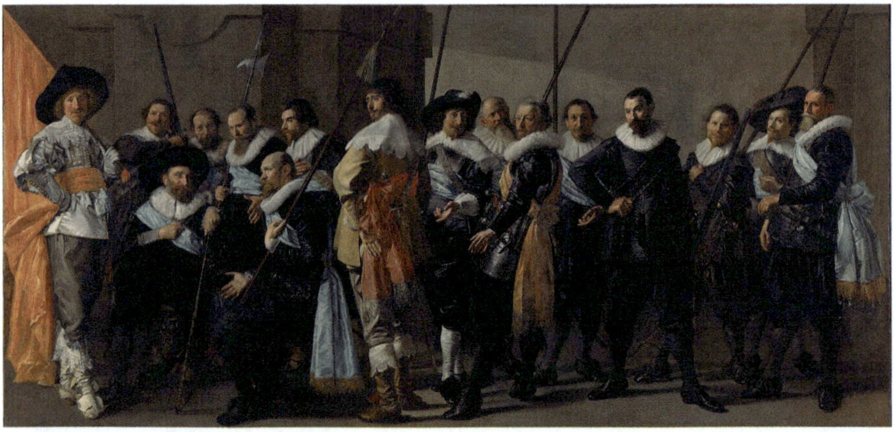

Fig. 6.6 Frans Hals and Pieter Codde, *Militia Company of District XI under the Command of Captain Reynier Reael, Known as 'The Meagre Company'*, 1637, oil on canvas, 209 × 429 cm. (Rijksmuseum, Amsterdam) (sn.pub/43xin5)

examined and a small-scale guard scene by Codde was analysed for comparison (Fig. 6.7). The stylistic and technical analyses were related to seventeenth-century art theory and primary sources in order to better understand the works in their original context. As the study yielded a very large amount of data, advanced digital tools were created to facilitate the interpretation and comparison (https://images.erdmann.io/Draper/?collection=/NICAS/Frans_Hals).

Specifically for this study, the six large civic guard portraits were examined in the galleries both during and after opening hours.[4] The public MA-XRF scanning was a first both at the Rijksmuseum (January–February 2017, Gallery of Honour) and at the Frans Hals Museum (March–October 2017) (Fig. 6.8).[5] As the five reference works at the Frans Hals Museum contain a total of 68 portraits, this study provided a unique opportunity to explore the consistency and evolution in Hals's style, technique and workshop practice.

[4]The Rijksmuseum's technicians constructed an elevation for the MA-XRF scanner, allowing it to scan at different heights, in collaboration with Sapmetaal B.V. Beverwijk.

[5]For a short introduction to the scanning project see https://www.youtube.com/watch?v=G5VDHIztv4U and https://www.youtube.com/watch?v=gwHRH0YPPqI. The latter includes also an introduction to the NICAS *21st Century Connoisseurship* Project. In 2016, two other paintings by Frans Hals had already been scanned *in situ* at the Frans Hals Museum, in a gallery that was temporarily closed off to the public. In 2017, the public could witness the scanning in the large civic guard gallery, view the results in real time on the computer screen, and was informed about the research by an introductory video and wall text.

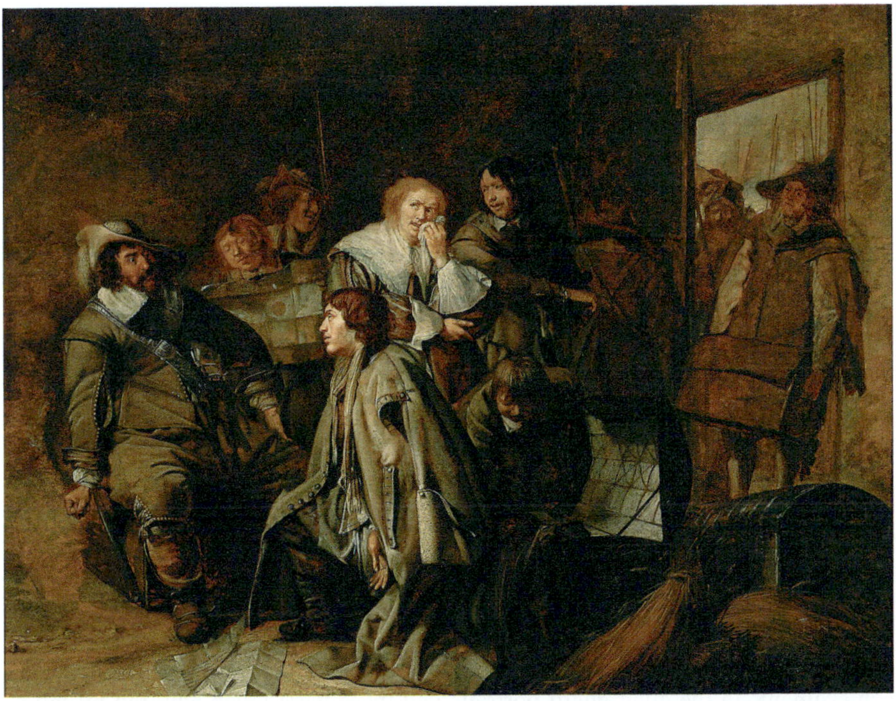

Fig. 6.7 Pieter Codde, *Plundering Soldiers in a Barn*, c. 1635, oil on panel, 35 × 44 cm. (Frans Hals Museum, Haarlem)

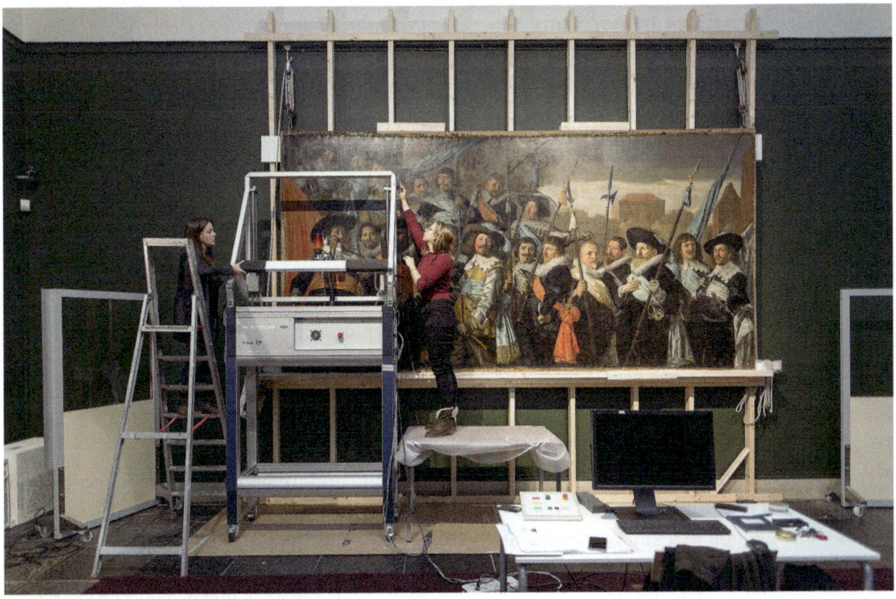

Fig. 6.8 MA-XRF scanning at the Frans Hals Museum by Nouchka De Keyser and Jennie Allred, Oct 2017

6.1 'Rightly Admired by the Greatest Masters'

It is hard to overstate the importance of Hals's civic guard portraits. According to Hals expert Seymour Slive, Hals's earliest work of this type announced the arrival of the golden age of Dutch painting like a cannon shot (Slive (ed.) 1989–1990, p. 15). His *Banquet of the Officers of the St George Civic Guard* (1616) is one of the most important group portraits of the Dutch seventeenth century—the dynamic, natural grouping of the figures, the true-to-life characterisation of the individual faces and the lifelike colours, lighting and shadows give it an unprecedented reality effect (Frans Hals, *Banquet of the Officers of the St. George Civic Guard*, 1616, oil on canvas, 175 × 324 cm. (Frans Hals Museum, Haarlem)). Whereas earlier civic guard portraits look rather contrived in the way the figures are grouped, and the expressions and movements seem somewhat wooden, Hals manages to give the impression that we are encountering this group of men at a fleeting moment in time, as if we have just walked in on them at their banquet (compare with Fig. 6.9) (see Tummers (ed.) 2013a, p. 14 ff).[6] This is no mean feat, and it seems no coincidence that this milestone was reached in Haarlem—at the time the leading centre of the arts in the Northern Netherlands, the 'Florence of the North'.[7] After a visit to Haarlem in October 1616, the English Ambassador in the Netherlands, Sir Dudley Carleton, wrote that the painters were the city's 'chiefest curiosity'.[8]

Already in the seventeenth century, the city of Haarlem took great pride in Hals's achievements in the civic guard portraits and these were much admired by fellow artists. The early Haarlem city descriptions by historians Samuel Ampzing (1628) and Theodoor Schrevelius (1648) praise Hals for the vitality of his work and for his unique, assured painting style, referring only to his civic guard paintings as specific examples. According to Ampzing, Hals's *Banquet of the Officers of the St George Civic Guard* (1627) was painted 'very boldly after life'.[9] Schrevelius explained that Hals 'through an exceptional manner of painting that is entirely his own, surpassed almost all, for there is in his paintings such vigour and life that he even seems to defy nature itself with his brush, to this testify all the likenesses he made, and these are incredibly numerous, which are painted in such a way that they seem to breathe and to live.'[10] According to a 1660 poem by Herman Frederik Waterloos of Amsterdam, Haarlem 'boasted' about Hals's 'early masterpieces', no doubt refer-

---

[6] See also cat. no. 25 in (Tummers (ed.) 2013b) on the importance of Cornelis van Haarlem's civic guard portraits in their own right and contemporary praise for these by Karel van Mander.

[7] See also below and note 14 on the importance of Hals's earliest civic guard portrait for the city of Haarlem. On the importance of Haarlem painting in the Netherlands, see (Van Bueren 1991).

[8] In a letter to John Chamberlain (see Gibson 2000, p. 115).

[9] 'Daer is van Franz Hals een groot stuck schilderije van enige Bevelhebbers der Schutterije in den Ouden Doelen ofte Kluyveniers, seer stout naer't leven gehandeld' (Ampzing 1628, p. 371).

[10] '[Hier can ich ook met stille swijghen niet verbij gaen, Frans ende Dirck Hals Gebroeders, van de welcke d'eene, die] deur een ongemeene manier van schilderen, die hem eyghen is, bij nae alle overtreft, want daer is in sijn schildery sulcke forse ende leven, dat hy te met de natuyr selfs schijnt te braveren met sijn Penceel, dat spreecken alle sijne Conterfeytsels, die hy ghemaeckt heeft, onge-

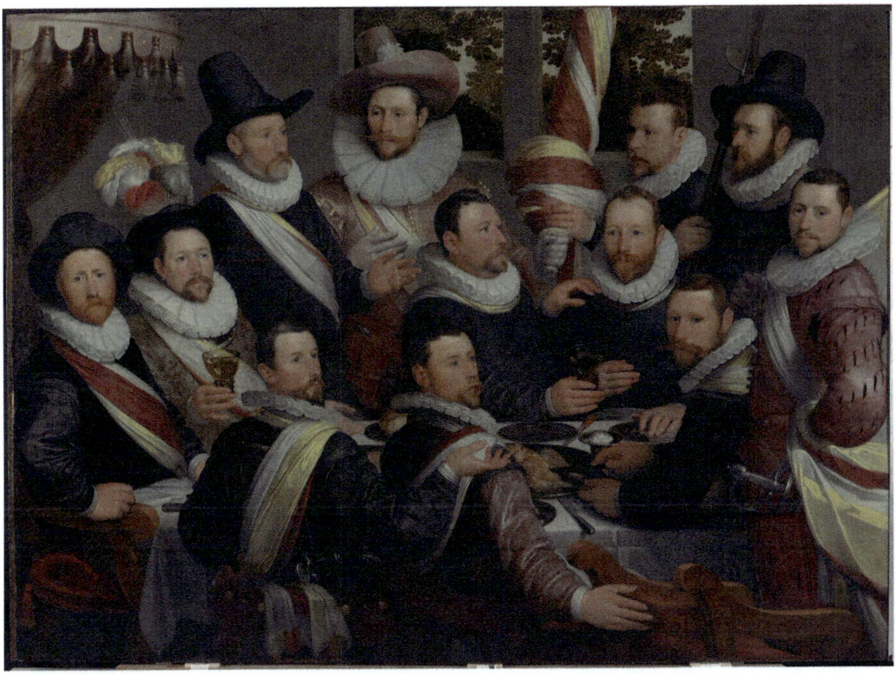

Fig. 6.9 Cornelis van Haarlem, *Banquet of the Officers and Subalterns of the Calivermen Civic Guard*, 1599, oil on canvas, 169 × 223.5 cm. (Frans Hals Museum, Haarlem)

ring to these large group portraits.[11] Significantly, the French diplomat Balthasar de Monconys, who visited the Haarlem militia headquarters in August 1663, noted in his diary that he had seen the civic guard portraits 'among others by Hals, who is rightly admired by the greatest painters'.[12]

Around 1687, the Haarlem burgomasters even decided to move Hals's earliest civic guard portrait (1616) from the militia headquarters to the city collection at the Prinsenhof, which had served as a gallery of honour for famous artists since the very beginning of the seventeenth century.[13] The Prinsenhof functioned as a

looflijcke veel, die soo gecoloreert zijn, dat se schijnen asem van haer te gheven, ende te leven' (Schrevelius 1648, p. 383). With thanks to Eric Jan Sluijter for his advice on the translation.

[11] 'stoft Haarlem op uw kunst, en jonghe meesterstukken', (Tummers and Gration 2013, p. 75 and note 9, p. 146).

[12] 'Il y a [à Haarlem] une maison nommé le *Doul* … et il y a force grands portraits de ces Messieurs [officers] assemblez, et entre autres d'Als, qui est avec raison admiré des plus grands peintres.' (De Monconys 1665–1666, vol. 1, p. 159). See also (Slive (ed.) 1989–1990, Hals Doc. 173; Tummers (ed.) 2013a, pp. 13 and 37, and notes 1 and 48, pp. 144–145).

[13] In 1687 the painter Jan de Bray was paid 275 guilders to paint a replacement portrait of the militiamen, which in the end he never finished. This is the earliest indication that Hals's civic guard portrait has been moved to the Prinsenhof (Köhler and Levy- Van Halm 1990, p. 18 and note 10, p. 19; Van Thiel-Stroman 2006, p. 118).

museum-avant-la-lettre and could be visited free of charge by art lovers.[14] It was the first time that a group portrait was added to the collection; almost every other picture there was a history painting. The choice for Hals's civic guard portrait cannot have had much to do with the people portrayed, dignitaries of a bygone generation. Rather, the inclusion of Hals's earliest civic guard portrait in the Prinsenhof collection underlines the extraordinary esteem in which Hals's work was held and, in his wake, the growing appreciation of portraiture as an art form. Although traditionally considered the lowest type of painting, Hals had succeeded in establishing portraiture as an important art form in its own right. Interestingly, in 1678, the painter and art theorist Samuel van Hoogstraten described portraiture as the lowest form of art *unless* the painter managed to capture the soul of the person portrayed.[15]

Hals's earliest biographer, Arnold Houbraken, emphasised that the appreciation for Hals's art extended beyond the Netherlands (just like the French diplomat Balthasar de Monconys had suggested when he wrote that Hals was admired by the greatest painters). According to Houbraken (1718–1721, vol. 1 pp. 92–93), the famous painter Anthony van Dyck (1599–1641) was an admirer of Hals and tried unsuccessfully to persuade him to go to England to work for King Charles I (1600–1649) (see also De Clippel and Vermeylen 2013, pp. 47–50; Tummers 2013a, p. 37). Although the passage was long dismissed as improbable in the Hals literature, research in preparation for 2013 Hals exhibition proved the contrary: not only did Van Dyck visit the Netherlands in the 1630s and pay homage to Hals in one of his portraits, but another less talented Haarlem artist, Hendrick Pot (1580/81–1657), left Haarlem to paint for Charles I around the same time (Tummers 2013a, p. 37). After his visit, Van Dyck was said to have remarked on many occasions that if Hals 'had blended his colours a little more delicately or thinly, he would have been one of the greatest masters. For his equal was not to be found regarding control of the brush in that, once he had applied the underpainting of a portrait, he could give the characteristic features, highlights and shadows their proper place with one brushstroke, without tempering or change'.[16]

The high esteem in which Frans Hals was held is also demonstrated by the fact that he was commissioned to paint the Amsterdam civic guard portrait known as *The Meagre Company*. Hals is one of only two non-Amsterdam artists to receive such a prestigious commission in the seventeenth century, the other being Paulus Moreelse, who painted a civic guard portrait in 1616 (Rijksmuseum, inv. no. SKC-623) (see

[14] On the use of the Prinsenhof as a museum-avant-la-lettre, see (Tummers and Gration 2013, p. 73; Van Bueren 1993).

[15] 'Jae de konterfeyters, die al reedelijke gelijkenissen maeken, en oogen, neuzen, en monden al fraeit jes naevolgen, wil ik zelfs niet buiten, of booven den eersten graet stellen, ten zyze haere tronyen met de gemelde hoedanigheyt van de verstandelijke ziele overstorten' (Van Hoogstraten 1678, p. 87).

[16] 'indien hy in zyne vermenginge iets meer van het teere, of dunne gehad had, hy een der grootste meesters zouden hebben geweest; want dat hy zyn weerga niet kende, die't penceel zoo tot zyn wil had, dat hy, na hy een Pourtret had aangeleid, de vaste wezenstrekken, hoogsels, diepsels met een penceelzet, zonder verzagtinge of verandering zoo hun behoorlyke plaats wist te geven' (Houbraken 1718–1721, vol. 1, pp. 92–93). With thanks to Eric Jan Sluijter for his advice on the translation.

Bikker (ed.) 2007, p. 181). Both Hals's reputation as a painter and the fame of his earlier pictures of this type, painted in Haarlem, must have made him an attractive candidate. By the time he began working on this group portrait, Rembrandt had already moved to Amsterdam and had created his well-known group portrait *The Anatomy Lesson of Dr. Nicolaes Tulp* (1632). But when it came to group portraits of civic guards no one enjoyed a greater reputation than Hals at the time. Significantly, the Amsterdam artist Nicolaes Eliasz Pickenoy borrowed several motifs from Hals 1632 portrait of the Amsterdam civic guard, *Banquet of Civic Guardsmen of the Company of Captain Jacob Backer and Lieutenant Jacob Rogh*, including the arrangement of the group of figures around a central figure carving a bird, and the gesture of a man holding his glass upside down (as a sign that he wants it refilled) (see Atkins 2013, pp. 60–62). Given the prestige of Hals's Amsterdam commission, however, it is all the more surprising that Hals never finished the work.

6.2 A Commission Gone Awry

The exceptional commission for *The Meagre Company* and the subsequent conflict between Hals and his patrons is well documented, allowing us to reconstruct the situation in great detail. The earliest document relating to the conflict is dated 19 March 1636 (Appendix I) (Slive (ed.) 1989–1990, Hals Doc. 73).[17] In the presence of the notary Frans Bruijningh, the alderman and captain of the civic guard Reynier Reaal (the seated figure on the left) and his lieutenant Cornelis Michielsz. Blaeuw (possibly the seated figure to his left) declared –on behalf of the entire civic guard– that they had commissioned Frans Hals three years before to paint a group portrait of all the officers of their civic guard. Hals, however, had defaulted: 'the aforementioned painting should have long been entirely completed and finished, as he [Hals] agreed and promised to do and accomplish, on St John's day of the previous year [i.e. 24 June 1635]' (Appendix I). But he had still not done so: 'In spite of several interpellations, both verbal and written, addressed to him', he had painted the picture 'only partially'. They therefore demanded that Hals come to Amsterdam within fourteen days to complete painting the piece and add the missing parts 'in proper form' (*in behoorlijcke forme*). If he still failed to comply, they would 'have another good master here complete the same piece' (*'t selve stuck alhier door een ander goet meester sullen laeten voltrecken*), and would demand the return of all previous payments made to Hals as well as all other expenses related to the commission 'until the very end of the case' (*tottet uyteynde van de saecke toe*).

The next day, 20 March 1636, the statement from captain Reynier Reaal and lieutenant Cornelis Michielsz. Blaeuw was read out by the Haerlem notary Egbert van Bosvelt to Frans Hals, who was bedridden with a bad leg (*die sieckelyck aen een*

[17] Municipal Archive Amsterdam GAA NA 833 (notary Frans Bruijningh), fol. 110v–111. The documents pertaining to the conflict were first published by Abraham Bredius (Bredius 1913, pp. 81–84; Carasso-Kok et al. 1988, cat. no. 194, pp. 383–384).

quaet been te bedde lach, see Appendix II) (Slive (ed.) 1989–1990, Hals Doc. 74).[18] Hals replied that he had agreed to make the piece in Haarlem, not Amsterdam; only afterwards had he 'conceded to begin the heads in Amsterdam and work these up in Haarlem' (*bewillicht dat hij de troinges tot Aemstelredamme soude beginnen ende tot Haerlem voorts opmaecken*), which he had already begun and would have finished if he had been able to gather the people involved, which he had not been able to do, resulting in him being away from home in Haarlem a lot and spending a lot of money in an Amsterdam inn, although he had been promised that these expenses would be paid for him. Nevertheless, Hals indicated that he would still like to do what he had accepted and promised: if the officers would kindly pose for him in Haarlem he would 'promptly take up the work and would finish it without further ado and in such a way that it would do him credit' (*'t werk datelyk bij der hant [zou] nemen ende sonder vertouven affmaecken sal ende zijn eere daarin betrachten*). He indicated that he would rather do this in Haarlem than in Amsterdam, as he would then be at home with his own people and could also keep an eye on them [i.e. his workshop] (*overmits hij dan binnenshuys ende bij zijn volck zijnde 't oge oock daer op mach hebben*).

On April 29th 1636, captain Reynier Reaal and lieutenant Cornelis Michielsz. Blaueuw, on behalf of the entire civic guard, objected to the reason Hals had given for the delay, calling his answers 'frivolous and beside the truth'(*frivole en onwaerachtige andtwoorde*, see Appendix III) (Slive (ed.) 1989–1990, Hals Doc. 75).[19] They explained that they had indeed originally agreed 'that he [Hals] would start the heads in Amsterdam and work these up in Haarlem' (*dat hij de troniges alhier ter stede soude beginnen en tot Haerlem voorts opmaecken*) and that they would each pay him 60 guilders for their portrait. However, they had subsequently agreed with Hals that for an additional six guilders, so a sum of 66 guilders per person (making a total of 1056 guilders for the entire painting), Hals would paint 'both the bodies and the heads of the figures properly and work these up entirely in Amsterdam, not in Haarlem' (*alhier ter stede, en niet tot Haerlem, de personagien, soowel van lichamen als tronigens, ende sulck als behoort soude schilderen en volcoomentlijck opmaecken*), as he had already begun to do for some of the figures (*gelijck hij oock alreede eenige personagien hier ter stede alsoo heeft beginnen te doen*). Once again, the officers gave Hals a deadline: he had to come to Amsterdam within 10 days in order to continue working on the painting and to work it up properly and finish it (*voors. Stuck schilderije voorts te vervolgen ende naer behooren op te maecken ende voltrecken*). This time they demanded a simple 'yes' or 'no' answer to the question of whether or not he intended to finish the painting as agreed.

It took their spokesman almost two months to deliver the message to Frans Hals. It was not until 26 July that their statement was read to the painter, who stood by his earlier answer and suggested a compromise (Appendix IV) (Slive (ed.) 1989–1990,

[18] Municipal Archive Haarlem GAH NA 63 (notary Egbert van Bosvelt), fol. 80r.

[19] Municipal Archive Amsterdam GAA NA 833 (notary Frans Bruijningh), fol. 143v–144v.

Hals Doc. 78).[20] Given not only the prestige of the commission but also the large sum he would receive for the painting if he had completed it –1056 guilders equaled three and half times the annual salary of a skilled craftsman at the time[21]– he could not have been keen to lose the commission altogether. Hals declared that he would be happy to transport the painting directly from Amsterdam to his home in Haarlem in order to first complete the unfinished clothing. Once that was done, he would paint the heads of those who were willing to come to Haarlem—given the short time needed, he did not think anyone would mind. However, if six or seven persons were unwilling or unable to come to Haarlem, he would take the nearly finished painting back to Amsterdam and finish it there properly (*naer behooren voltrecken*).

There are no other known documents relating to this conflict. According to the date inscribed on the painting it was finally completed in 1637 (*A° 1637*). In a 1653 list of the portraits at the Crossbowmen's civic guard headquarters in Amsterdam by Gerard Schaep, the painting is described as '*Captn Reynier Reael, Lutn^t^ Cornelis Michielsz. Blau A° 1637 door Francois Hals begonnen en door Codde voorts opgemaeckt*', thus: 'begun by Frans Hals and subsequently worked up by [the Amsterdam artist Pieter] Codde' (Schaep Pietersz 1630–1653, no. 123, p. 134). It remains unclear if Hals continued to work on the picture after his last statement (on 21 July 1636). The officers of the civic guard had threatened to reclaim all their previous payments to Hals, but seem not to have followed through. He may have placated them somewhat. They did, however, ask 'another good master here' to complete the work as they had announced.

Pieter Codde was a reasonably successful local painter, known for his portraits and small-scale scenes of daily life, including so-called '*kortegaerdjes*' (guard scenes) (Fig. 6.7) (Rosen 2020). As I.H. van Eeghen discovered in 1974, he lived in exactly the same district as the officers of the civic guard (Wijk 11) and might have had a personal connection to them. He certainly had a common acquaintance: the notary, Frans Bruijningh, who drew up the document summoning Hals on behalf of the guard, had also drawn up a divorce document for Pieter Codde a few months earlier (Van Eeghen 1974).[22] Codde may even have been a member of the civic guard himself—all men of the appropriate age and able to afford their own outfit and weapons were –in principle– required to join. It was also quite common for civic guards to have themselves portrayed by a painter in their midst.[23] Hals, for

---

[20] Municipal Archive Haarlem GAH NA 165 (notary Jacob van Bosvelt), fol. 265r.

[21] A skilled craftsman earned about 250 guilders per year at the start of the seventeenth-century, rising to about 400 guilders at the end of the century, according to Marten Jan Bok (see Rietveld in Tummers (ed.) 2013b, cat. no. 26, note 7). A master-carpenter or master-mason earned around 300 guilders at the time of the commission, according to Prof. L. Noordegraaf (see Slive (ed.) 1989–1990, cat. no. 43, p. 256 and note 1, p. 257). The total sum for the painting was thus substantial; compare also the price Rembrandt fetched in 1642 at the height of his fame for *The Night Watch*: 1600 Guilders.

[22] The 1636 inventory list of Pieter Codde's household goods is quoted in full in (Rosen 2020, Appendix II).

[23] See (Slive (ed.) 1989–1990, cat. no. 43) for Amsterdam examples. See also (Middelkoop 2019).

example, served as a musketeer for the St George Civic Guard in Haarlem from 1612 to 1624 and portrayed them three times. Admittedly, the inventory related to Codde's divorce agreement of 1636 does not mention a guardman's outfit or weapon. However, it was not uncommon for a divorce agreement to exempt personal clothing and paraphernalia; no garments or jewelry are mentioned either.

While the primary documents give us a great deal of information about the context of the commission, the exact extent of Hals's and Codde's involvement in the painting is not specified in detail. Indeed, the question of who exactly completed what has given rise to considerable debate. Before looking into the different views on the attribution and analysing the civic guard paintings in detail, some background information on seventeenth-century art theory and workshop practice is required.

6.3 Seventeenth-Century Views on Authenticity

The brief description of the portrait by Gerard Schaep in 1653 'begun by Francois Hals and worked up by Codde' is more significant than it might at first appear. In the seventeenth century, artists and art experts did not always care to distinguish different hands in group portraits or collaborative works. Successful master painters often ran relatively large workshops and used advanced assistants and journeymen in the production of their works. As long as the overall quality of the work was sufficient, masters were entitled to sign it and sell it as a work 'by their hand' (Tummers 2011b, ch. 3). They would usually supervise these works and retouch them where necessary (*retokkeren*), to ensure the final quality. Master painters also deliberately produced works of varying levels of quality and adjusted their prices accordingly;[24] individual painters varied the price level of their works according to the painting technique and materials they used and/or the size of their paintings, while painters who ran larger workshops also took into account the relative contribution of workshop assistants in their works. Rubens, for example, distinguished five different levels of quality in the paintings produced in his workshop (see Rosenberg (ed.) 1881, p. 42 ff; Tummers 2011b, pp. 93–94). As we saw in chapter four, Hals also signed a work that he did not literally paint himself, a work that was not of the highest quality but apparently still acceptable—witness his monogram. For prestigious commissions for public display (such as civic guard portraits) artists would strive for the highest quality; yet in other types of pictures, the quality was less important or was adjusted to the price—although the art theorist Samuel van Hoogstraten (1678, p. 235) urged painters to always produce works of high quality, even if these were more sketchy in their execution (and thus more affordable) (see also Tummers and Gration 2013, p. 73).

When judging the attribution of seventeenth-century paintings, it is important to keep in mind that from a seventeenth-century perspective, the question is not

[24] This practice harkens back to the fifteenth century. One of the first portrait specialists, Hans Memling, produced both costly, elaborately executed portraits and more affordable and quickly executed likenesses. See the excellent analysis by Holger Borchert in (Borchert et al. 2005).

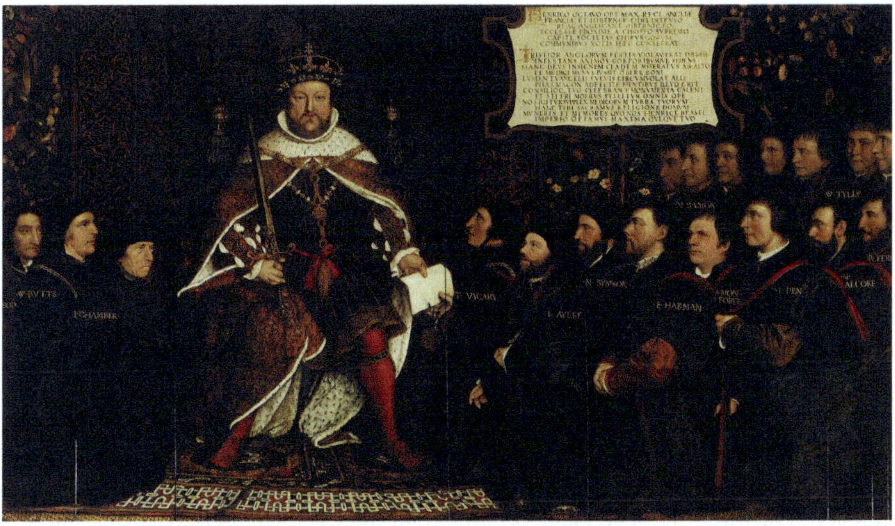

Fig. 6.10 Hans Holbein, *Henry VIII and the Barber Surgeons*, begun 1541–1543, oil on panel, 180.3 × 312.4 cm. (Hall of the Barber-Surgeons Guild, London)

whether or not other hands had contributed to the painting but whether or not the overall quality of the work was good enough to be considered the master's. A laudatory poem on Karel van Mander's well-known *Schilderboeck* of 1604 is revealing in this respect. It compares the book to a portrait of Pictura, the personification of the art of painting. According to the poet, this metaphorical portrait had benefitted from many contributions by others. However, Van Mander deserved to sign the work as an 'original' (*principael*) since he created the most impressive part, a crown of pearls on Pictura's head. Therefore, according to the poet, it would be a great mistake not to sign it with his name.[25]

Karel van Mander explains the attribution practice of early seventeenth-century artists and art experts even more clearly in his book when he discusses a group portrait by Hans Holbein, *Henry VIII and the Barber Surgeons* in London (Fig. 6.10). There was some debate at the time as to whether or not Holbein had finished the portrait himself or if it had been completed after his death. We now know that the latter was the case, as some of the sitters were appointed after Holbein's death (Rowlands 1985, cat. no. 78, pp. 148–49). According to Van Mander, however, this made no difference to art experts. It was done so well, he stated, that 'no painter or

[25] 'Waer in hy dan maer een groot faut bedrijvet,/Soo hy sich self daer onder niet en schrijvet / als principael: want hebben sy dit Beeldt,/ Elck nae sijn macht, verciert end'bejuweelt,/ Déen met een eringh, en dánder met een keten,/ Dees met een bagg', en die, ist wel te weten/ Met wat ghesteent: soo heeft van Mander haer / Versorght een Croon van enckel peerlen claer.' *Ode, op het Schilder-Boeck van den Const-rijcken Carel van Mander* by A.V.M. (see Van Mander 1604, fol. *7r; see also Tummers 2011b, p. 88).

art expert' ('*schilder of konstverstandige*') would distinguish different hands here (Tummers 2011b, pp. 107–108).

In the case of a deliberate collaboration between well-known masters with different specialisations, it was of course a different matter. In such cases, art experts were indeed quite keen to identify the different hands (see Honig 1995; Newman and Nijkamp (eds.) 2021; Tamis 2016).[26] In Codde's case, however, it is clear that he was trying to complete the work in Hals's style as best he could, painting on a much larger scale than he was used to and with much looser brushwork. The fact that Codde's share is mentioned so explicitly by Schaep suggests that Codde's contribution was not just the addition of a relatively small part in Hals's style; the wording suggests that his share was in fact more substantial.

6.4 A Contested Attribution

When the historian Jan van Dyck described all the paintings in the Amsterdam city hall in 1758, he mentioned 'one piece by Frans Hals, painted in 1637, in his best period' (Van Dyck 1758, p. 30). In his opinion, it was 'of a very different taste' than all the other works there (*van geheel anderen smaak als alle de overige*). He explained that from close up it looked as if it had been painted in one go, without 'dead colouring' (*dootverwen*), i.e. without the usual underpainting.[27] He added that it was 'painted very quickly' (*recht vlugtig gepenceelt*), 'drawn well' (*frai getekent*), that the figures were 'delightfull in their pose and relation to one another, also good in their general composition' (*de Beelden heerlyk van stand en houding, ook goet van Ordonnantie*). Moreover, they were all so 'dry and slender'(*dor en rank*) that one could rightfully call them the 'meagre Compagny' (*magere Compagnie)* in his view.[28] His nickname stuck; the picture has been called *The Meagre Company* ever since. His attribution, however, did not.

It was not until municipal archivist Pieter Scheltema published Schaep's description of the painting (*geschildert* [sic] *bij Hals en door Codde voorts opgemaeckt*) in 1885, that art historians began to question the attribution to Hals and to try to distinguish who had done what (Scheltema 1885, p. 121 ff, esp. pp. 134 and 141).[29]

[26] In case of Dirck Hals (specialist in small-scale figures) and Dirck van Deelen (specialist in architectural interiors) three collaborative works are known. Interestingly, two were signed by Van Deelen and one by Hals; possibly, it was the master who sold the work, who signed in this case (see Tummers 2011a, cat. no. 17, pp. 88–89).

[27] *Als men 't zelve van naby beziet, is men overtuigt dat op eenmaal zonder doorverwen geschildert is.* (Van Dyck 1758, p. 30).

[28] *Want 't is recht vlugtig gepenceelt, frai getekent, de Beelden heerlyk van stand en houding, ook goet van Ordonnantie, 't aardigste is datze alle zo dor en rank zyn, dat men ze met recht de magere Compagnie zoude kunnen noemen.* (Van Dyck 1758, p. 30). With thanks to Eric Jan Sluijter for his advice on the translation.

[29] Scheltema mistakenly transcribes 'begonnen' as 'geschildert'.

Abraham Bredius was the first to express his opinion in his book *Die Meisterwerke des Rijksmuseum zu Amsterdam* in 1890. In his opinion, only the gentleman in yellow in the middle was by Pieter Codde (Bredius 1890, p. 102). A more extensive analysis followed shortly afterwards.

In 1893, Jan Six described the main difference between the two painters in this work as follows: 'Codde hatches with the brush, Frans Hals moves his brush in every direction required by the form, with an ease and assurance that no one has ever equalled'.[30] He believed that the seven figures on the left and the ninth head from the left were by Hals. He added that the two heads at the back beside to the standard bearer, Nicolaes van Bambeeck were not quite finished. He also attributed all the hands in the left part of the painting to Hals, except from the right hand of the seated lieutenant. As for the clothing, the situation was more complex. On the left, the standard bearer's outfit and the clothing in the upper part were by Hals; in the lower part the clothing was, however, by Codde, including the blue sash of the lieutenant.

As for the nine standing officers on the right: the figure in the centre was entirely done by Codde, but most of the other heads were strongly reminiscent of Hals, yet different in their execution. Therefore, the most likely solution was that Hals had laid in the heads (*gedoodverfd*), working them up somewhat less than the two unfinished heads on the left, and that Codde had finished them: hence the strong relief and less vigorous brushwork. The figure with the reddish face on the far right, however, was entirely Codde's as were the cloths and hands throughout the right-hand side of the painting, according to Six.

When the archival documents relating to the conflict between Hals and his patrons were discovered and published by Abraham Bredius in 1913, they did not directly affect the attribution of the painting (Bredius 1913). It is noteworthy that the term 'dead colouring' (i.e. laying in the design and distribution of light and dark) is not used at all in the documents.[31] Instead, the documents mention the terms 'beginning' (*beginnen*), 'working up' (*opmaecken*) and 'finishing' (*voltrecken*). Perhaps Hals did not use this term (*dootverwen*) or did not use it with patrons; as we will see, he certainly did not underpaint his entire work in blacks, whites and greys as some theorists recommended, but he usually built it up in stages (Hendriks et al. 1991, 'Underpaint', pp. 26–31; Miedema 1987, p. 142).

The next in-depth analysis of the attribution of *The Meagre Company* was written the conservator Maurits van Dantzig in 1946 (Van Dantzig 1946). As was his wont, Van Dantzig related the painters' styles to their presumed characters (see also above, Chap. 2). Hals, in his view, was the greater master, far more lively, spontaneous and outward-oriented than Codde. By contrast, Codde was more hesitant, restrained and inward-focused. In the painting, the difference was evident in the

[30] 'Codde arceert met het penseel. Hals beweegt zijn penseel, onverschillig in welk richting zoo als de vorm dat meebrengt, met een gemak en een zekerheid door geen ander ooit geëvenaard.' (Six 1893, pp. 96–104, esp. pp. 102–104).

[31] The absence of the term deadcolouring (*dootverven*) was first noted by Koos Levy-Van Halm in (Carasso-Kok et al. 1988, p. 382).

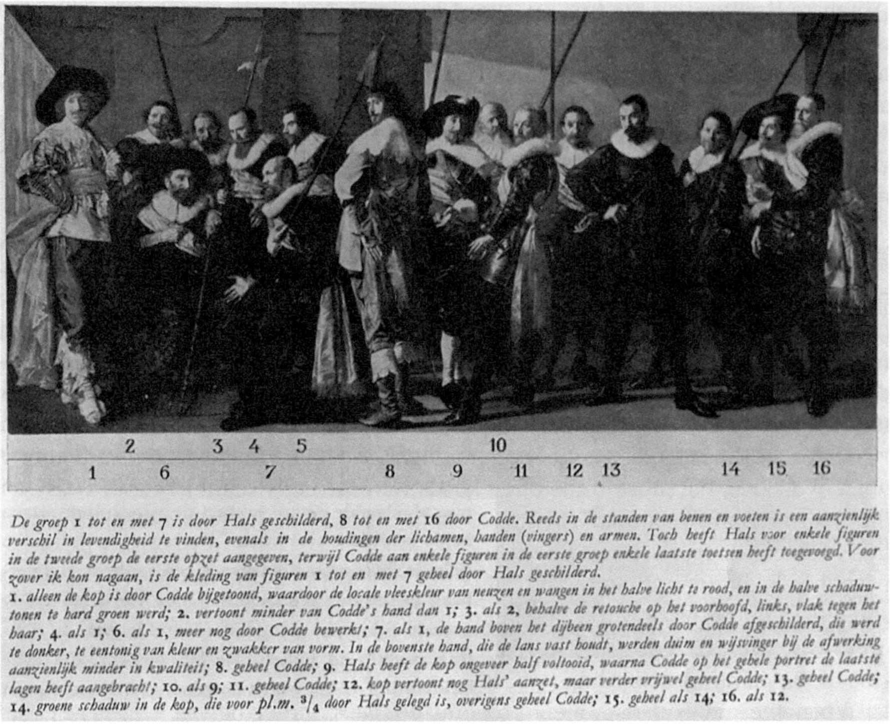

Fig. 6.11 Van Dantzig's attribution of *The Meagre Company* in. (Van Dantzig 1946)

shape of their strokes and their use of colour and tone—Codde's strokes were more uniform and repetitive, his colours less varied and often lacking sufficient mid-tones for an animated rendering. For example, Hals's contours were 'powerful and agile, emphasising the ever-changing directions of the different planes', whereas Codde's were 'dull and evenly smooth' by comparison.[32] The amount of effort involved was also revealing, according to Van Dantzig (1946, p. 9): 'Codde achieves little by doing a lot. Hals achieves a lot by doing a little.'

Like Six and Bredius, Van Dantzig attributed the seven men on the left almost entirely to Hals (See Figs. 6.6 and 6.11). Unlike Six, he believed that the clothing of these figures was also entirely by Hals, including the blue sash of the seated luitenant. In the faces of these figures, however, he detected some final touches by Codde: middle tones in the noses and cheeks (making them appear too red) and in the half shadows (making them appear too green), especially in the standard-bearer, the fourth standing figure from the left and the two seated men. Furthermore, the right hand of the seated lieutenant was largely finished by Codde as were the thumb and index finger of this figure's left hand (Van Dantzig 1946, p. 5).

In Van Dantzig's opinion, the figures eight to sixteen were mainly done by Codde, although Hals had laid in the first sketch for some of these figures. He

[32] 'saai en gelijkmatig glad', (Van Dantzig 1946, p. 8).

pointed out that the positions of the legs and feet were less animated in this part of the painting, as were the postures of the bodies, hands and arms. He attributed the man in yellow entirely to Codde, as well as the officer with the orange sash and the officer in black with the bandolier (the thirteenth person from the left). The heads of the other figures were, in his opinion, based on initial outlines by Hals: the ninth and tenth heads were half done by Hals and then finished by Codde; the fourteenth and fifteenth heads showed a green shadow in the face that was largely (3/4) done by Hals; and the twelfth and sixteenth were started by Hals and subsequently finished almost entirely by Codde.

In his elaborate three volume monograph on Hals (1970–1974), Seymour Slive attributed *The Meagre Company* somewhat differently. He was struck by the work's similarity to Hals's 1633 civic guard portrait (Frans Hals, *Officers and Subalterns of the Calivermen Civic Guard*, 1633, oil on canvas, 207 × 337 cm. (Frans Hals Museum, Haarlem); Frans Hals and Pieter Codde, *Militia Company of District XI under the Command of Captain Reynier Reael, Known as 'The Meagre Company'*, 1637, oil on canvas, 209 × 429 cm. (Rijksmuseum, Amsterdam)). He therefore believed that the design of the painting was 'entirely the master's invention' (Slive 1970–1974, vol. 1, p. 137). According to Slive, Hals painted most of the left half of the composition, although he also saw 'traces of his hand' on the right side in the portraits of the third, fifth and seventh men from the right. He suspected that Codde had touched passages on both the right and the left of the painting to unify the portrait, and emphasized that Codde did not manage to equal Hals's touch, witness the difference in quality between the standard bearer and the man in yellow.

In the context of a large survey exhibition in 1989–1990, *The Meagre Company* was restored and re-examined through a series of technical analyses. Conservator Martin Bijl published the new findings and his interpretation in the catalogue accompanying the exhibition (Bijl 1989–1990, pp. 103–108). He noted that the painting was done on a single piece of canvas –like Hals's Haarlem civic guard group portraits– and speculated that he may have ordered it from an Amsterdam primer (Bijl 1989–1990, p. 104).[33] As for the use of pigments he found nothing unusual except the use of two different types of white: a finely ground lead white in the skin colours (*'loodwit'*) and a coarser lead white mixed with chalk elsewhere (*'schelpwit'*) (Bijl 1989–1990, p. 108; Van de Graaf 1961). The most revealing new evidence for the attribution, however, came from large radiographs that were made by the Belgian KIK-IRPA. These revealed an underlying sketch in what he believed to be grey oil paint, visible in places where the original design departed from the final execution (Bijl 1989–1990, Fig. 4, pp. 106–107).[34] For example, the contours of the jacket of the thirteenth figure from the left differed from the final version (as could already be seen with the naked eye). In addition, a cursory zigzag line indicated boots similar to those worn by the standard bearer, which were never executed.

Some sketched lines and underpainted areas were easier to identify and interpret than others. The eight figure from the left had been significantly altered by Codde,

[33] It was only stretched once, not restretched after priming.

[34] The order of the radiographs in the catalogue is mistaken; the second one should be switched with the fourth. On the sketch see also below 'New light on the attribution of *The Meagre Company*'.

Bijl noted. The figure was originally turned further away from the viewer and held his left hand to his chest. Instead of a sash, he wore a bandolier with powder charges just like the second and thirteenth figure from the left, and possibly a rapier as well, depicted at the level of his current hand. A few folds in his pants and his boots could also be distinguished. Furthermore, Hals had worked up the face and collar of the eighth figure from the left to a 'very advanced stage',[35] according to Bijl (1989–1990, p. 105); therefore, it remained unclear why Codde had reworked this figure so heavily. He speculated that figure 9 had been as elaborately laid in as figure 8. At the end of the essay, however, he concluded that the head and costume of figure 9 were near finished when Codde took over, while presumably only a rough sketch for figure 8 had been on the canvas (thus contradicting his earlier analysis somewhat).

Bijl also suspected that a large part of the background had been sketched too. The radiographs showed a building in the background and close to the fourteenth figure from the left, a layer of green could be distinguished underneath the grey top layer. Moreover, he saw a first indication of all the collars below all current ones.

In short, Bijl agreed with Six's attribution of Hals's and Codde's share in the painting and elaborated on it. In his view, Hals had worked from left to right, leaving just a few details for a later stage, such as the right hand of the lieutenant and the final reflection lights (*reflectielichten*). Also, he left the two heads in the background on the left unfinished. Codde had worked mostly in the middle (on the man in yellow, figure 8) and on the right (from figure 10 onwards); the heads completed by Codde at the right showed 'sharp contours of opaque light coloured paint near the edges of the foreheads' and 'eyes drawn with greenish paint' (Bijl 1989–1990, p. 108).

While the Hals survey exhibition was in full swing, Hals specialist Claus Grimm published a different opinion. In his view, the division of work was more nuanced (Grimm 1990/1989, pp. 70–84, esp. pp. 82–83). Grimm recognised Hals's characteristic diagonal brushstrokes in all of the faces depicted, also on the far right. He also saw Hals's sharp-edged umbral shadows on some of the collars and, in some cases, his typical highlights (developed as if it were from the brush movement). However, the hair and beards of the nine figures on the right were predominantly done by Codde and easily recognisable, according to Grimm, in their cautious execution. Similarly, the difference between Hals's confident tonal depiction of the collars and Codde's rather dull rendering of the collars was not difficult to discern. He did not see any additions by Codde in the faces themselves and suspected that he had avoided touching them out of respect for Hals. The stylistic variations, especially the contrast between the sharp-edged brushwork in the faces of the standard-bearer and the two seated figures, on the one hand, and the smoother brushwork with more subtle nuances and softer transitions in the faces in the background and on the right, on the other hand, were, according to Grimm, due to Hals's stilistic development between 1634 and 1636.

He suspected that Hals had painted his portraits in a centrifugal manner directly on the canvas starting with the faces and collars, and subsequently modeling the position of the shoulders, arms and hands. The only figure Hals appeared to have

[35] In Dutch: *ver gevorderd stadium*.

finished completely was the standard-bearer on the far left. The other figures on the left were merely finished as 'busts' in his view—only the heads and collars had been completed when Codde took over. Grimm had the impression that the definitive position of the legs had not been indicated unambiguously when Hals had already modelled the twists in the upper bodies of the figures. This would explain why – especially on the right– the figures appeared excessively long and slumped back in their poses. He referred to earlier observations of a few sketch lines visible in the radiograph, which indicate the body postures of the seated captain and the figure in yellow in the centre of the painting to support his theory about Hals's working process (Levy-Van Halm 1988, p. 382; Grimm 1990/1989, note 53, p. 81). His conclusions about the attributions of the different parts of the finished painting were predominantly based on his stylistic analysis (Grimm 1990/1989, p. 83).

As we have seen in Chap. 2, a fierce public debate followed the sharp contrast between Slive's and Grimm's insights into to Hals's oeuvre, and led to an extensive technical assessment of many works present in the exhibition. *The Meagre Company* and the civic guard portraits were not included, however, presumably because of their size.

Since then, no other scholar has published a new in-depth analysis of the attribution of the entire work. Nevertheless, some valuable insights have been gained from the study of specific parts. Margriet van Eikema-Hommes has studied the use of indigo in early seventeenth-century painting and analysed Hals's characteristic use of the pigment in great depth (Hendriks et al. 1998; Van Eikema Hommes 2004, pp. 104–109).[36] In the blue sashes in *The Meagre Company*, she recognised Hals's characteristic underpainting with its typical greyish-blue colour (which is characteristic of indigo paint that contains a lot of binding medium). The upper paint strokes were most likely applied by Codde, in her view, an artist who had presumably not worked with the pigment before: "the stiff, sometimes even scratchy paint handling differs greatly from the technique evident in Hals's Haarlem portraits where opaque and transparent, blended and unblended strokes alternate in a sophisticated manner" (Van Eikema Hommes 2004, note 80, p. 159). In 2013, Christopher Atkins nevertheless attributed the hanging sash on the left to Hals and the one on the right to Codde, while underscoring Grimm's analysis (Atkins 2013, p. 62 and note 14, p. 146).

Interestingly, many authors claim that the difference between Hals and Codde is substantial and therefore very clear; however, interpretations of the exact share of Hals and Codde still vary considerably from author to author and continue to change. In the Rijksmuseum's 2007 catalogue of paintings, Jonathan Bikker states that technical examination has confirmed the idea, derived from the primary documents, that Hals left seven heads incomplete. He identifies the seven [sic] rightmost portraits as the work of Codde, adding that Codde substantially altered the eight figure from the left and must also have added finishing touches to some of Hals's figures (Bikker (ed.) 2007, cat. no. 112, pp. 180–182, esp. p. 182).

---

[36] See also Chap. 5.

6.5 Hals's Design Process: Unique Sketches Discovered

One of the most spectacular results of the current research projects has been the discovery of various types of initial sketches that shed new light on Hals's working process. Hals's *Banquet of the Officers of the St. George Civic Guard* of 1627 proved particularly revealing in this respect. In the background, at the height of the curtain, the MA-XRF scan revealed traces of a rare initial sketch of a face in a calcium containing material, presumably chalk, as well as –in its vicinity– outlines of the banner (Detail of Fig. 6.2: visible light image and infrared reflectograph (IRR), in 'sync mode'). Moreover, IRR showed a very bold, cursory initial sketch in a carbon-containing wet material below the faces and collars of colonel Nicolaes Druyvesteyn and captain Nicolaes Verbeek (Detail of Fig. 6.2: visible light image and infrared reflectograph (IRR), in 'sync mode'), allowing us –for the first time– to see how Hals set up a face. Similar

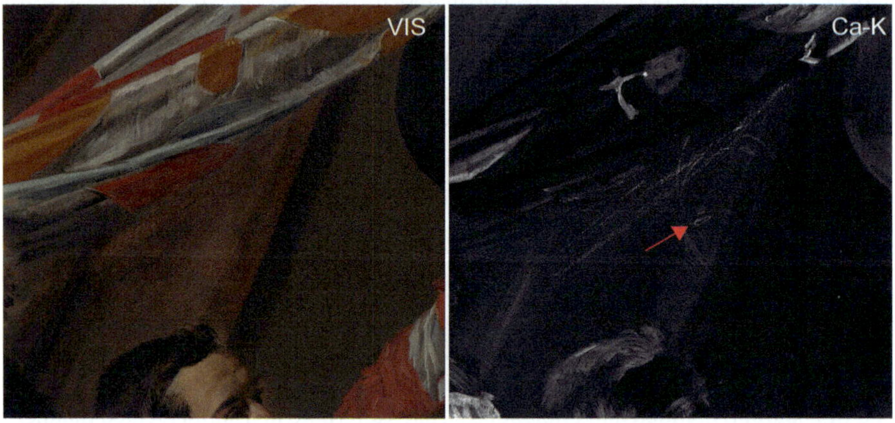

Detail of Fig. 6.2: visible light image and calcium (Ca-K) map, in 'sync mode' (sn.pub/a1g90i)

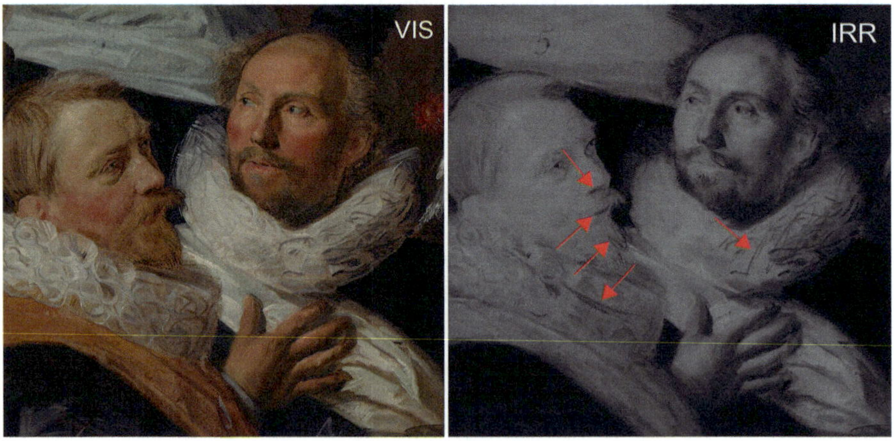

Detail of Fig. 6.2: visible light image and infrared reflectograph (IRR), in 'sync mode' (sn.pub/jjstm0)

carbon-containing sketch lines can also be seen below the outstretched arm of Michiel de Wael in the yellow coat, who initially raised his glass (Detail of Fig. 6.2: visible light image and IRR, in 'sync mode').[37]

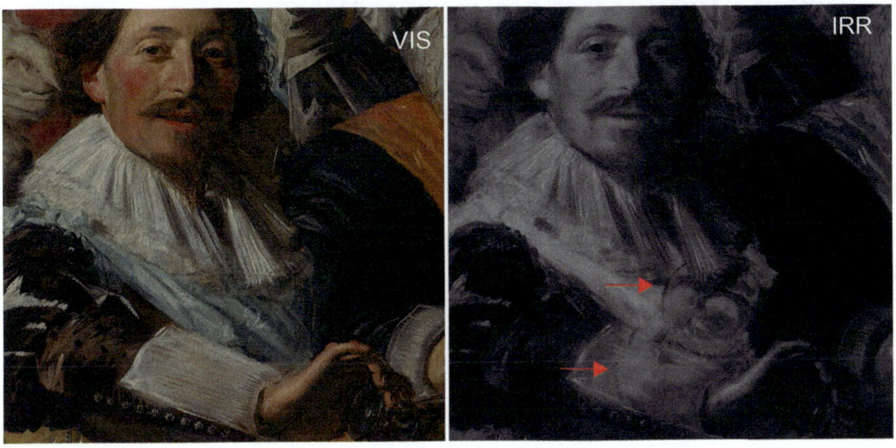

Detail of Fig. 6.2: visible light image and IRR, in 'sync mode' (sn.pub/qrzirl)

Previously, only some incidental traces of initial sketch lines in a dark, wet material had been observed, mostly underneath collars and around contours where Hals had deviated from his design (Hendriks et al. 1991, pp. 24–25).[38] The newly discovered sketch lines show very clearly that Hals designed these complex compositions directly on the primed canvas and that he had a characteristic, bold way of sketching. Moreover, Hals continued to sketch while setting up the picture in paint, resulting in other, often broader, types of sketch lines.[39] In the same painting, for example, such brushstrokes are visible in the MA-XRF map of manganese (and iron) which indicate sketch lines containing the dark brown earth pigment umber. They correspond to dark brushstrokes that are sometimes (partially) left open at the surface. These brushstrokes roughly indicate the outlines and/or the deepest shadows, such as in the hat and costume of the flag-bearer Dirck Dicx (who is standing behind captain Michiel de Wael).

[37] See also below: Revisions, inconsistencies, workshop practice and assistance.

[38] Based on research conducted by Prof. J.R.J. van Asperen-de Boer. See also (Packer and Roy 2021–2022; Tummers et al. 2019).

[39] On the different paint phases, see also below: *Life moved into the paint*: the challenge of creating lifelike portraits.

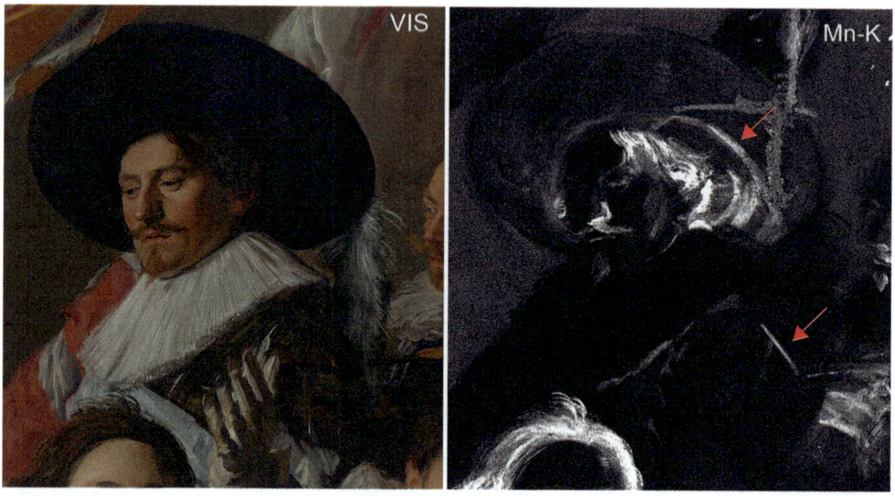

Detail of Fig. 6.2: visible light image and manganese (Mn-K) map, in 'sync mode' (sn.pub/nlnmvq)

Karel van Mander, who was in all likelihood Hals's teacher, as we have seen, explained in his *Schilderboeck* of 1604 that 'experienced painters with a steady hand [...] worthy of their master's title because of their art, ably draw freehand on their panels what they have first painted in their mind's eye' (*eenighe wel gheoeffent expeerdich,/En vast in handelinghe cloeck beraden, [...] om hun Const zijn Meesters name weerdich,/Gaen toe, en uyt der handt teyckenen veerdich/Op hun penneelen, t'ghene nae behooren/In hun Ide' is gheschildert te vooren*) (Van Mander 1604, Grondt XII, 04).[40] Subsequently, they 'begin directly, without worrying much, with their brush and paint, and a bold mind' (*vallender aen stracx, sonder veel quellen, / Met pinceel en verw', en sinnen vrymoedich*) (Van Mander 1604, Grondt XII, 05).[41] This is probably exactly how Frans Hals set about his work.

When artists started to work on coloured grounds in the second half of the sixteenth century, it made sense to make the initial sketch of the composition in white chalk. It would be visible against the coloured background but would become invisible when covered with oil paint. Chalk in oil has a translucent light beige tone with little colour strength—which was a great advantage compared to underdrawings in graphite or black chalk (Nicolaus 1990/1980, p. 77; Ward (ed.) 2008, pp. 721–722). Indeed, we know from various seventeenth-century depictions of artists at work, such

[40] Three seventeenth-century sources that identify Van Mander as Hals's teacher are cited in (Tummers (ed.) 2013, p. 16 and note 13, 14 and 15, p. 144). See also Chap. 4, note 30.

[41] In the margin, Van Mander summarized the passage as follows: 'designing directly on the panel, Master's work' (*Stracx eerst op penneel te stellen, Meesters werck*).

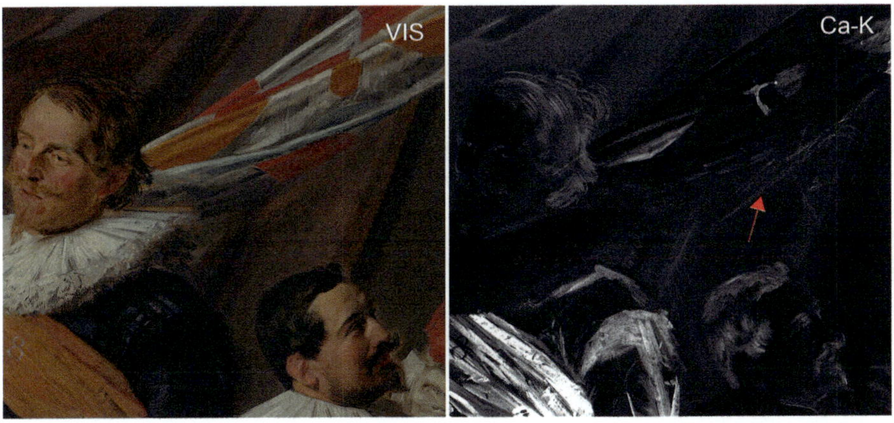

Detail of Fig. 6.2: visible light image and calcium (Ca-K) map, in 'sync mode' (sn.pub/a1g90i)

as Vermeer's famous *The Art of Painting* (c. 1666–1668) and a *Portrait of a Young Artist,* created in Hals's surroundings, that painters used white chalk for their initial design.[42] However, until now no concrete evidence of such an initial sketch in chalk had been found (which is not surprising given that chalk normally thus becomes translucent in oil paint and is invisible in traditional imaging techniques such as IRR and radiography). During the investigation of Hals's civic guard portraits in Haarlem, the sketch lines showed up in the MA-XRF calcium map (Detail of Fig. 6.2: visible light image and calcium (Ca-K) map, in 'sync mode'). Observations in the gallery with a handheld microscope confirmed that the grainy calcium rich lines in the MA-XRF scan correspond to an underlying white material, that is now partially visible at the surface (Fig. 6.12). The rare visibility of the white calcium rich sketch is presumably related to the very thin surface paint layer and the fact that the face and lower part of the banner were not placed here in the final design, which makes it possible to visualize it with MA-XRF.[43]

The sketch lines give us a unique glimpse into Hals's initial working process. Much like the artist depicted in Vermeer's *Art of Painting*, Hals appears to have focused initially on outlines, witness the contours of the eyes, the face, hair and the top and bottom of the banner pole. The upper outline of the banner can be seen below the hair of the standard-bearer, Boudewijn van Offenberg,

---

[42] For images of both paintings, see: https://artsandculture.google.com/story/KQWhTLphPokuKg; https://commons.wikimedia.org/wiki/File:Portrait_of_a_painter_-_copy_after_Frans_Hals.jpg

[43] Further research is needed to confirm that the chalk was indeed applied as a dry material and not in a medium. The microscope used was a Dino-Lite Edge digital microscope. See also below on the reason why the sketch below the faces in a wet material is so clearly visible.

and its lower edge (which was originally lower than in the final design) near the face in the background to Van Offenberg's left. Various lines indicating the contours of the face and the lower edge of the banner indicate that we are witnessing Hals's initial thoughts: he was drawing freehand, presumably in a dry material. It is not a neat outline of just the final contour like we see in the *Art of the Painting*. Moreover, it does not seem to be a coincidence that we only see contours of a figure and the main attribute of another figure here. Presumably, Hals's initial sketch concerned only the placement of the figures and their main accessories. Similarly, the chalk sketch in Vermeer's painting does not specify any background details. According to the art theorists Willem Beurs (1656–1692) and Gerard de Lairesse (1640–1711), this is exactly how painters of figure scenes should begin their work. Whereas other types of pictures required the opposite approach (i.e. working from background to foreground), scenes in which figures were the most prominent, should be started where one would see the greatest strengths: one should design and paint the figures first and then work towards the back, according to De Lairesse (1707, vol. 1, p. 12). Beurs specified that:

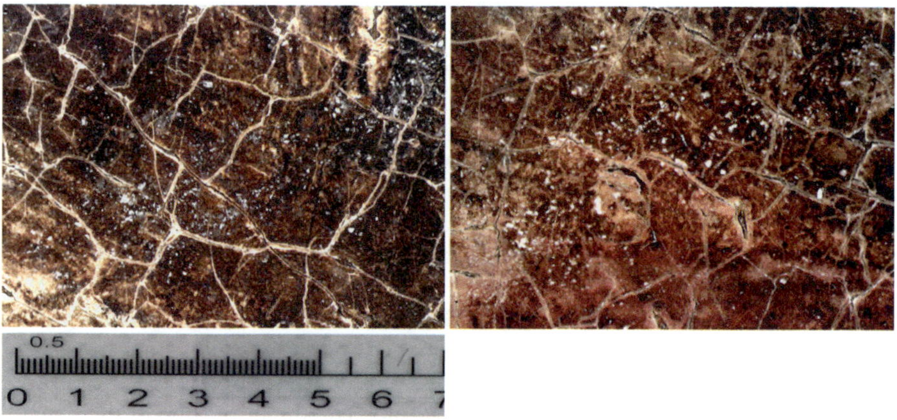

Fig. 6.12 Detail of Fig. 6.2, taken with a dino-lite microscope (x65), showing the calcium rich white material that is visible at the surface

> The figures (to begin with) the painter must apply first. Both in a room, a landscape or in another meeting place of people. In a room it is not indelicate to paint in the background, which recedes most, at the end. But in a landscape, after [first painting] the figures (whether they are painted herein because of a history [i.e. narrative] or [just] as ornament), [one has to paint] the sky, the mountains and the ground from the background to the foreground (Beurs 1692, pp. 51–52).[44]

Hals presumably both sketched and began to paint his figures first, just like the artist in Vermeer's painting. The clearest evidence of this is the halo effect around the heads of standard-bearer Jacob Cornelisz Schout in Hals's earliest civic guard portrait and of the hall steward Willem Ruychaver in the 1627 Calivermen's portrait. The heads were painted first and built up in several layers; the banner and background vista were clearly added later (Detail of fig. 6.1; detail of fig. 6.3). In Hals's case, the design phase consisted of several stages and involved different materials.[45] Like the first white outlines, the subsequent sketches are remarkably free and summary in their execution. As we will see, their cursory character had implications for his working process, as the lines offered little guidance to others.

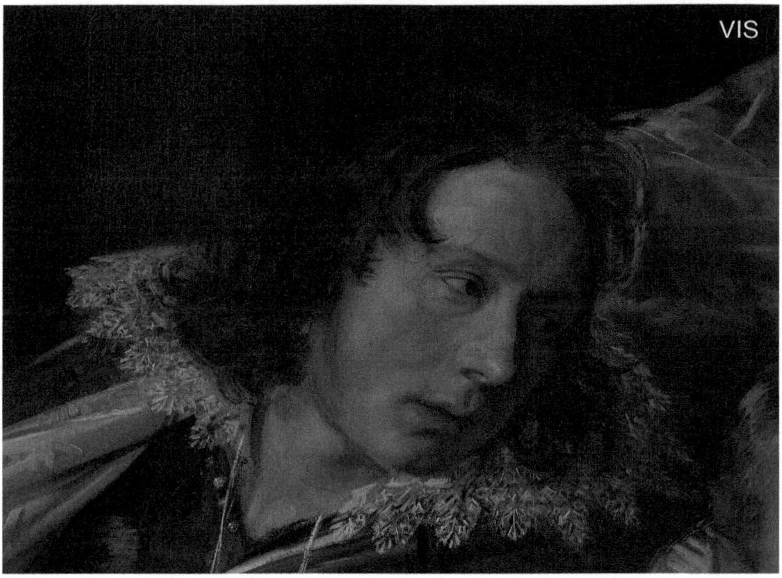

Detail of Fig. 6.1 (sn.pub/9tweuj)

[44] The translation is taken from (Van Eikema Hommes 2004, p. 14).

[45] See also below on traces of the initial sketch below Michiel de Wael in os I-110: *Revisions, inconsistencies, workshop practice and assistance.*

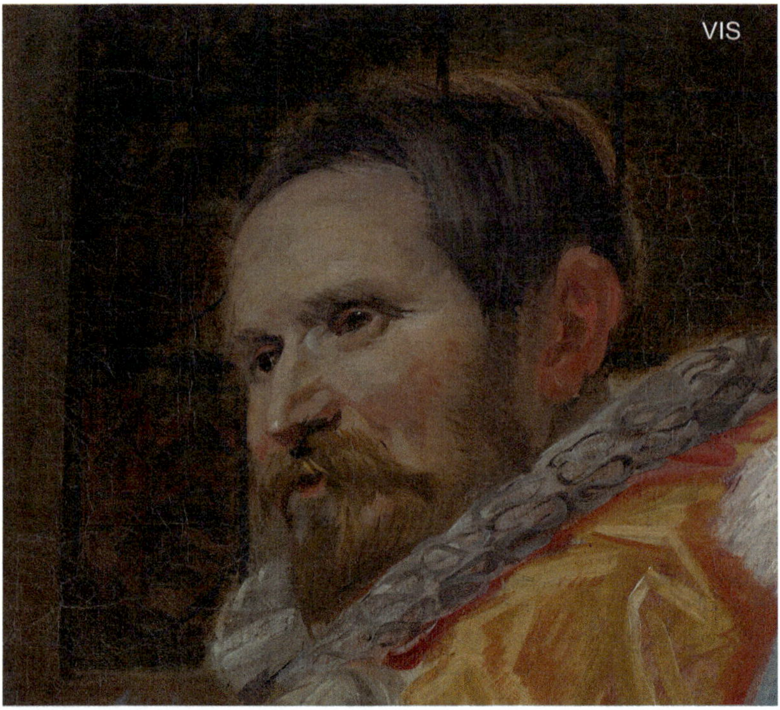

Detail of Fig. 6.3 (sn.pub/g7kvro)

The black carbon-based sketch in Hals's design process is visible in the infrared reflectogram (IRR) of the painting, showing black paints both on the surface and in deeper layers. In particular, an exceptionally loose sketch can be seen underneath the faces and collars of colonel Nicolaes Druyvesteyn and captain Nicolaes Verbeek (Detail of Fig. 6.2: in IRR). With just a few touches, Hals indicates the position of the beard, the middle of the mouth, the underside and top of the nose and presumably the position of the eyes (the top layer also contains black here, making it hard to distinguish). He does not bother to describe the forms in much detail; it is merely a shorthand notation for himself, a quick reminder of where to put what. Note, for example, how simply and effectively he indicated the shadowy part of colonel Verbeek's collar with just one swift curly line. There was no need for specifics at this stage.[46]

---

[46] It is important to keep the very cursory character of the newly discovered sketches in mind when interpreting technical evidence relating to Hals. Based on an infrared reflectogram of *The Laughing Cavalier*, Lelia Packer and Ashok Roy concluded in their recent essay on the painting's genesis that Hals did not plan to paint the cavalier's right nostril. However, the newly discovered sketches show that Hals certainly did not outline every single element he was about to paint—just a few key elements sufficed. Given the direction and angle of the face, it seems unlikely that Hals would have

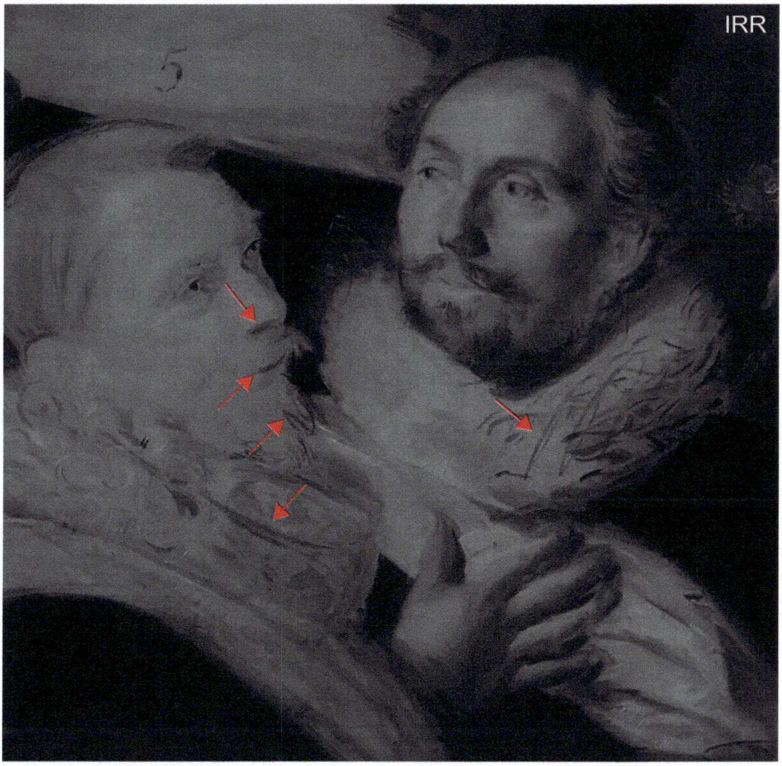

Detail of Fig. 6.2: in IRR (in 'sync mode': sn.pub/tcpe3m)

The reason that the underlying sketch is so clearly visible here is that these parts were painted more thinly than the rest of the painting, possibly because of the time the sitters were available. The underpainting below the faces is missing here, as observations made with a hand-held microscope confirmed. In fact, the paint layer is so thin that small abrasions directly show the canvas support, while the faces in the rest of the painting are painted more thickly (compare Figs. 6.13 and 6.14).[47] The collars of the two sitters make heavy use of the warm beige ground as a midtone in their build-up.

Just how unique Hals's spontaneous and assured working process is, becomes clear when we compare his sketch to a similar sketch by Haarlem's second most well-known portrait painter at the time: Johannes Verspronck (1600–1662).[48] Underneath his paintings quite a few sketches have been found by Ella Hendriks and Anna

intended to skip the nostril. He presumably simply did not need more specifics (see Packer and Roy 2021–2022, pp. 81–91, esp. p. 90).

[47] On Hals's usual underpaint stage in flesh colours and drapery see (Hendriks et al. 1991, p. 50).

[48] Compare also the underdrawings under different types of seventeenth-century paintings at the National Gallery of Art: https://www.nga.gov/features/underdrawings-revealed.html

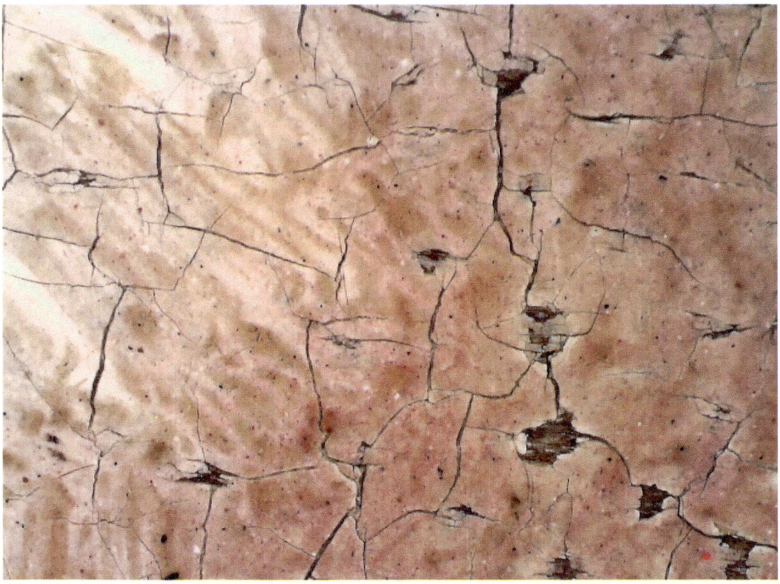

Fig. 6.13 Detail of Fig. 6.2, taken with a dino-lite microscope (x65), showing the thin paint layer in the face of colonel Aernout Druyvesteyn

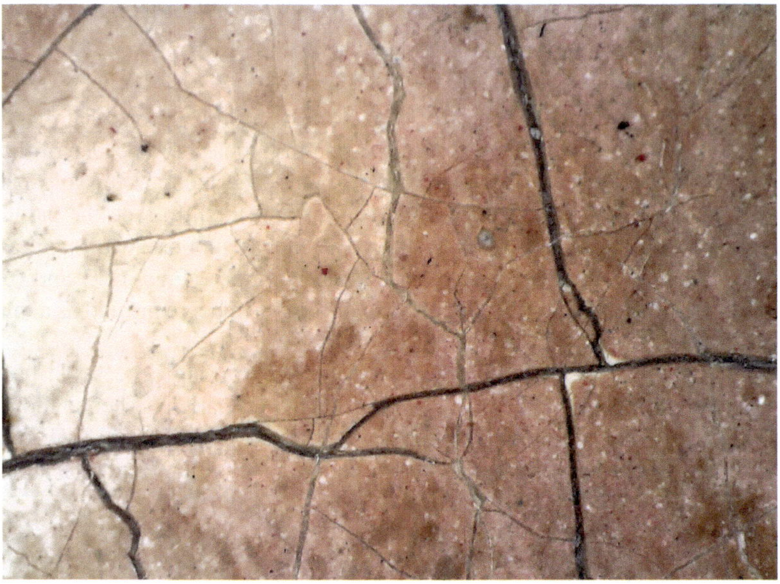

Fig. 6.14 Detail of Fig. 6.2, taken with a dino-lite microscope (x65), showing the thicker paint application in the face of captain Nicolaes Lefubre on the right

Krekeler (Hendriks 1998; Krekeler et al. 2014). A beautiful and entirely characteristic sketch can be seen in the infrared reflectogram (IRR) of the *Portrait of a Man*, dated 1646, at the Rijksmuseum (Fig. 6.15). While Hals's sketch is very bold, fluent and cursory, Versponck's is more elaborate and descriptive. We see both a man's and –on the right– a woman's face, sketched with elegant thin lines. Presumably, the picture was originally intended as a woman's portrait. Her outlines are particularly clear as no portrait was painted on top: the upper contour of her eyes, her iris and pupil, the underside of her nose and nostrils, her lips and the shadow beside her mouth are all neatly indicated. Occasionally, though, Verspronck is searching for the correct shape—witness the many contours of the man's face (right), hair (top left) and hat. By contrast, there is no sign of Hals searching or hesitating anywhere; his touches are swift, confident and remarkably loose.

It brings to mind an anecdote recounted by Hals's earliest biographer Arnold Houbraken. As we have seen, according to Houbraken, the well-known painter Anthony van Dyck had great respect for Hals's art. Even after he had unsuccessfully tried to persuade Hals to come work at the British court, Van Dyck reputedly remarked on many occasions that Hals had no equal in his control of the brush, that

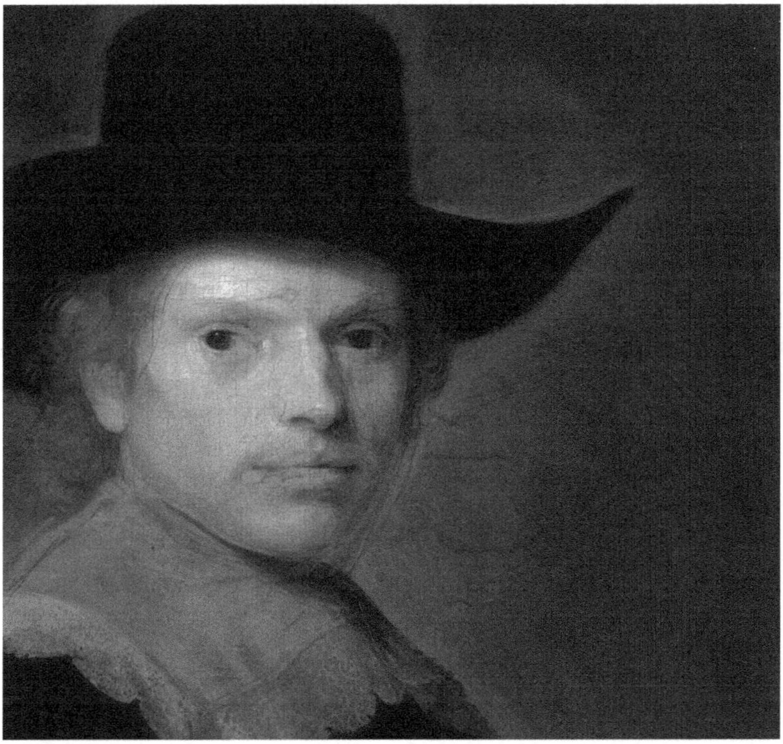

Fig. 6.15 Infrared reflectogram (IRR) of Johannes Verspronck, *Portrait of a Man*, 1646, oil on canvas, 86 × 65.5 cm. Rijksmuseum, Amsterdam. (Credits: A. Krekeler, Conservation&Science, Rijksmuseum)

once he had set up the underpainting he could portray the sitter's essential features, the highlights and shades with one touch of the brush without needing any corrections.[49] Indeed, from start to finish, Hals painted with a rare ease and accuracy.[50]

No preparatory drawings for paintings or partial studies in oil for larger works by Hals are known, and this may well be due to his *virtuoso* working process. Indeed, both the documents pertaining to his conflict with the Amsterdam guardsmen and the presence of the different sketches make it unlikely that he used such studies. He needed the guardsmen to be present so he could paint their portrait directly on the large canvas in either Haarlem or Amsterdam (see Slive (ed.) 1989–1990, Hals Docs. 74 and 78). Moreover, the sketch lines indicate that before working up the individual portraits, he worked out the sketch for the entire design for the picture also directly on the canvas, just as Van Mander had recommended.

In the early seventeenth century, painters and art theorists were keenly aware of the importance of a good design. In fact, Karel van Mander introduces a lot of new terms to describe the 'design' (*ontwerpsel, bewerp, beworp, voor-beworp*), which do not occur in earlier art theoretical treatises—as Hessel Miedema (2013, p. 23 and note 79, p. 29) discovered. He also occasionally used the term 'composition' (*compositie*), although he mostly uses the more common terms for 'design': *byeenvoeghing* or *ordonnantie* (which literally mean: 'putting together' and 'arranging') (Van Mander 1604, de Grondt, ch. V; Miedema 2013, p. 23 and note 81, p. 29). As to how to create a successful design, Van Mander is rather brief, mentioning just a few possibilities, such as creating a clear focal point (*scopus*) for the main protagonist by placing this figure on a slightly higher plane, grouping the other figures in circular fashion around the main protagonist, and introducing witnesses to an event much like a stallholder cunningly arranges his wares on high shelves, down either side and at the bottom (see Tummers 2011b, pp. 204–208, esp. p. 206). He added that it could be very pleasant to have one figure looking out of the picture at the viewer—a popular pictorial device since the Renaissance (Tummers 2011b, p. 208).[51] Although Van Mander appreciated 'abundance' (*copia*) and 'variety' (*varietas*) in designs (characteristic of the mannerist fashion in his time), he also mentioned that 'good masters of originals often avoid abundance or *Copia* and rejoice in achieving quality in

[49] '*Men zegt dat Van Dyk veel moeite deed om hem mee te troonen naar Engeland, maar hy wilde daar niet naar luisteren [...] Egter behield hy groote achting voor des zelfs Konst, gelyk hy ook naderhand dikmaals gezegd heeft: Dat, indien hy in zyne vermenginge iets meer van het teere of dunne gehad had, hy een der grootste meesters zoude hebben geweest; want dat hy zyn weerga niet kende, die 't penceel zoo tot zyn wil had, dat hy, na hy een Pourtret had aangeleid, de vaste wezenstrekken, hoogsels, diepsels, met een penceelzet, zonder verzagtinge of verandering zoo hun behoorlyke plaats wist te geven*' (Houbraken 1718–1721, vol. 1, pp. 92–93). See also above: '*Rightly Admired by the Greatest Masters*'.

[50] In art theory, Michelangelo was praised in a similar way. According to Vasari and Van Mander, his hand was so steady and experienced that he could draw directly from his mind without any preliminary sketches (see Miedema 2013, p. 21 and note 58, p. 29). Van Mander also noted that Titian worked directly on the surface without drawing the composition first so that he painted with 'wet good judgement'(*nat met goet oordeel*) (Van Mander 1604, fol. 174; see also Atkins 2003, pp. 289–290 and note 49, p. 305).

[51] Alberti already recommended this pictorial device in the Italian renaissance.

simplicity'.[52] He also emphasised that good design was to be learned in practice, a view shared by all seventeenth-century art theorists (Tummers 2011b, pp. 204, 207 and note 103, p. 292). Recommendations in the written literature were therefore scarce. One other piece of design advice worth mentioning in connection to Hals's civic guard portraits comes from the painter Pieter de Grebber (c. 1600–1652/53), who wrote down a set of *Rules of Art* in 1649. De Grebber (1649, Rules IV, VIII and VIIII) strongly advised against placing the heads of flanking figures at the same height and warned that, in order to avoid confusion, figures should not overlap too much or be cut off by the picture frame.

As we have seen above, Hals's early civic guard portraits were highly appreciated as art, not only for Hals' bold, individual manner of painting but also for their overall liveliness and reality effect. Their designs contribute significantly to their effect. While Hals inspired the grouping of the officers around a banquet table on earlier examples by his well-known predecessors Cornelis van Haarlem and Hendrick Goltzius, he introduced a new sense of dynamism evoked by pictorial devices called 'joining/connecting' (*koppeling*) and 'leap' (*sprong*) in the seventeenth century (Tummers 2011b, p. 210; Van Hoogstraten 1678, p. 193). The powerful diagonal lines that quite a few art historians observed in these paintings are in fact the result of a careful spatial arrangement. Hals 'joined' diagonal directions and movements in the painting (*koppeling*), guiding the viewer's eye through the picture and creating a convincingly natural illusion of space, contrasting the officers in the foreground with a view into the distance through a window (*sprong*) (compare Fig. 6.16 with Frans Hals, *Banquet of the Officers of the Calivermen Civic Guard*, 1627, oil on canvas, 183 × 266.5 cm. (Frans Hals Museum, Haarlem)).[53]

[52] '*[...] goede meesters van den principalen/ D'overvloet oft Copia veel vermijden/ En in't weijnich eensaem/ weldoen verblijden*' (Van Mander 1604, fol. 17v).

[53] On Hals's design, see also (Liedtke 2011, pp. 18–19; Slive 1970–1974, p. 43 ff; Rietveld in Tummers (ed.) 2013, cat. nos. 26, 27, 28, 29, 30 and 31).

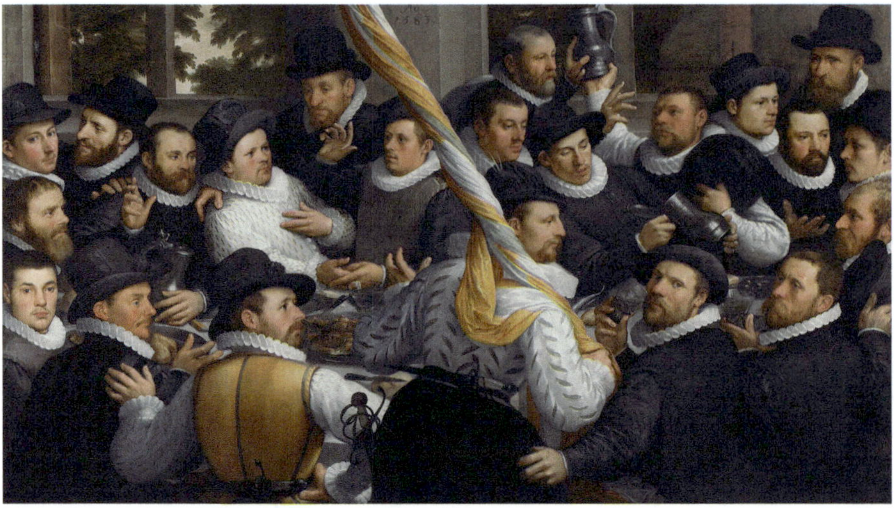

Fig. 6.16 Cornelis van Haarlem, *Banquet of the members of the Calivermen Civic Guard*, 1583, oil on canvas, 135–233 cm. (Frans Hals Museum, Haarlem)

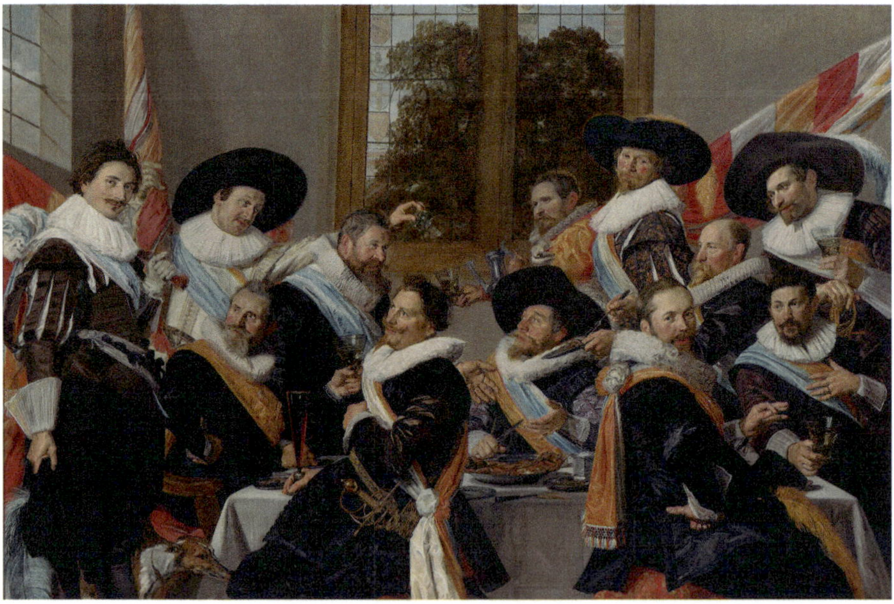

Frans Hals, *Banquet of the Officers of the Calivermen Civic Guard*, 1627, oil on canvas, 183 × 266.5 cm. (Frans Hals Museum, Haarlem) sn.pub/2ma71r

The powerful suggestion of reality in these paintings is often referred to as a 'snapshot' effect. However, unlike a casual photographic snapshot, these portraits have been well thought out in their design and adjusted to their specific context. Both the strict hierarchy within the civic guard and the portrait's specific location on the wall had a considerable impact on the works. As Seymour Slive discovered, Hals depicted the officers in his three banquet scenes in strict hierarchical order: the colonel, the highest in rank, seated at the head of the table next to the provost (or fiscal). The others are placed from left to right according to their rank: first three captains (one of which had the privilege of carving the meat) and then three lieutenants. The three ensigns or standard-bearers were the lowest in rank and are therefore depicted standing (Slive 1970–1974, vol. 1, pp. 42–44). Moreover, it was tradition for the standard-bearers to present themselves during the celebratory meal and ceremoniously hand in their banners (which gave the artist the opportunity to depict them while carrying their banners) (Carasso-Kok et al. 1988, p. 203; Van Asch van Wijck 1848–1851, p. 159). The later guard portraits show most officers standing, and have a somewhat more mixed order, though still carefully thought out. In the 1639 portrait, for example, sergeant Nicolaes Jansz. van Loo is placed behind his father, colonel Johan Claesz. van Loo, and faces the exact same direction, which nicely emphasises the family resemblance (Detail of fig. 6.5, nos. 1 and 16; the third figure from the left in the front row and the fifth from the left in the back row, both dressed in black) (Slive 1970–1974, vol. 1, pp. 138–140).

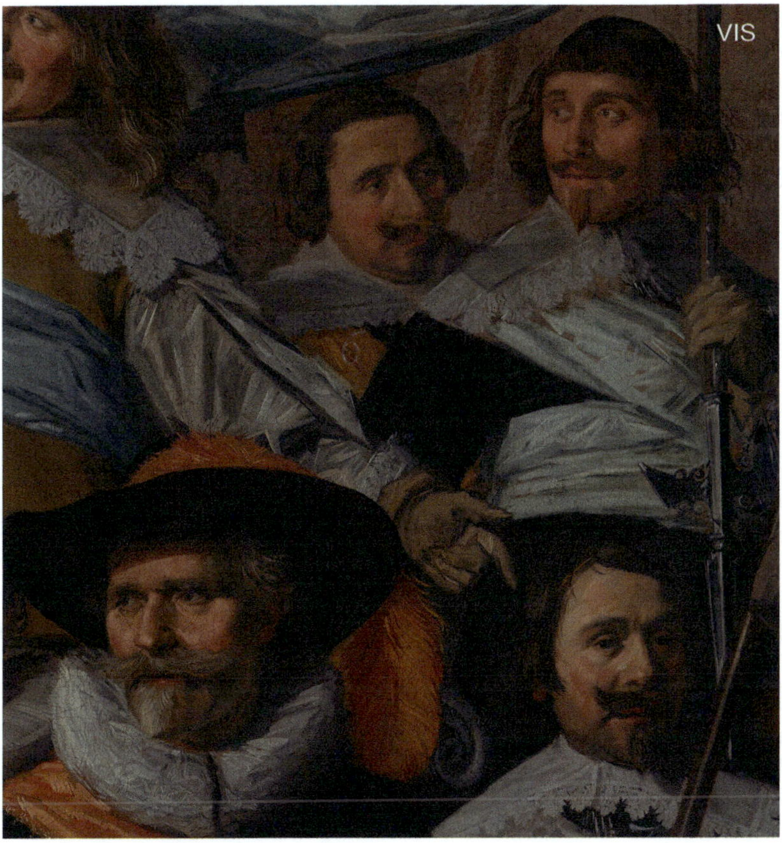

Detail of fig. 6.5 (sn.pub/696qrr)

The intended location of the portrait was also an important factor to consider. In the early seventeenth century, the earliest civic guard portrait created in the city by Cornelis van Haarlem (1583) occupied the place of honour above the chimney place in the Calivermen's headquarters (see Tummers and Gration 2013, p. 74). Artistically bolder than his later piece, it depicts the sitters in hitherto unprecedented 'lifelike' interactions, including the artist (upper left), his teacher (beside him) and his brother (lower left), resulting in an animated arrangement (see above Fig. 6.16).[54] It was later replaced by Hals's 1627 *Banquet of the Officers of the Calivermen Civic Guard*, which explains the latter's format, but also its particularly lively design, emulating Van Haarlem's earlier achievement. Hals's *Banquet of the officers of St. George Civic Guard* was similarly meant as a chimney piece for the headquarter's of Haarlem's other civic guard. By contrast, Hals's last Haarlem civic guard portrait of circa 1639 is much more static and traditional in its design—presenting the guardsmen in two rows (so-called *koppenrijen* or 'head-rows'). Again, the location is key here: it was meant for the long wall and must have been adjusted to an earlier painting hanging in the same location and showing the guardsmen in two neat rows (see Tummers and Gration 2013, p. 77).

In these large group portraits, the design depends almost entirely on the figures and their attributes such as the banners and arms. A we have seen, Hals's underlying sketch for the entire design did indeed focus on such elements, as seventeenth-century art theorists recommended. The rest was considered 'bywork' ('*bywerk*'), secondary parts (Tummers 2011b, pp. 103, 105, 232). The different postures of the figures, their interactions and their distribution across the canvas were key to the success of the overall design. Indeed, the lively placement of the figures, always at an angle with regard to the picture plane, their various, natural gestures and postures, and their subtle interactions with each other and with the viewer contribute greatly to the pictures' overall effect: their unprecedented sense of dynamism and lifelikeness.

6.6 *'Life Moved into Paint'*: The Challenge of Creating Lifelike Portraits

One of the greatest challenges for a seventeenth-century portrait painter was not only to capture the sitter's likeness, but also to make him or her appear 'alive'. According to seventeenth-century art theory, inner emotion was expressed through outer movement such as postures and facial expressions (Tummers 2011b, pp. 212–227). In addition, colour and brushwork had a huge impact in creating the illusion that the depicted scene was 'real' and in conveying characters and emotions

[54] Interestingly, Karel van Mander, who knew Van Haarlem well, mentions only this civic guard portrait by him (not the later one) in his *Schilder-Boeck*, praising it extensively (Van Mander 1604, fol. 292v). As almost all the sitters were members of the civic guard rather than officers and personal acquaintances of the artist, van Haarlem probably had much more artistic license in the way he depicted them then in his later piece. On the identification of the sitters, see (Niessen 2012–2013).

(see Tummers 2011b, p. 213 ff). Even the use of particular pigments could make or break the lifelike effect of a painting (see Van Hout 2008).

Achieving a convincing suggestion of reality was a *tour de force*, requiring a lot of talent, knowledge and experience. Much depended on the paint application. After the initial design had been completed, the paint was commonly applied in three phases, not unlike the three stages ('beginning' (*beginnen*), 'working-up' (*opmaecken*) and 'finishing' (*voltooien*) described in the legal documents about *The Meagre Company* (see also above and Appendices).[55] The first phase involved the division of light and shade and often also the first lay-in of the colours in the underpainting, usually referred to as 'dead- colour' (*dootverf*).[56] Once the dead-colour had dried, the second phase could start: the so-called 'working up' of the painting, in which the painter elaborated the colour, modelling and textural effects. In the third and final phase the strongest highlights and deepest shadows were added as well as various details. Like the overall design, Hals applied the different paint layers with a remarkable ease and accuracy, often integrating the different phases, which makes it hard to distinguish between the different stages.

In the seventeenth century Hals was praised both for the powerful reality effect of his pictures and for his distinctive personal style. As we have seen, Samuel Ampzing (1628, p. 371) commended Hals for painting 'very boldly after live' and according to Theodoor Schrevelius (1648, p. 383), the people portrayed by Hals even appeared to 'breathe and to live' (see above, notes 10 and 11). The playwright Govert Bidloo described this effect of Hals's paintings evocatively: *Where does […] Hals lead my attention?/ What are those faces, hands, covers, cloths?/ What master strokes are there in the velvet and in the pearls?/As if life itself, indeed, life was cheated and moved into the paint.*[57]

Although it may seem contradictory to praise Hals's brushwork and the lifelike effect of his paintings at the same time, according to seventeenth-century art theory, these features were actually related. As the Flemish art theorist Cornelis de Bie (1661, pp. 281–282) explained, Hals's paintings were intended to be viewed from a distance. As such, loosely applied brushwork often had a stronger reality effect than more finely painted works. It was not unlike recognising a friend at dusk, according to the painter and art theorist Samuel van Hoogstraten. Loosely applied brushstrokes produced great results when viewed from the right distance: then the paints seemed to coalesce a little because of the 'thickness of the air' and the beholder's eye filled

---

[55] On seventeenth-century sources about the three stages in the painting process, see (Van Eikema Hommes 2004, p. 12 ff).

[56] An example of a portrait in the dead colouring stage is Rembrandt's unfinished *Self-Portrait with Beret* of circa 1659 (Musée Granet, Aix-en-Provence, Inv. no. 860–1–17).

[57] 'Wat voord'Jordaens, en Hals mijn'aandacht met zich heen?/Wat zijn die troniën, die handen, hulsel, kleên?/Wat zijn daar in't fluweel, en paarlen meester streeken?/Als was het leeven, juist, het leeven afgekeken,/En in de verf verplaatst' (Bidloo 1719, pp. 173–185). Bidloo praised both Jacob Jordaens and Frans Hals in the same breath in a poem celebrating the paintings of the Amsterdam collector Filip de Flines written between 1683 and 1685 (see also Tummers 2013a, pp. 28–32; 2011b, pp. 230–234).

in the details (Van Hoogstraten 1678, pp. 27 and 264; see also Tummers 2013a, p. 32 ff). Omitting, not describing every detail, created space for movement; the figures appeared to come to life. It was the reason that only painters with a sure touch and a rather loose painting style could convincingly depict a person laughing or create the so-called 'snapshot' effect discussed above; too much detail would make the expression or scene appear to be 'frozen'. It was a notorious challenge for painters. Karel van Mander (1604, fol. 25v) reported that many artists had tried to depict a convincing laugh, but that in many cases it looked more like their figures were crying. Their smile had turned into a grin. In the case of the guardsmen, Hals of course paid heed to the rules of decorum or appropriateness; openly smiling was considered unseemly at the time. In the civic guard portraits Hals's loose brushwork complements the lively gestures and glances of the militiamen, making it seem as if some of the men have just turned to us, the viewers, and could move again at any moment.

Another notoriously difficult challenge for painters was to convincingly depict skin colours and their various tonal gradations. Skin colour was considered the single most important and complex colour to render successfully. In the words of the painter and art theorist Willem Beurs (who wrote extensively on the subject): "*Just as we humans consider ourselves foremost amongst animals; so, too, are we the foremost subject of the art of painting, and it is in painting human flesh that its highest achievements are to be seen, whenever a painter succeeds in rendering the diversity of colours and strong hues found in human flesh and particularly in faces, adequately depicting the intricacy of the diversity of people or their different emotions*."[58]

Karel van Mander (1604, fol. 49v) explained that painting faces required as many colours as landscapes, including green, blue and yellow. He recommended the use of vermillion as it created a 'glowing' effect (*gloeyendheyt*), a sense of translucency. Furthermore, one should avoid lamp black in shadowy parts of flesh tones in his view for it turned too dark; it was better to use umber, bitumen, Cologne earth or terra verde (*umbre*, *aspalten*, *Ceulsch' eerden*, en *terreverden*).[59] Moreover, the shadows should be 'flesh-coloured' (*vleeschachtig*) and the highlights 'skin-coloured' (*carnaty*); the latter should never be pure white, he warned, for he had seen many painters err in this respect (Van Mander 1604, fol. 49r).[60] The successful

[58] 'Gelijk wij menschen ons zelven het voornaamste der dieren stellen: zoo mede van de schilderkonst, en't was die tot een toppunt verheven te zien, als 'er een Schilder was, die alle verscheidentheid der verwen, en kragtige kolorijten, die in menschen vlees, en met namen in de tronien voorkomen, 't zywe de verscheide menschen besien, of haare verscheide hertstogten, behoorlijk genoeg in alle doorwrogtheid kan uitbeelden.' (Beurs 1692, p. 184). Translation taken from (Van de Wetering 1997, pp. 146–147 and note 27, p. 305).

[59] On Cologne earth pigments see (Languri 2004, pp. 73–115). On the term 'terra verde' and its different meanings, see (Van Loon and Speleers 2011).

[60] A similar warning was later repeated by Samuel van Hoogstraten: "'Ik meyne hier niet, datmen geverfde kleederen of iets dat van natueren bruin is, noch zelfs het blanke naekt, met witten of mastekotten moet ophoogen, want dat acht ik belachelijk.' (van Hoogstraten 1678, p. 268).

portrait painter also needed a ground or middle tone, preferably, in his view, a 'glowing' one. Other art theorists recommended somewhat different mixtures. Beurs described as many as thirteen different mixtures composed of nine pigments that were needed to successfully depict people of different ages and gender as well as their emotions: '*1. Light ochre and white 2. Light ochre vermillion and white 3. Light ochre vermillion [red] lake and white 4. Vermillion and white 5. [Red] Lake and white 6. Carbon black and white 7. Carbon black [red] lake and white 8. Umber and [red] lake 9. Umber, bone black and [red] lake 10. Terra verde and white 11. Terra verde, black and white, and [red] lake 12. [Red] lake and black, and lastly 13. Light Persian berry lake (schietgeel), [red] lake and brownred.*'[61]

The exact paint mixtures used by different artists have recently been shown to be much more characteristic of individual painters than art historians have long assumed. The pigments described above were widely available and thus not unique to a specific artist. However, their specific application could vary considerably and thanks to relatively new non-invasive techniques such as XRF we can now study the way in which painters such as Hals used their pigments in much more detail than ever before, how they reserved certain pigments for costly types of paintings and/or varied their use of pigments throughout their career.[62]

The large amount of data gathered in the context of the *Frans Hals/not-Frans Hals* and *21st Century Connoisseurship* projects allows us to study Hals's depiction of skin tones in great detail. Just like Beurs would later recommend, Hals used umber, (containing manganese (Mn) and iron (Fe)) in his facial shadows, and occasionally mixed in some bone black (containing calcium (Ca) and phosphorus (P)) for the darkest tones. For example, the shadow on the cheek of standard-bearer Boudewijn van Offenberg in Hals's earliest civic guard portrait is indicated with umber while the mixture with bone black is reserved for the deepest shadows underneath his nose and in his eye sockets (Detail of Fig. 6.1: manganese (Mn-K) map, visible light image, calcium (Ca-K) map, in 'sync mode'). The reddish glow in these facial shadows suggests that he also mixed in a red pigment, possibly a red lake, which corresponds in the MA-XRF scans to potassium (K), originating from the lake substrate and/or manufacturing process. Indeed, red lake seems to have been used in the deepest shadows of the incarnates, such as the partition of the lips, the nostrils, the eyelids and the folds of the earlobes, as seen, for example, in the face of Captain Nicolaes Woutersz van der Meer, seated in front of the table in the foreground of *The Banquet of the Officers of the St George Civic Guard* from 1616 (Detail of Fig. 6.1: visible light image, potassium (K-K) map, in 'sync mode').

[61] '1. Ligten oker en wit 2. Ligten oker vermilioen en wit 3. Ligten oker vermilioen lak en wit 4. Vermilioen en wit 5. Lak en wit 6. Koolswart en wit 7. Koolswart lak en wit 8. Omber en lak 9. Omber, beenswart en lak 10. Terreverde en wit 11. Terreverde, swart en wit, en lak 12. Lak en swart, en ten laatsen of 13. Ligte schijtgeel, lak en bruinrood' (Beurs 1692, p. 186).

[62] See for example the excellent essay (Gifford and Glinsman 2017).

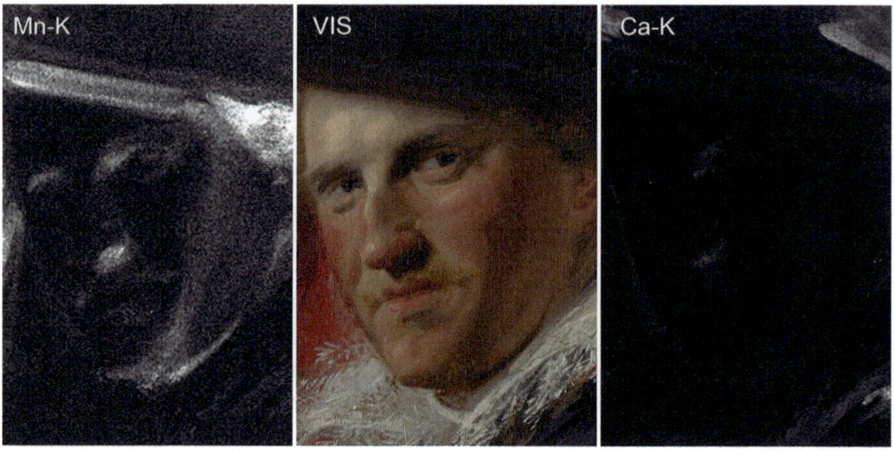

Detail of Fig. 6.1: manganese (Mn-K) map, visible light image, calcium (Ca-K) map, in 'sync mode' (sn.pub/p9nhja)

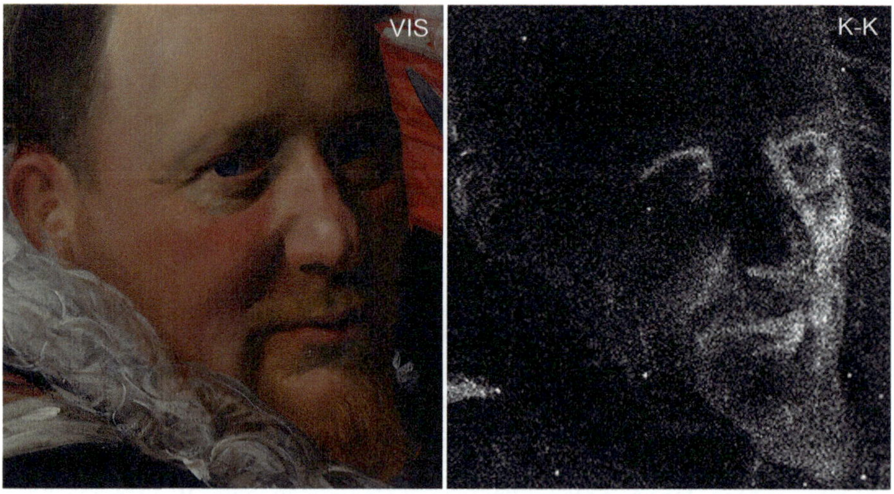

Detail of Fig. 6.1: visible light image, potassium (K-K) map, in 'sync mode' (sn.pub/bwf5ga)

As Karel van Mander had advised, Hals applied vermillion a lot, the pigment known for its 'glowing' effect. Especially in his earlier civic guard pieces, Hals used vermillion abundantly and consistently; the mercury maps of these group portraits show occasional changes in the position of hands, faces or the direction of a glance very clearly, as we will see. In only a few cases, vermillion is absent in a depicted hand: in the hand of the standard bearer Jacob Matham on the far left in the 1627 calivermen civic guard portrait and in the hand of Luitenant Cornelis Boudewijnsz on the far left in 1627 St George civic guard portrait, and in the hands of the standard-bearer Jacob Steyn and luitenant Nicolaes Olycan in the 1633 portrait. In the 1639 civic guard portrait only two hands contain vermillion, while the three other hands lack it: the hands of figures 3, 5 and 7 (the figures surrounding captain Florens Pietersz van der Hoeff with the orange sash). In a general sense, Hals appears to have employed the pigment more sparingly for skin tones as time progressed. We already saw that in the Berlin *Portrait of a Woman* Hals used vermillion in the face but not for the hands (see Chap. 3).

His selective use of the pigment presumably relates to its price. Vermillion was a costly pigment and some seventeenth-century recipe books therefore recommended painting skin colours with red earth first and only adding vermillion in the top layer to save costs (Van Eikema Hommes 2004, p. 12). When seen from this perspective, it is striking how often Hals actually used vermillion for both faces and hands in his civic guard portraits, which confirms the special status of these paintings. These were large works intended for semi-public display, and it was common for well-known artists to reserve their finest materials and best manner of painting for such works.[63]

Van Mander's remark that a range of pigments including red, yellow and blue, was needed to depict skin colours, is also largely consistent with Hals's painting practice. Admittedly, in his earliest civic guard portrait he did not employ any blue pigment for the skin tones but from at least the late 1620s until the late 1630s he commonly mixed in a copper containing blue: presumably azurite, visible in the Cu maps.[64] He used it to create a greenish grey shadow middle tone for the half shadows, while reserving brown tones containing umber (Mn) and bone black (Ca) for the deeper shadows (Detail of Fig. 6.2: copper (Cu-K) map, visible light image, manganese (Mn-K) map, calcium (Ca-K) map, in 'sync mode'). In Hals's last civic guard portrait (1639), however, his use of copper in skin tones is sparse; only captain Quirijn Jansz Damast, the third figure from the right, has an abundance of grey tones in his face that contain copper (Detail of Fig. 6.5: visible light image, copper (Cu-K) map, in 'fade mode'), and it has not been found in any later paintings by Hals.[65]

---

[63] On the art theory about adjusting one's manner of painting to the price or prestige of a painting, see (Tummers 2011b, p. 134 ff). On the selective use of expensive materials, see (Gifford and Glinsman 2017).

[64] The presence of azurite was confirmed by observation through the microscope in the Berlin *Portrait of a Woman* (see Chap. 3); before, it had been observed in a sample taken from *Cornelia Vooght* (1631, Frans Hals Museum, os I-118) in the lower layer of the flesh tone and in observations though the microscope of Hals's 1639 civic guard portrait, witness the report by Karin Groen of 22 December 1984, Frans Hals Museum, Archives of the Conservation Department.

[65] It was not used in the skin colours of the Hals's regent group portrait of 1641 and the 1663–64, as research with MA-XRF in relation to their recent conservation treatment confirmed, and neither does it occur in Hals's *Portrait of a Man, possibly a Clergyman* of c. 1657–1660 at the Rijksmuseum (inv. no. SK-A-2859).

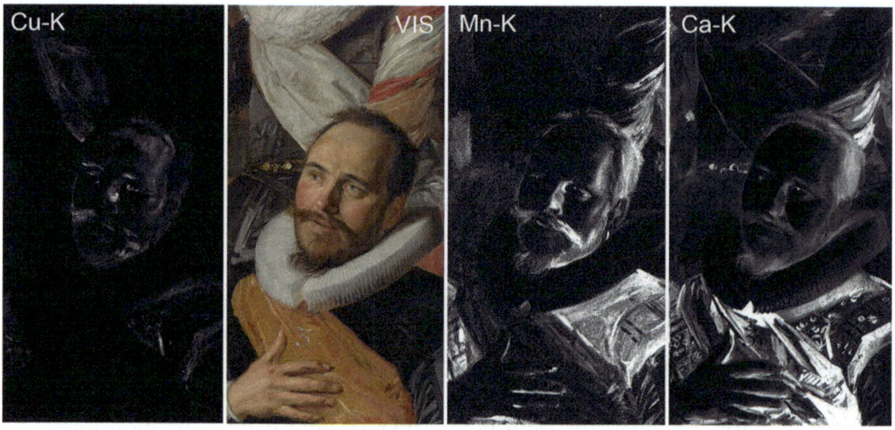

Detail of Fig. 6.2: copper (Cu-K) map, visible light image, manganese (Mn-K) map, calcium (Ca-K) map, in 'sync mode' (sn.pub/khr0by)

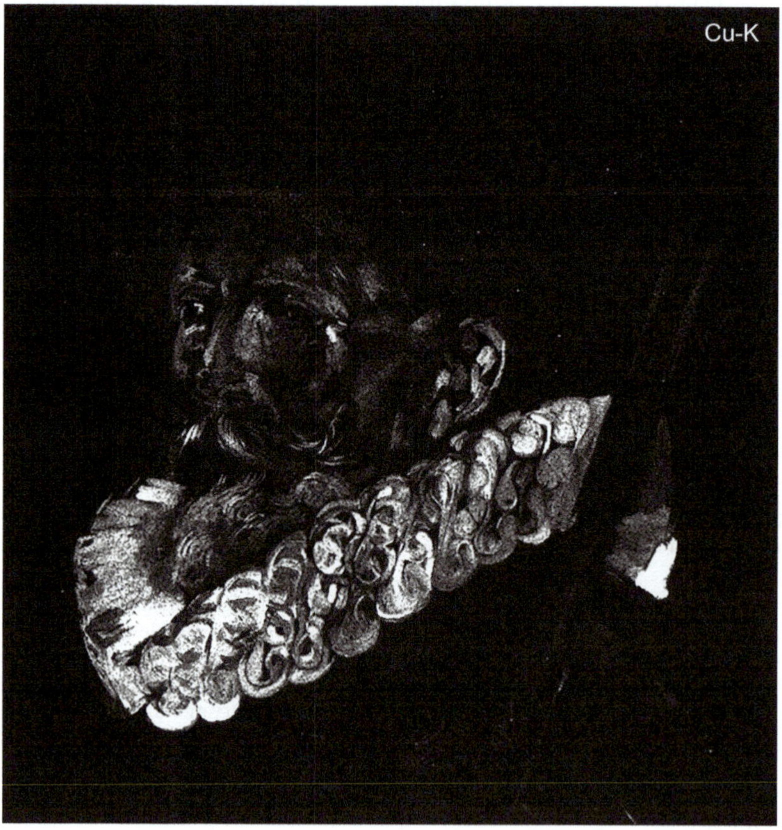

Detail of Fig. 6.5: copper (Cu-K) map (in 'fade mode' sn.pub/o1xkv1)

As we have seen, Hals's last civic guard portrait is the least innovative of his Haarlem civic guard pieces in terms of its overall design. It also differs somewhat from his earlier works in terms of the materials used; relatively costly pigments such as vermillion are applied more sparingly in the flesh tones.[66] Moreover, the execution also falls short of his earlier standard: the use of colours and tones is somewhat less balanced than in his earlier group portraits. Furthermore –as will be discussed below-workshop assistance can very clearly be seen in this painting.

Nevertheless, in this portrait, as in the earlier ones, Hals must have accurately captured the likenesses of the various guardsmen. The large civic guard portraits created in Haarlem actually provide us with a unique opportunity to judge some of the likenesses ourselves, since five of the guardsmen portrayed appear twice in the paintings. As early as 1690, lists were drawn up to preserve the names of all the guardsmen depicted for posterity, which greatly facilitates their identification (see Köhler and Levy- Van Halm 1990, p. 8 and note 2, p. 13).[67] The standard-bearers Boudewijn van Offenberg and Cornelis Schout appear both in the 1616 portrait and again some 15 years later in 1627 (Details of Fig. 6.1 and Fig. 6.2, in 'gallery mode'; Details of Fig. 6.1 and Fig. 6.2, in 'gallery mode'). Captain Michiel de Wael appears in 1627 and again, as fiscal, in 1639 (Details of Fig. 6.2 and Fig. 6.5, in 'gallery mode'). Lastly, the painter and luitenant Hendrick Gerritsz Pot and the colonel Johan Claesz van Loo appear both in the 1633 and the 1639 group portraits (Details of Fig. 6.4 and Fig. 6.5, in 'gallery mode'; Details of Fig. 6.4 and Fig. 6.5, in 'gallery mode').

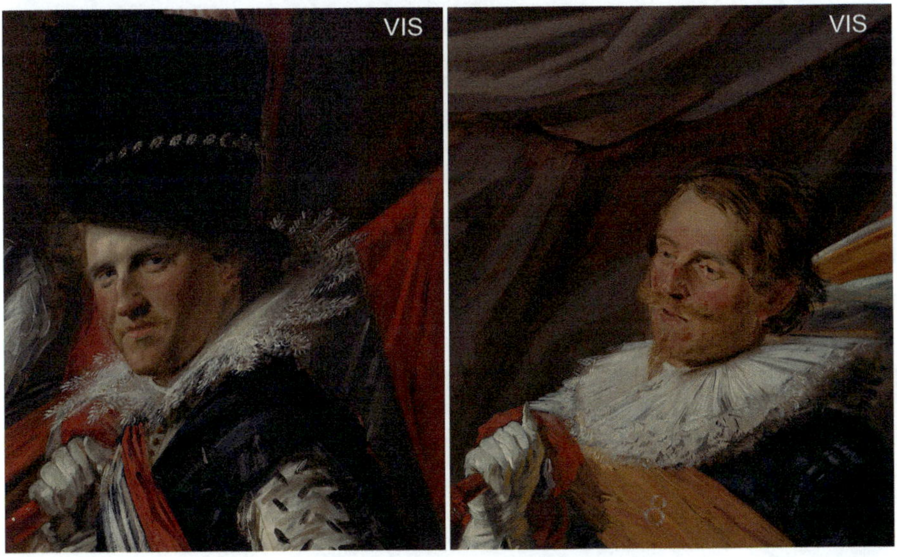

Details of Fig. 6.1 and Fig. 6.2, in 'gallery mode' (sn.pub/ixqdar)

[66] The prices for pigments varied greatly at the time; earth pigments such as ochres were relatively inexpensive, but the price of bright yellow, red and blue pigments could be considerable, see (Van Eikema-Hommes 2004, p. 12; Krekel and Burmester 2003; Henny 1994).

[67] Municipal Archive Haarlem, *Oud Archief der Schutterij*, inv. 23–2, 2 Oct 1690.

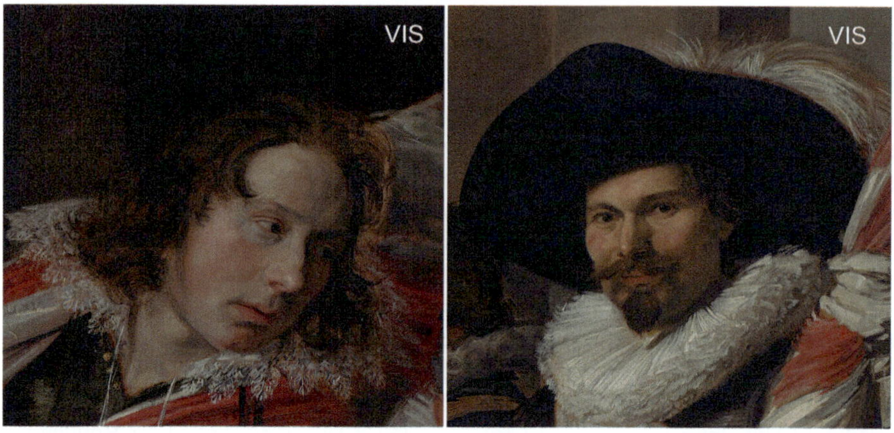

Details of Fig. 6.1 and Fig. 6.2, in 'gallery mode' (sn.pub/ziyfnv)

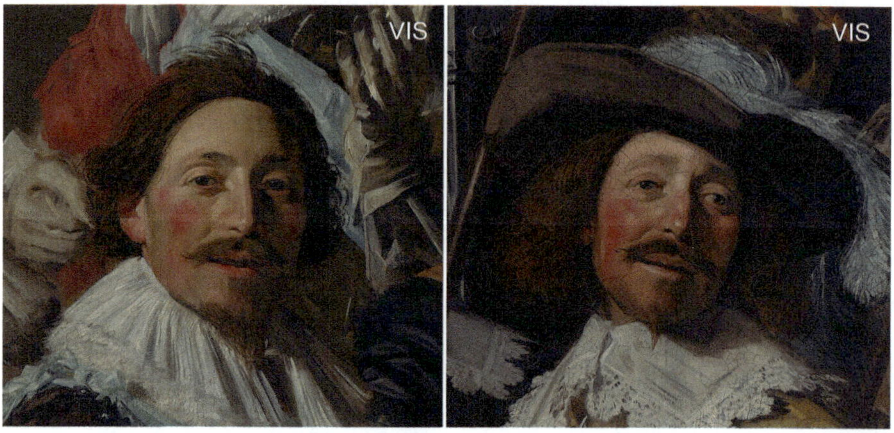

Details of Fig. 6.2 and Fig. 6.5, in 'gallery mode' (sn.pub/7fry92)

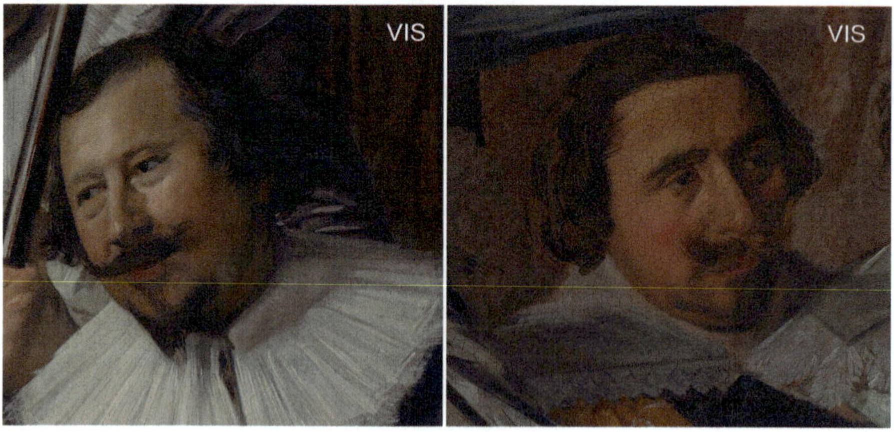

Details of Fig. 6.4 and Fig. 6.5, in 'gallery mode' (sn.pub/9bjotz)

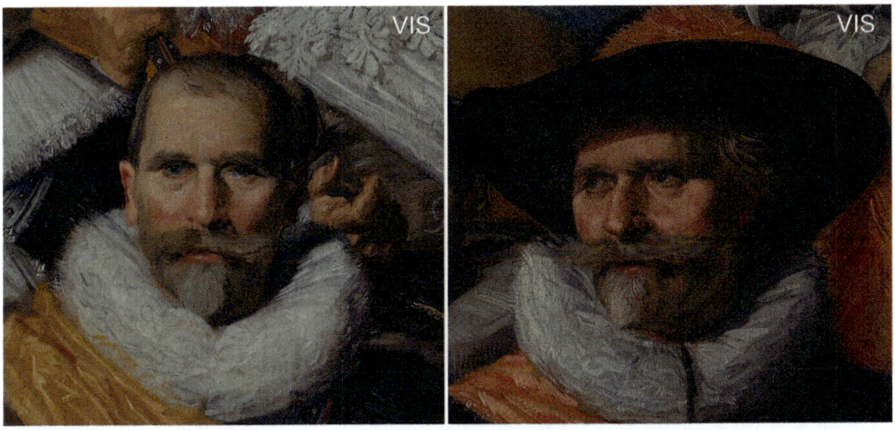

Details of Fig. 6.4 and Fig. 6.5, in 'gallery mode' (sn.pub/uxmm8e)

6.7 Revisions, Inconsistencies, Workshop Practice and Assistance

The painter and art theorist Karel van Mander (1604, fol. 46v) advised skilled painters with a steady hand and an abundance of ideas to 'act like the bold and correct a mistake here or there' (*die overvloedich / In't inventeren zijn, doen als de stoute, / En verbeteren hier en daer een foute*). It was a sign of virtuosity, of true mastery, when painters were able to design their compositions directly on the canvas or panel; such bold painters simply corrected their work where necessary during the painting process (Van Mander 1604, fol. 46v).[68]

The single most striking characteristic of all the different technical data gathered in the context of the recent research projects is how few changes these show. It confirms Hals's reputation of having a remarkably steady hand and a sure touch. We do not see him correcting actual mistakes in the placement of, for example, a contour, an outline or a loose accent. Instead, we only see deliberate changes such as a different gesture, attribute or interaction between the figures. Often his first sketch is spot-on and he continues to work it up at exactly the same place, which can make it hard to distinguish the underlying sketch or initial outline. However, there are a number of significant changes and some interesting inconsistencies in his large civic guard pieces that shed new light on Hals's working process as well as on his collaboration with workshop assistants.

Compared with what was previously known about Hals's working process based on the study of the radiographs of these large works, the new analyses with IRR and MA-XRF yielded many new insights. In some cases, this clarified a previously observed change. For example, the radiograph of Hals's earliest civic guard portrait

[68] In the margin: "Stracx eerst op penneel te stellen, Meesters werck."

showed a change in the position of the hands of captain Jacob Laurensz (Detail of Fig. 6.1: visible light image and x-radiograph, in ‘sync mode’). It has been suggested that the captain originally held both hands to his chest, which would have been a rather odd posture for a gentleman (Köhler and Levy- Van Halm 1990, pp. 18–19). However, the element maps of iron (Fe) and lead (Pb), both present in the skin colour, showed that in fact not one but two hands had changed position, and made the change more understandable: Hals had simply reversed the gesture. Originally, captain Laurensz held his right hand to his heart and made a speaking gesture with his left hand raised. In the final version, he holds his left hand to his heart while making the speaking gesture with his right hand, creating a more close-knit visual unity with the colonel beside him, to whom he is speaking (Detail of Fig. 6.1: visible light image, iron (Fe-K) map and lead (Pb-L) map, in ‘sync mode’).[69]

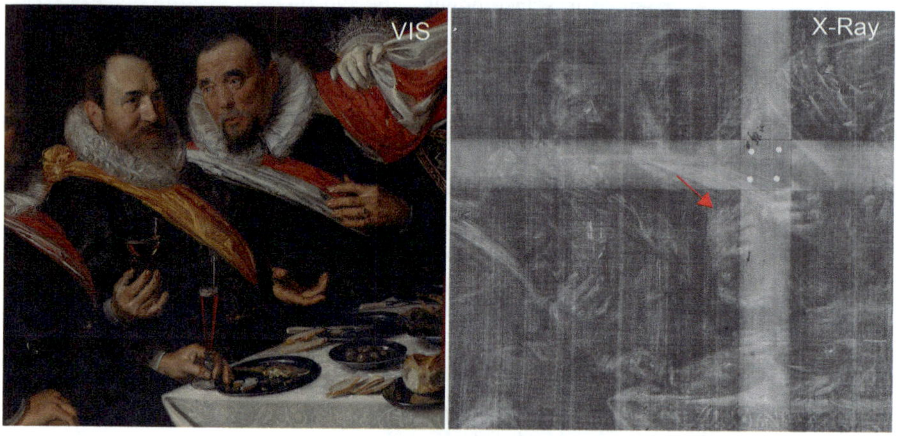

Detail of Fig. 6.1: visible light image and x-radiograph, in ‘sync mode’ (sn.pub/upbj3c)

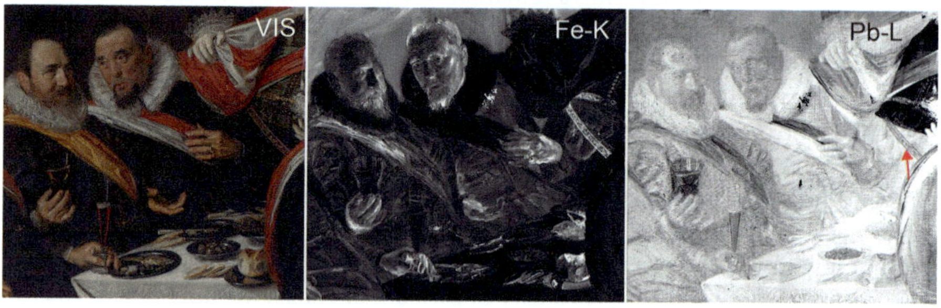

Detail of Fig. 6.1: visible light image, iron (Fe-K) map and lead (Pb-L) map, in ‘sync mode’ (sn.pub/xtdeqv)

[69] The changes are more visible in the elemental maps because the separate elements in the hands are mapped; by contrast the radiograph is less legible, specifically at the height of the second hand where the signal is overshadowed by the presence of the stretcher bar.

Many of these changes seem intended to strengthen the visual unity of these complex designs by emphasising the connection between adjacent shapes and directions (*koppeling*) or by removing distracting elements. In the 1616 painting, apart from the change discussed above, a hand gesture of lieutenant Hugo Mattheus Steyn was eliminated after it was first painted. Steyn originally held his right hand on his heart, as can be seen in the vermillion map (Detail of Fig. 6.1: visible light image and mercury (Hg-L) map, in 'sync mode'). On second thoughts, Hals removed the hand, presumably to avoid '*haspeling*' ('reeling') or '*sparteling*' ('flounder') as it was called in art theory: It was well-known that shapes in close vicinity influenced one another and that this could have an unintentional chaotic effect (Detail of Fig. 6.1: visible light image and mercury (Hg-L) map, in 'sync mode').[70]

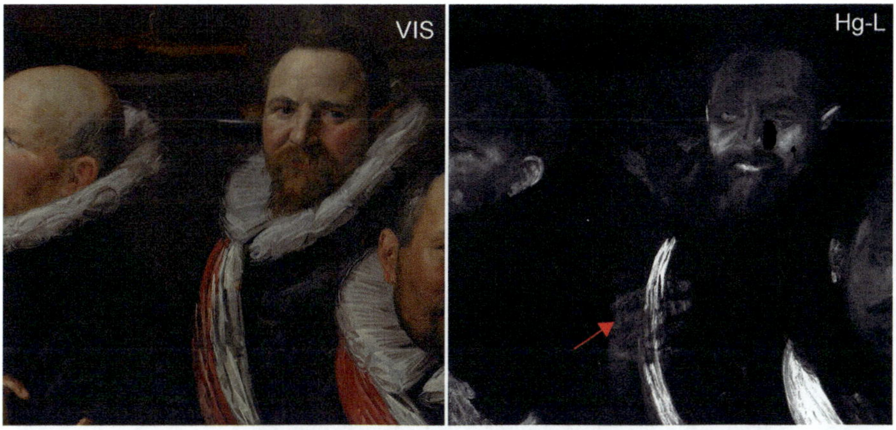

Detail of Fig. 6.1: visible light image and mercury (Hg-L) map, in 'sync mode' (sn.pub/tj14dq)

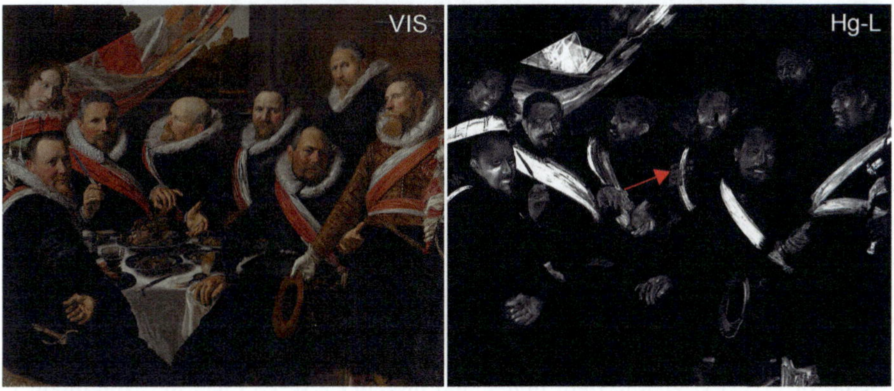

Detail of Fig. 6.1: visible light image and mercury (Hg-L) map, in 'sync mode' (sn.pub/mr26qp)

[70] Karel van Mander (1604, fol. 18v) and Pieter de Grebber (1649) warned painters to avoid 'haspeling'; Gerard de Lairesse (1707, vol. 1, pp. 236 and 263–264; vol. 2, p. 256) discussed specifically how neighbouring shapes impacted one another and could cause 'sparteling' (see De Vries 2004; Tummers 2011b, pp. 207 and 210, and notes 101 and 114, p. 202).

Furthermore, Hals reversed the orientation of the sash of the standard-bearer Gerrit Cornelisz Vlasman, which now creates a powerful diagonal thrust together with the standard worn by Boudewijn van Offenberg. In seventeenth-century terms, such a 'joining' of adjacent shapes and directions was called a *'koppeling*', as we have seen, and Hals used this pictorial device to create a refined coherence in these rather complex designs. The above-mentioned line of sight echoes the strong diagonal thrust at the left of the painting where the direction of Jacob Cornelisz Schout's banner points at the highest official at the table: colonel Hendrick van Berkenrode (with the orange sash)—a diagonal orientation that is also echoed in the green curtain above Van Berkenrode's head. The change in the orientation of Vlasman's sash occurred early in the design process: the sash was never painted in the other direction, witness the lead and vermillion maps; it is only visible in the IRR (Detail of Fig. 6.1: visible light image, IRR, mercury (Hg-L) map and lead (Pb-L) map, in 'sync mode').[71]

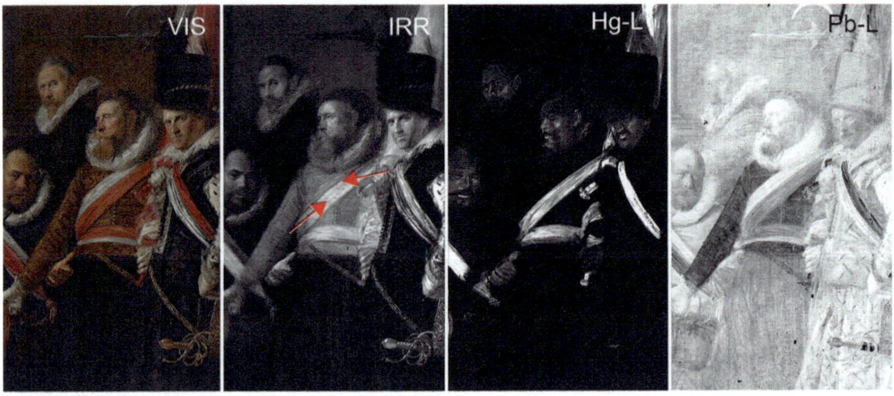

Detail of Fig. 6.1: visible light image, IRR, mercury (Hg-L) map and lead (Pb-L) map, in 'sync mode' (sn.pub/1fqtkf)

Also, some minor adjustments have been made: the trees in the vista below Jacob Schout's banner have been extended, partially overlapping a bright patch of sky in the background and thus integrating the background a little more into main design (Detail of Fig. 6.1), and the direction of captain Vechter Jansz van Teffelen's glance has been altered. He is the captain who had the so-called privilege of the knife, i.e. he was allowed to carve the bird at the ceremonious meal. Initially, he looked at the bird he is about to carve; in the final design, however, he faces the viewer instead (Detail of Fig. 6.1: visible light image and mercury (Hg-L) map, in 'sync mode'). Possibly, the last change was made at the request of the sitter, although Hals himself could also have wished to strengthen the connection with the viewer.

[71] Interestingly, only in this civic guard piece do two of the standard bearers wear their sash over their left shoulder. In all the other civic guard painting by Frans Hals the sashes are worn exclusively over the guardmen's right shoulders or over their bellies. Possibly, the rules for wearing the sashes changed somewhat over the course of the years.

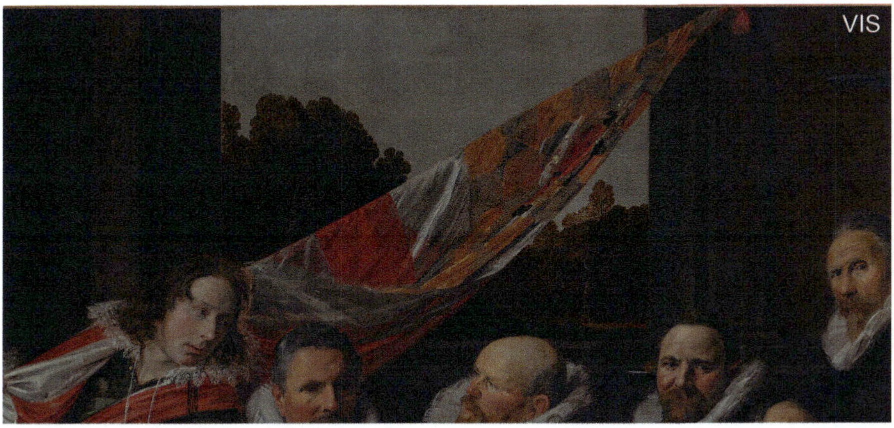

Detail of Fig. 6.1 (sn.pub/9wnvb0)

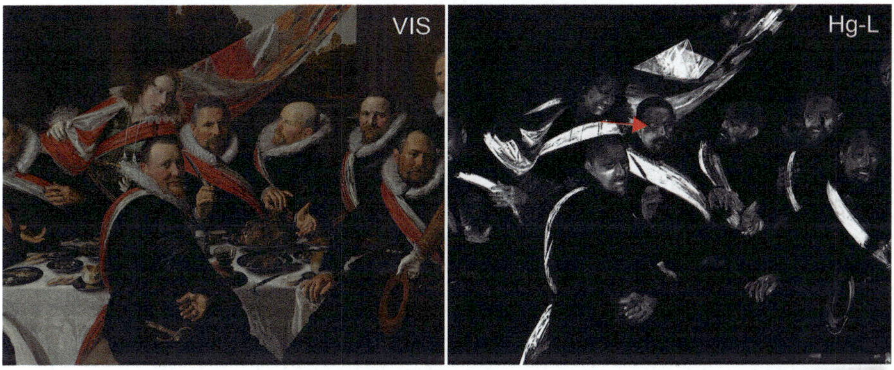

Detail of Fig. 6.1: visible light image and mercury (Hg-L) map, in 'sync mode' (sn.pub/bh3us6)

Similar compositional changes can be detected in the 1627 St George civic guard portrait. The interaction between several guardsmen has been altered, witness the earlier direction of their faces and glances that can be seen in the mercury element map (showing areas containing vermillion, a pigment Hals used in the skin colour) (Detail of fig. 6.2 in visible light and mercury map).[72] Originally, captain Michiel de Wael and lieutenant Frederick Coning faced each other (Coning's head was originally depicted a bit higher too), while in the final design they look out at the viewer. Especially captain Michiel de Wael with his central position, flushed cheeks and parted lips (as if he were speaking) greatly enhances the connection with the viewer. Indeed, he is a favourite in promotional material of the Frans Hals Museum, such as book covers and bags.[73]

[72] Two of these three changes in the direction of the faces were noticed before in the radiographs, and there has been some speculation that the change in figure 9 (see Fig. 6.2) could be related to a different guard member occupying the position. Now that the change(s) can be seen in much greater detail there is no reason to assume that the head itself was changed; just its orientation was altered (see Köhler and Levy- Van Halm 1990, pp. 28–29).

[73] Among other things, Michiel de Wael graces the cover of the museum's latest publication of its highlights: (Erftemeijer et al. 2014).

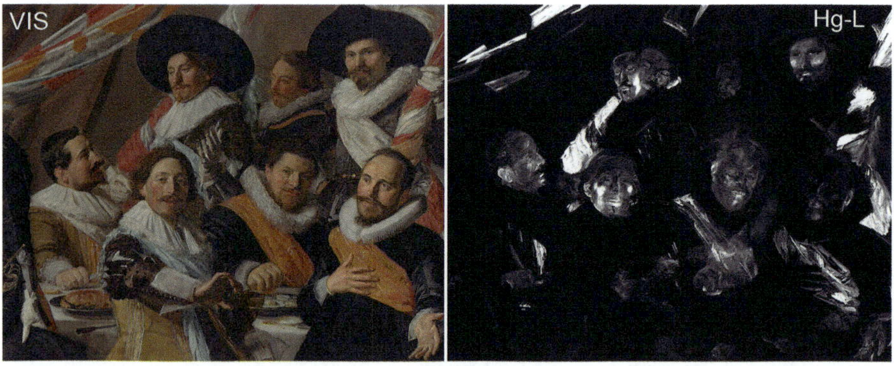

Detail of Fig. 6.2: visible light image and mercury (Hg-L) map, in 'sync mode' (sn.pub/w1gsk8)

The infrared reflectogram shows that Hals altered De Wael's hand gesture too. While facing Frederick Coning, De Wael initially held out his glass towards the viewer, as we can see in the sketch lines in the infrared reflectogram (IRR), the green glass in the copper map and in the subsequent paint layer in flesh colour that contains mercury (Hg) (Detail of Fig. 6.2: visible light image, IRR and mercury (Hg-L) map, in 'sync mode'). When Hals turned De Wael's face and made him appear to address the viewer, he also had De Wael turn his glass upside down– a sign at the time that he wanted his glass to be refilled (Detail of fig. 6.2 in visible light and infrared reflecogram). Possibly, Hals wanted to avoid the suggestion that De Wael was involving the viewer in a toast. Raising the glass in the so-called '*heilsdronk*' or prosperity toast was a privilege reserved to the highest officer at the meal (Schama 2001/1987, pp. 188–189). Therefore, in this painting, like in the earlier group portrait of the St George civic guard, it is the colonel (recognisable by his orange sash) who raises his glass. Just behind Michiel de Wael, standard-bearer Dirck Dicx originally had his head turned more towards the right and looked in the direction of the table, while in the final design he exchanges looks with lieutenant Jacob Olycan, strengthening the internal cohesion of the group.

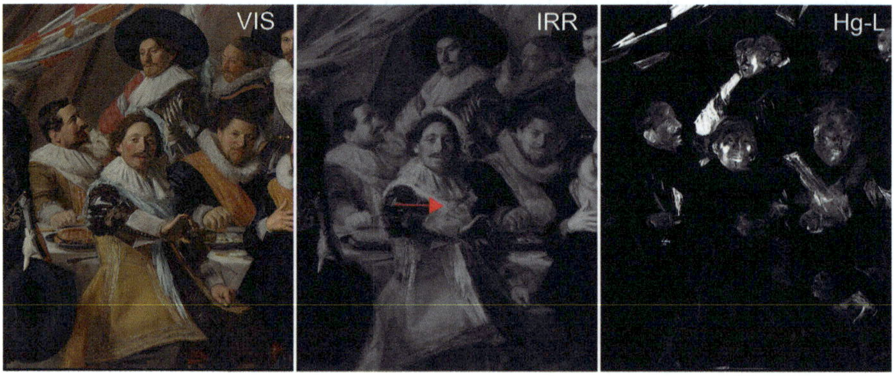

Detail of Fig. 6.2: visible light image, IRR and mercury (Hg-L) map, in 'sync mode' (sn.pub/mbyj1u)

Changes such as these give us a unique insight into the design process; we can see how Hals worked out his thoughts for the composition, strengthening the unity in the design and the connection with the viewer, while paying heed to the decorum and hierarchy that was characteristic of the civic guard. Having been a member of the St George civic guard himself from 1612 until 1624, as we have seen, Hals was of course familiar with the traditions and hierarchies of the guards. In these two early civic guard portraits, the changes in the design are the most substantial. In the other 1627 civic guard portrait, only the position of one head was altered: hall warden Willem Ruychaver was originally depicted a little closer to standard-bearer Michiel Ramp (Detail of Fig. 6.3: visible light image and IRR, in 'curtain mode'). Note how one of the main diagonals in the composition here runs from Ramp's banner along Ruychaver's collar, his outstretched hands with the pewter and *roemer* glass towards the highest ranking officer present: colonel Willem Claesz Vooght (with the orange sash), who raises his glass in a toast while looking at standard-bearer Adriaen Matham. The latter has taken off his hat to present himself (as was customary for the standard-bearers during the festive meal, as we have seen), and makes the connection to the viewer by looking straight out of the picture. Once again, Hals created a powerfully unified design, caused by an exquisite '*koppeling*' (joining) and '*sprong*' (spatial leaps), in which also the guard in the back with his outstretched arm and the collars of several guardsmen play a role.

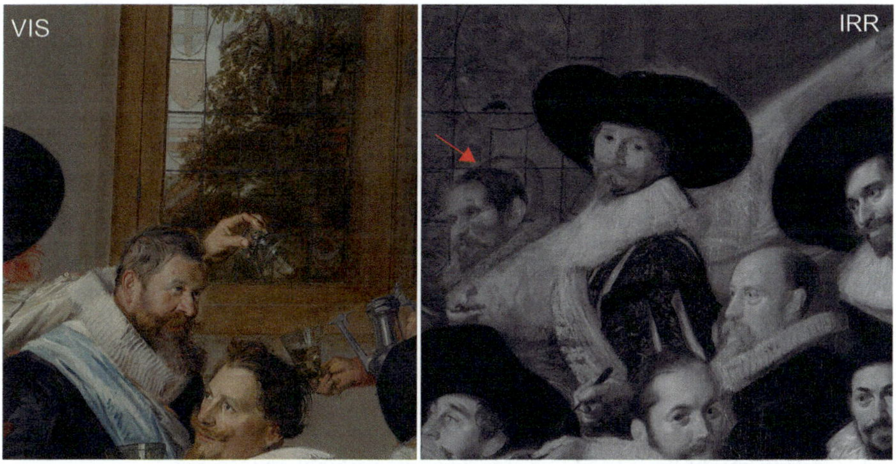

Detail of Fig. 6.3: visible light image and IRR, in 'curtain mode' (sn.pub/wurxmu)

In terms of their inventions, the two later Haarlem civic guard portraits show more subtle alterations. Presumably, Hals's increased experience in designing such large, complex works, made him ever more efficient in working out his ideas. His 1633 outdoors portrait of the Calivermen's civic guard is unique in that it also includes sergeants (i.e. sub-alterns). The outdoors setting enabled Hals to include the different arms of the guardsmen in his design. The captains hold their partisans (some of which are decorated with tassels); the sergeants their halberds. The only change visible in the infrared reflectogram concerns captain Johan Schatter in the yellow coat, who tilts his head slighty further in the final version (mostly visible upon moving the curser: (sn.pub/f2zn8o)).

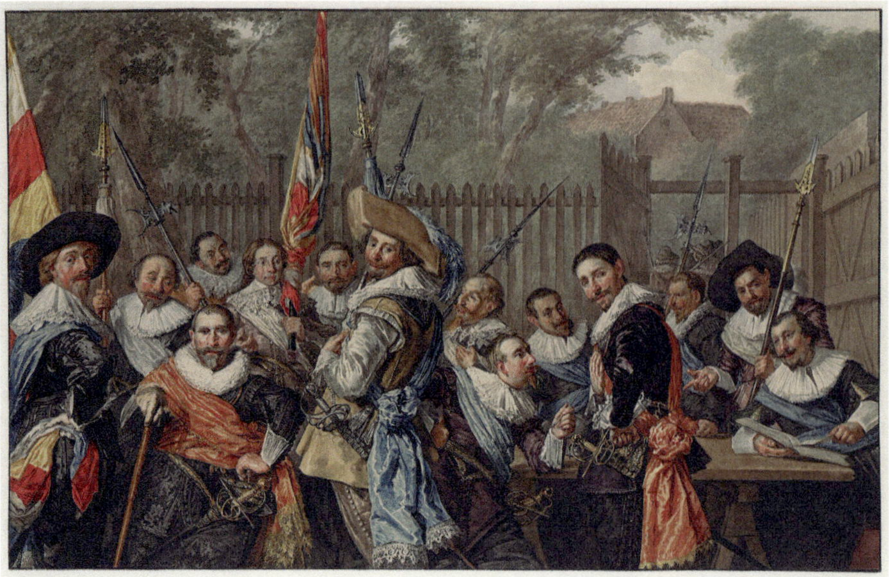

Fig. 6.17 Wybrand Hendriks (1744–1831), after Frans Hals, *Officers and Subalterns of the Calivermen Civic Guard*, copy in watercolour, undated, 33.3 × 51.4 cm. (Teylers Museum, Haarlem)

In fact, the most noticeable changes in this painting are unintentional: the background landscape has darkened substantially, making it hard to read. An eighteenth-century copy in watercolour by Wybrand Hendriks (1744–1831) presumably gives an impression of the vibrant colours and sense of spaciousness in the original picture (Fig. 6.17). It should not be taken as an exact record of what its colours looked like in Hendriks's time, however, for it is done in a different medium, known for its light, translucent colours, and it was a common eighteenth-century practice to correct the effects of age when copying older paintings; Hendriks himself noted on the back of one his light and bright watercolour copies of a civic guard portrait that it was done after an original that had become completely black and unclear (Van Eikema Hommes 2004, pp. 153–154 and notes 282 and 283, p. 169). Possibly, the 1633 portrait had already darkened substantially in the first century of its existence and Hendriks might have tried to reconstruct its original appearance.

With regards to the original colours and tones in Hals's civic guard portraits, it was not only the background landscape in the 1633 painting that was affected. In Hals's case, particularly the blacks in the different costumes and hats have lost their subtle variations. Van Gogh famously distinguished 27 different shades of black in Hals's paintings; in his time, the subtle differences must have showed much more clearly (Van Gogh 1953, vol. 3, 428 (Oct 1886), p. 74).[74] Indeed, Hendriks's copy shows a lot of nuance in these parts. As Hals used distinctively different paint mixtures for various blacks, reserving bone

[74] Van Gogh observed the same range of blacks in paintings by Rembrandt and Velazquez.

black for instance for the deepest darkest passages, some of the nuances can still be seen very clearly in the MA-XRF element maps (see for example the velvet decoration pattern in captain Johan van Loo's costume in the calcium (Ca) map, indicative of bone black: (Detail of Fig. 6.4: visible light image and calcium (Ca-K) map, in 'sync mode')).

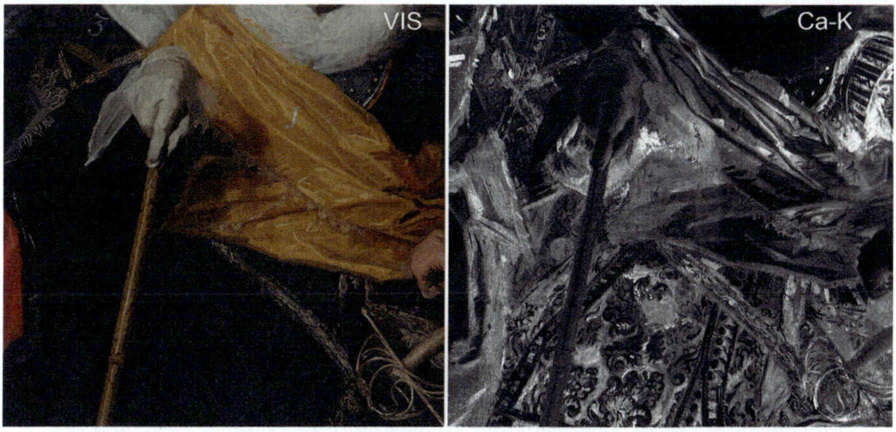

Detail of Fig. 6.4: visible light image and calcium (Ca-K) map, in 'sync mode' (sn.pub/fwb71q)

Furthermore, quite a few other colours lost some of their original appearance and/or intensity: indigo blue turned paler and somewhat greyish over the years (Hendriks et al. 1998; Van Eikema Hommes 2004, pp. 91–169); the purple and brown costumes must have darkened substantially as these appear almost black nowadays; the notoriously unstable blue pigment smalt (containing cobalt (Co), nickel (Ni), and potassium (K)) faded, making the blue pillow decorated with gold thread on which captain Nicolaes Woutersz van der Meer is seated in the 1616 portrait, appear rather grey (Detail of Fig. 6.1: visible light image, nickel (Ni-K) map and potassium (K-K) map, in 'sync mode'). Besides, the verdigris (copper green) glazes in the green curtain and background landscape in the earliest civic guard portrait have turned brownish -just like in the 1633 portrait- and were partially rubbed off during cleaning (witness a much greener area in the curtain, which is better preserved because it was covered by the frame for a long time and shows a much higher copper concentration in the Cu element map: (Detail of fig. 6.1).[75]

[75] On this ageing effect, see also (Van Eikema Hommes 2004, p. 75).

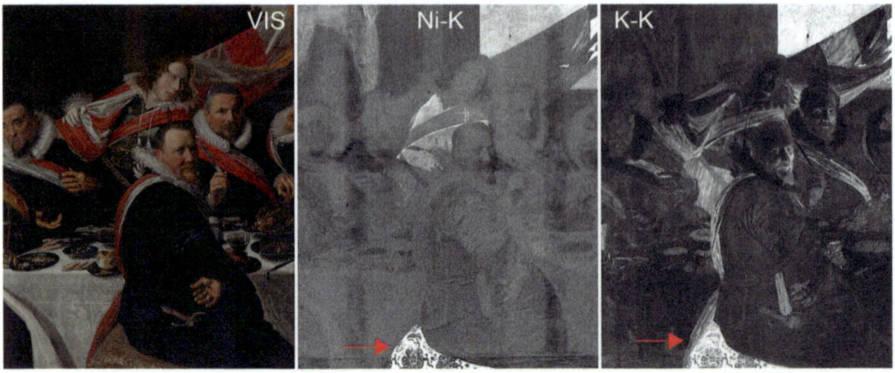

Detail of Fig. 6.1: visible light image, nickel (Ni-K) map and potassium (K-K) map, in 'sync mode' (sn.pub/vql2vt)

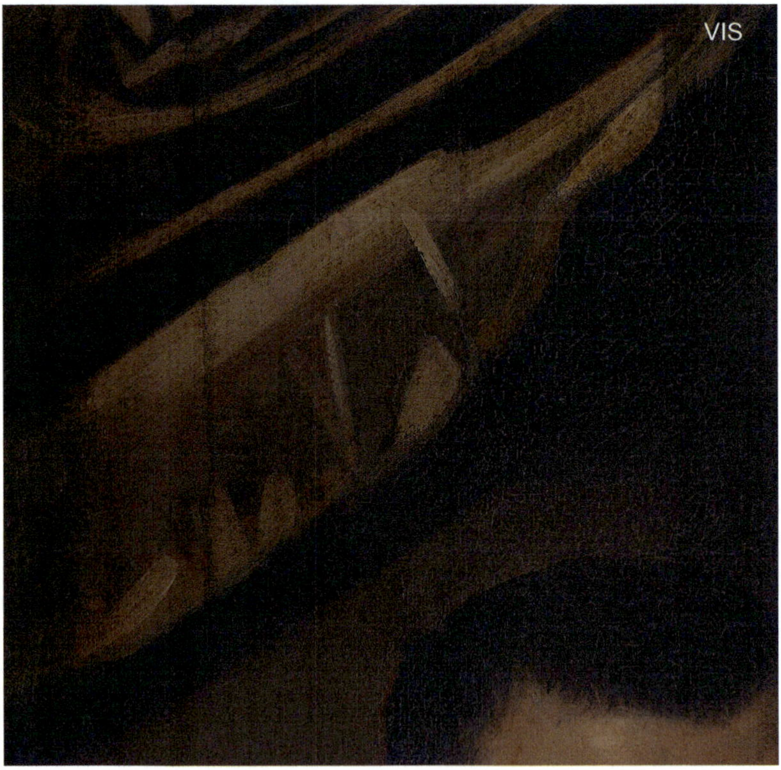

Detail of Fig. 6.1 (sn.pub/336u1h)

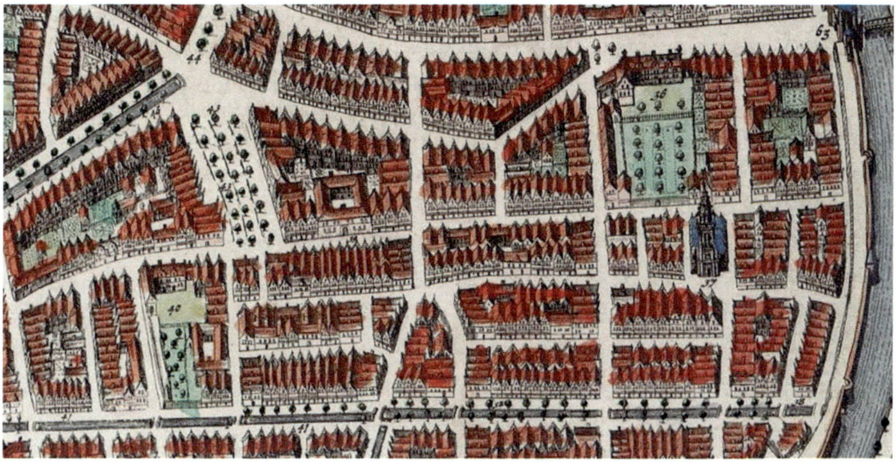

Fig. 6.18 Detail of Pieter Wils, *Map of Haarlem*, 1646, 56 × 55 cm. (Private collection)

In his last civic guard portrait of 1639 Hals used different pigments for the greens in the background landscape, but unfortunately, this mixture also proved unstable. He presumably used a mixture of yellow and blue. The high amount of calcium could come from the substrate of a now discolored yellow lake, while the overlapping of iron and phosphorus suggests the presence of the iron containing blue pigment vivianite here, known as 'Haarlem blue' or 'Haarlem ultramarine'.[76] However, the blue pigments in the greens have seriously discoloured and as a result these passages have lost most of their suggestion of three-dimensionality and shape, turning into a dull and uniform yellow-brown (Detail of Fig. 6.5: iron (Fe-K) map and visible light image, in 'curtain mode'). Moreover, the painting was altered somewhat in the early eighteenth century, notably the building in the centre of the background and the surrounding sky (Detail of Fig. 6.5). The paint layer covers a substantial earlier damage in the sky above the building (visible in the radiograph: Detail of Fig. 6.5: visible light image and x-radiograph, in 'curtain mode') which

[76] A sample taken analyzed by Karin Groen on 22 December 1984 contained P, Pb, Ca, Fe and some Si and Co; vivianite had not been identified at that time yet. The presence of vivianite is difficult to determine with optical microscopy, especially in green mixtures; it only really becomes apparent when analyzed by SEM-EDX. Vivianite was first discovered in paintings by Aelbert Cuyp, where it similarly discoloured (see Spring 2001–2002, pp. 66–67). During the recent restoration of Hals's three regent group portraits, it was discovered in his 1641 portrait of the Regent of the St Elisabeth Hospital (see Abrahamse 2018, p. 45).

may have been the reason for the early repainting; furthermore, the discolouration of the trees may have made the background look incoherent already in the early eighteenth century, which could explain the specific repainting of the building. According to a document in the Haarlem municipal archive, colonel Pieter Schatter commissioned the painter Dirk Maas (1656–1717) to 'repaint' ('*verschilderen*') this piece in 1706.[77] Originally, the building in the background had a saddle roof, much like a building close to militia hall (Fig. 6.18), which makes it likely that the backdrop initially depicted the grounds where the guardsmen practiced their skills.[78]

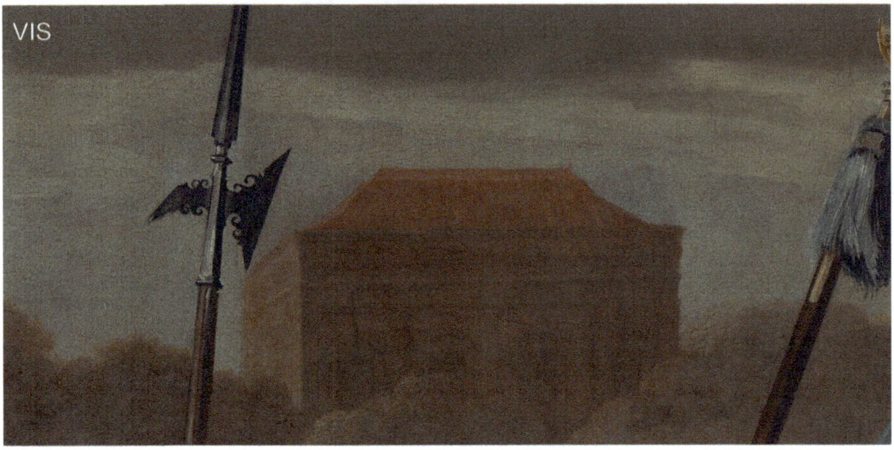

Detail of Fig. 6.5 (sn.pub/nbcbvn)

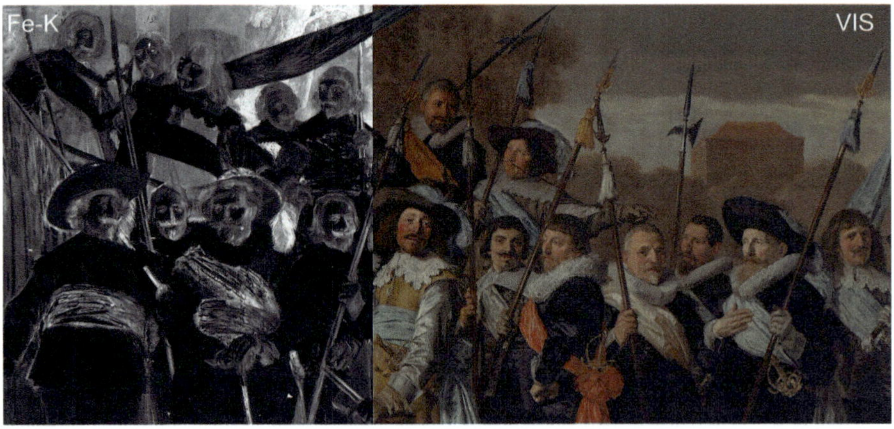

Detail of Fig. 6.5: iron (Fe-K) map and visible light image, in 'curtain mode' (sn.pub/sjeck3)

[77] Maas was paid for his work in 1709 Municipal Archive Haarlem, *Oud Archief der Schutterij*, inv. 49–3, 1709. See also (Köhler and Levy- Van Halm 1990, p. 43 and note 6, p. 45).

[78] There is no reason to believe the background was not initially intended as an actual location, as was suggested in the past (see Köhler and Levy- Van Halm 1990, p. 43).

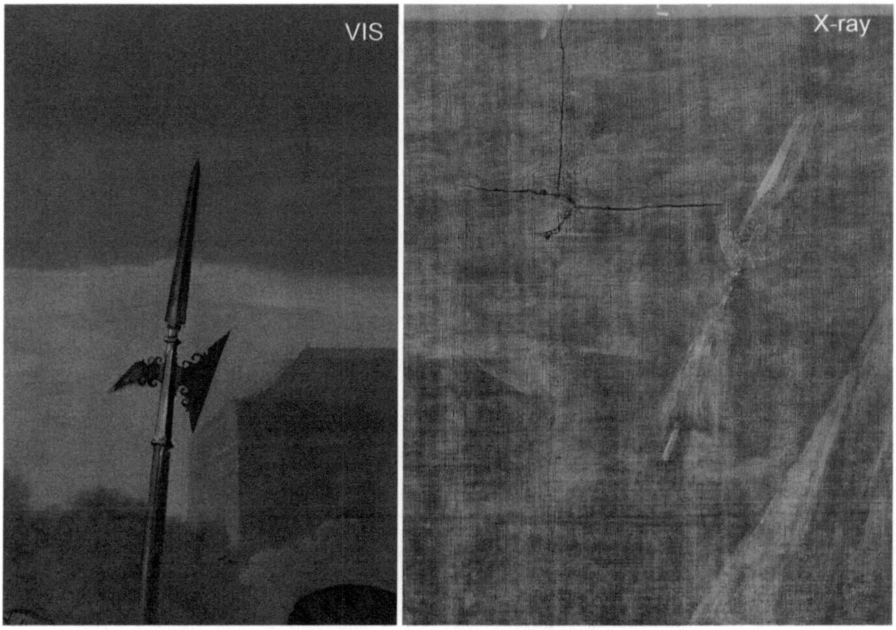

Detail of Fig. 6.5: visible light image and x-radiograph, in 'curtain mode' (sn.pub/5khpvs)

This last civic guard portrait also shows some interesting changes done by Hals himself and gives us a unique insight into his workshop practice. Captain Michiel de Wael originally wore a more old-fashioned, ruffled collar (like most of the elderly men in this painting), which is clearly visible in the radiograph (Detail of Fig. 6.5: visible light image and x-radiograph, in 'sync mode'). In the final version he wears a more fashionable, flat collar—an update he most likely requested himself (Slive 1989–1990, pp. 87–103). Likewise, sergeant Hendrick Coning must have requested an update of his weapon; he was promoted from sergeant at the St George civic guard to lieutenant at the Calivermen civic guard in 1639 and his attribute was changed accordingly: from halberd to partisan (Detail of Fig. 6.5: visible light image, iron (Fe-K) map, lead (Pb-L) map and x-radiograph, in 'sync mode'). The iron (Fe) map shows that a part of a tree was originally painted at the location of the current partisan; while the deeper laying lead-map (Pb-L) and the radiograph show the contour of the halberd.[79] Furthermore, colonel Johan Claesz van Loo was given a hat on second thoughts, after his hair had already been painted (Detail of Fig. 6.5: visible light image, copper (Cu-K) map, phosphorus (P-K) map and iron (Fe-K)

[79] The painting shows guardsmen from both groups as these had formally become one in 1632. This change explains why the formation shows four instead of three lieutenant and five instead of six sergeants (see Kurtz 1979, pp. 27–28).

map, in 'curtain mode'). In the 1633 group portrait, colonel van Loo is depicted without a hat and apparently, Hals initially intended to do the same here. He may have added the hat at Van Loo's request, or else it must have been a compositional choice. Either way, he balanced the hat out with two similar black hats at the right of the composition and two somewhat lighter, brown hats in the centre, thus creating a visual rhythm.

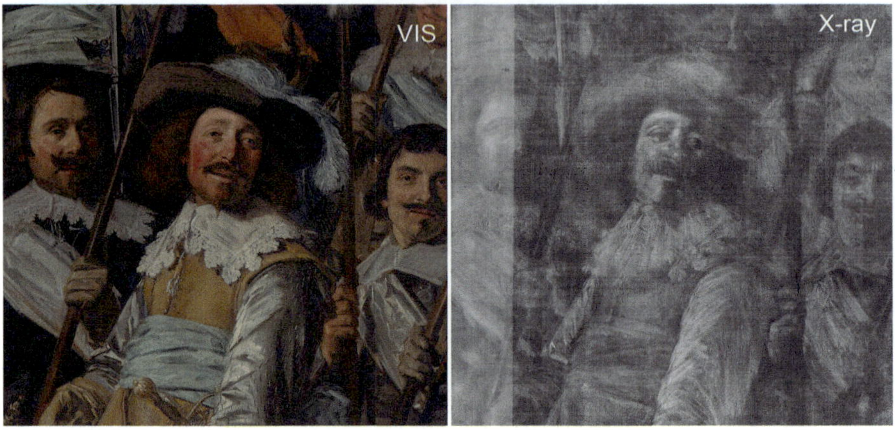

Detail of Fig. 6.5: visible light image and x-radiograph, in 'sync mode' (sn.pub/93fhrm)

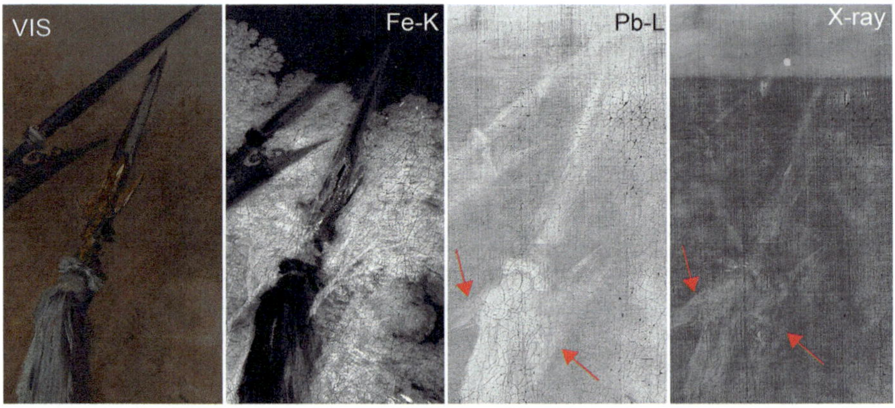

Detail of Fig. 6.5: visible light image, iron (Fe-K) map, lead (Pb-L) map and x-radiograph, in 'sync mode' (sn.pub/i67528)

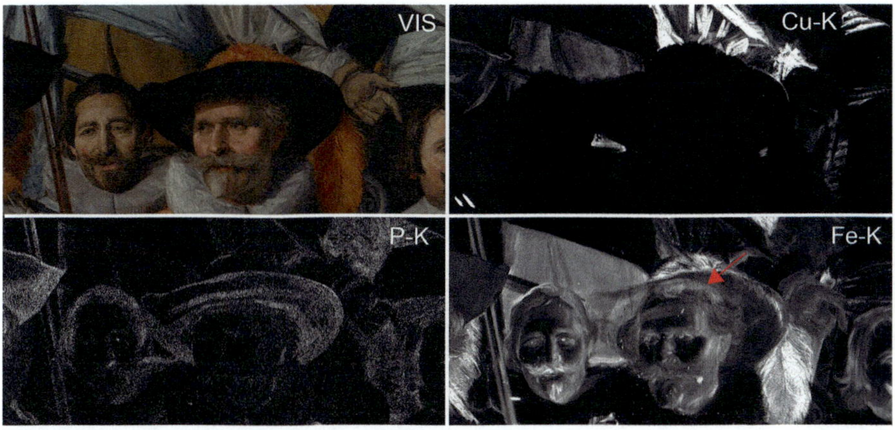

Detail of Fig. 6.5: visible light image, copper (Cu-K) map, phosphorus (P-K) map and iron (Fe-K) map, in 'curtain mode' (sn.pub/qepzti)

As to Hals's workshop practice, the MA-XRF copper map gives an illuminating insight into his painting process, and a mistake that has thus far gone unnoticed, provides a clear insight into the involvement of his studio—a topic that has caused much speculation in the past (see Chap. 2). The copper map shows the areas where the scanner detected copper, present in azurite and verdigris, blue and green copper-containing pigments. Hals used azurite sometimes for skin colours, as we have seen, but also for greyish tones in white fabrics, and verdigris both for green colours and in mixed colours that do not look green (Groen and Hendriks 1989–1990, p. 119). Sometimes, he also mixed in a copper-containing pigment with one of his blacks, presumably to speed up the drying process, as recommended in seventheenth-century paint recipe books.[80] In this case he mixed it in with a black he used for dark accents such as the final contour and darkest tones in the hat of standard-bearer Pieter Schout (Detail of Fig. 6.5: visible light image and copper (Cu-K) map, in 'sync mode'). The copper element map thus provides a composite image of various coppers present in the painting but is nonetheless revealing since Hals used little copper in this particular work.

[80] Seventeenth-century painting recipes recommended adding different siccatives to speed up the drying for different black pigments: ivory or bone black was best combined with verdigris (copper green); lamp or smoke black with umber, according to a workshop handbook *Le Petit peintre de Mr. de St. Jehan*, part of the De Mayerne manuscript (see Van Eikema Hommes 2004, p. 11). On Hals's use of verdigris mixed in with blacks, see (Groen and Hendriks 1989–1990, p. 119).

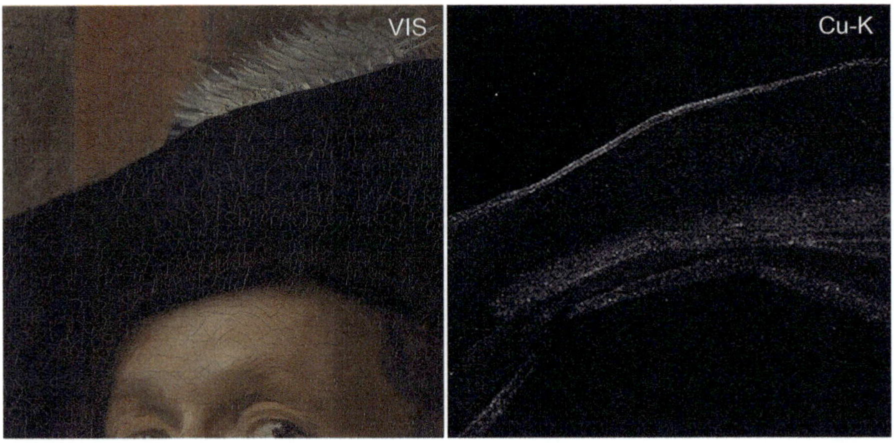

Detail of Fig. 6.5: visible light image and copper (Cu-K) map, in 'sync mode' (sn.pub/qx5k1z)

As discussed above, only one face in this painting contains a substantial amount of copper in the skin tones, the face of the elderly captain Quirijn Damast (Fig. 6.3). It is therefore likely that he was painted with a separate palette, presumably during a separate session. Interestingly, only six of the different white collars, one white sash, and two white sleeves contain copper. It is present in different shades of grey, which makes it likely that Hals worked up these particular parts of the clothing in one session. As we know from the documents about *The Meagre Company*, Hals could work up the clothing without the sitters being present (see the beginning of this chapter and Appendices II and IV). He could even complete the clothing before working on the faces apparently, something that some seventeenth-century art theorists warned against, as it was notoriously hard to place the head correctly on the shoulders if it was painted last (see Kirby-Talley 1981, pp. 53, 165, 275 and 359–374).

In the case of captain Damast's collar, we can see that Hals sketched its bottom edge very summarily with two curvy green(!) strokes at the height of Damast's chest and also indicated the lower edge of the collar behind Damast's shoulder with similar bold and summary green strokes (Detail of Fig. 6.5: visible light image and copper (Cu-K) map, in 'sync mode'; Detail of Fig. 6.5: visible light image and copper (Cu-K) map, in 'sync mode'). As he only used this green for the handles of the banners in this painting, he may have sketched the position of Damast's collar very summarily just after painting those.

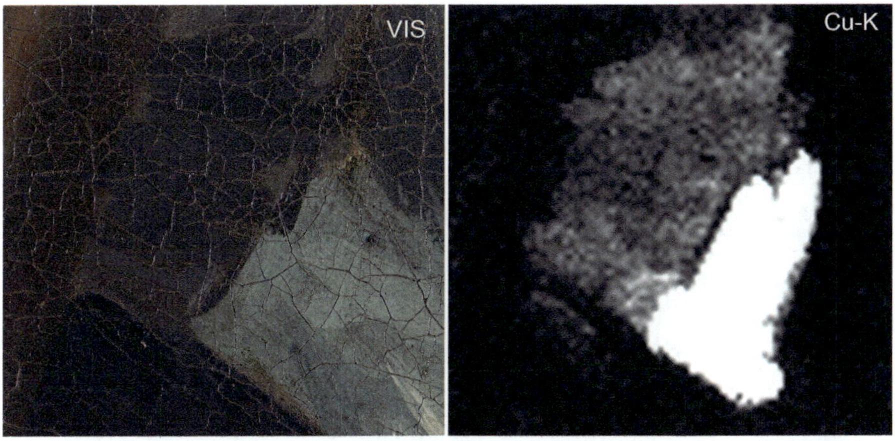

Detail of Fig. 6.5: visible light image and copper (Cu-K) map, in 'sync mode' (sn.pub/6lrz0t)

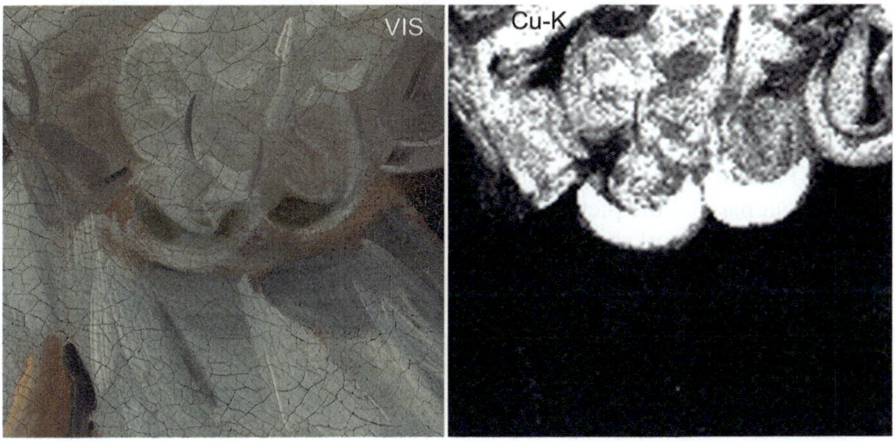

Detail of Fig. 6.5: visible light image and copper (Cu-K) map, in 'sync mode' (sn.pub/oub7rk)

After these green strokes had dried, the collar was worked up in greys and whites, and once that part was dry, the hairs of the beard were painted on top. So we can distinguish various stages in the paint application here, which is good to keep in mind when considering Hals's workshop practice. As we have seen several times now, Hals's first sketch was much like a shorthand notation for himself—it did not give a possible assistant much to go on when working up a

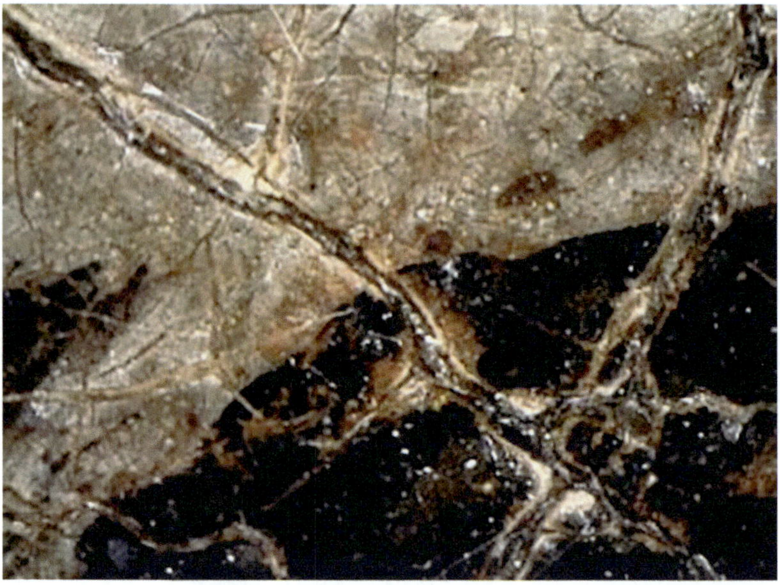

Fig. 6.19 Details of Fig. 3.62, taken with a dino-lite microscope (x65), confirming that the spiky hair is part of the original paint layer

passage. Parts of the final stage, such as adding hairs or beards and thus integrating different parts of the picture, were often less complex and could therefore be more easily left to workshop assistants.

Indeed, a rather clumsy mistake and oddly placed dark accents betray the hand of a workshop assistant near Damast's head and collar. The hairdo of captain Florens Pieters van der Hoeff (Fig. 6.4) was worked up by someone who did not understand it: it is fairly long in the front, like the hairdo of lieutenant Cornelis Coning to his right, but extremely short and spiky at the top (Detail of fig. 6.5). It makes captain van der Hoeff look somewhat like a hedgehog. Observations with a hand-held microscope confirm that the spiky hairs are part of the original paint layer, not a later restoration (Fig. 6.19). Clearly, the incoherent hairdo is a mistake; the initial sketch was apparently not sufficient here for the person executing this part to fully understand the shape to be depicted.

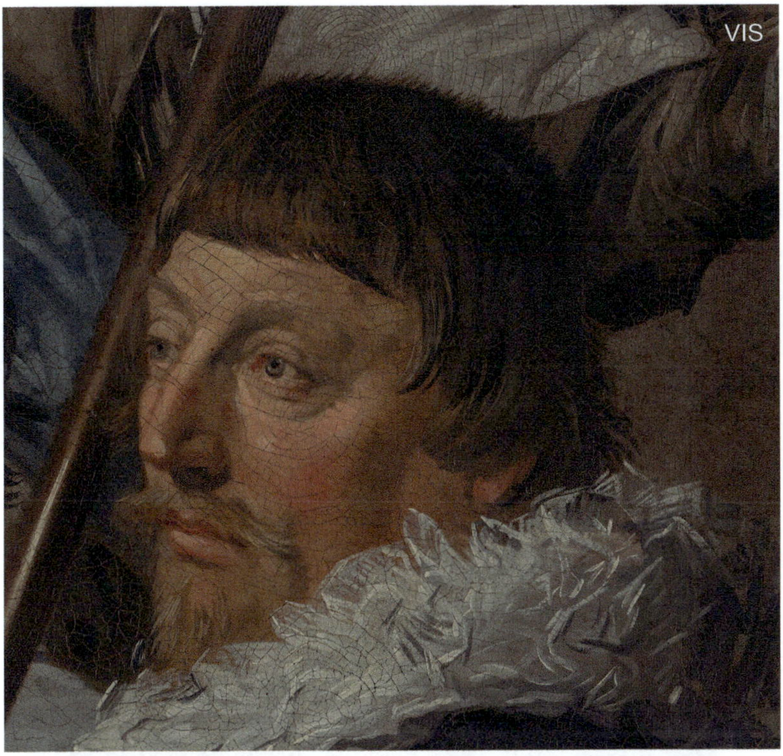

Detail of Fig. 6.5 (sn.pub/k4bzl9)

Interestingly, Wybrand Hendricks corrected the mistake when he copied the painting, erasing the spikes as if it were (Fig. 6.20). However, it is more likely that the spiky hair on top was actually correct and the longer bangs in the front were in fact the mistake. In an earlier Haarlem civic guard portrait we can see that captain van der Hoeff had short hair nine years before (Fig. 6.21). Presumably, he still had a similar hairdo when he was portrayed by Hals, and not medium-long hair like lieutenant Coning beside him.

In the collars of the two men depicted in between captain van der Hoeff and captain Damast we can also see evidence of workshop assistance (Detail of fig. 6.5, p. 211). The overall execution of the collars is rather stiff, and especially in the right collar the loose grey accents are not at all economic and efficient, like Hals's (Detail of fig. 6.5). Rather than enhancing the pictorial illusion these accents diminish it. As we have seen above, the different pigment mixture here suggests that these collars were not painted during the same session as the six collars discussed before.

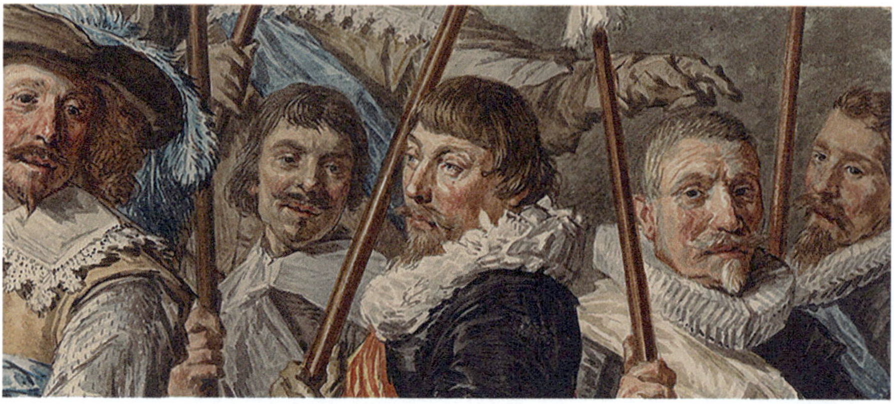

Fig. 6.20 Detail of Wybrand Hendriks, after Frans Hals, *Officers and Subalterns of the St. George Civic Guard*, copy in watercolour, 1979, 30.3 × 59.7 cm. (Amsterdam Museum)

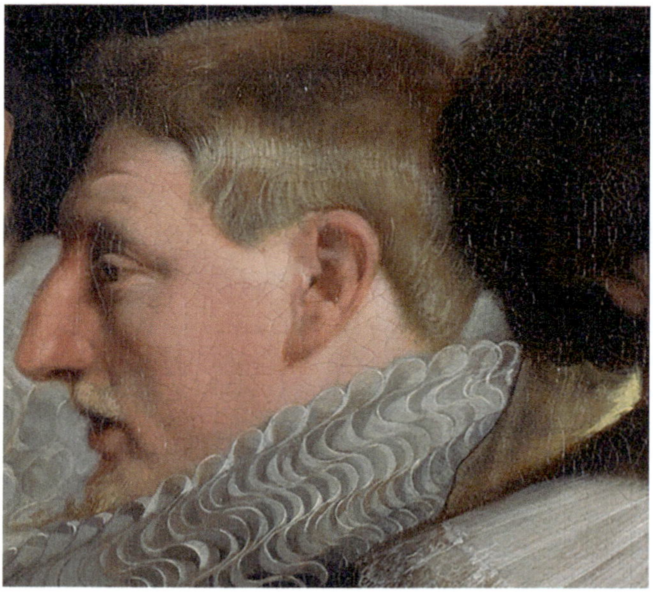

Fig. 6.21 Detail of Haarlem School, *Officers and Subalterns leaving the Calivermen's Civic Head Quarters*, 1630, oil on canvas, 214 × 285 cm. (Frans Hals Museum, Haarlem)

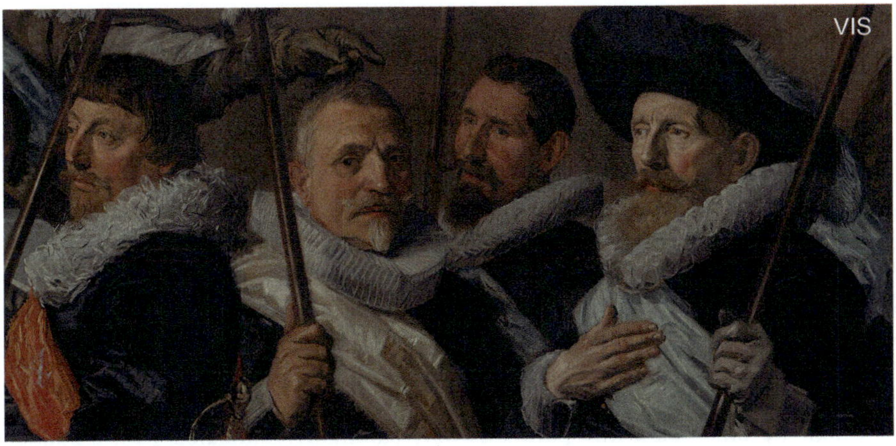

Detail of Fig. 6.5 (sn.pub/7ni20j)

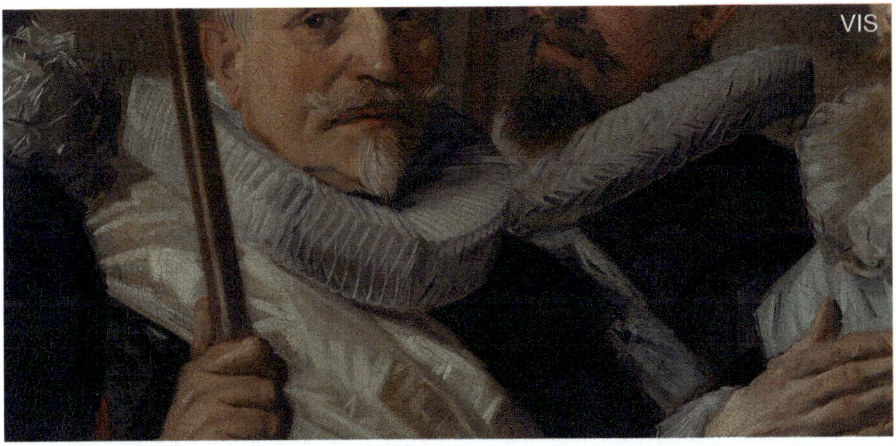

Detail of Fig. 6.5 (sn.pub/8ku1tz)

Before delving into the exact attribution of *The Meagre Company*, there are a few last striking features of the 1639 portrait worth noticing: a sash left unfinished and a quick indication of a background behind a figure. Like the collars, Hals presumably worked up sashes of the same colour largely in the same session. Although painted rapidly, the sashes could be very time-consuming to paint, especially the blue ones. These were painted with a rather unusual pigment, indigo, as Ella Hendriks and Margriet van Eikema Hommes have shown. Hals was one of the first artists in the Netherlands to use indigo in his oil paintings, a pigment utilized mostly in the cloth dying industry at the time. Hals built up his blue sashes using a lot of different glazes on top of one another; in some cross samples as many as six different paint layers could be distinguished, three of which contained indigo (Hendriks et al. 1998; Van Eikema Hommes 2004, pp. 104–109). In the 1639 portrait he worked up the indigo sashes rather late in the painting process. The indigo paint clearly overlaps the tip of captain Damast's middle finger (which technically lays on top of the sash) (Detail of Fig. 6.5), and the indigo of lieutenant Coning's sash partially overlaps his collar (which should of course lay on top): (Detail of Fig. 6.5). Sergeant Nicolaes Jansz van Loo's indigo blue sash is even left unfinished, leaving large areas of the ground exposed (Detail of Fig. 6.5). The intricate edge of Van Loo's collar had already been fully worked up and possibly Hals did not want to get too close to its lace edge. Furthermore, of the three blue banners in the picture, only two were painted with indigo. The right-most banner of which we only see a tiny part, was only summarily indicated with a copper containing blue: azurite. Hals thus varied his level of finish and use of pigments for similar colours somewhat.

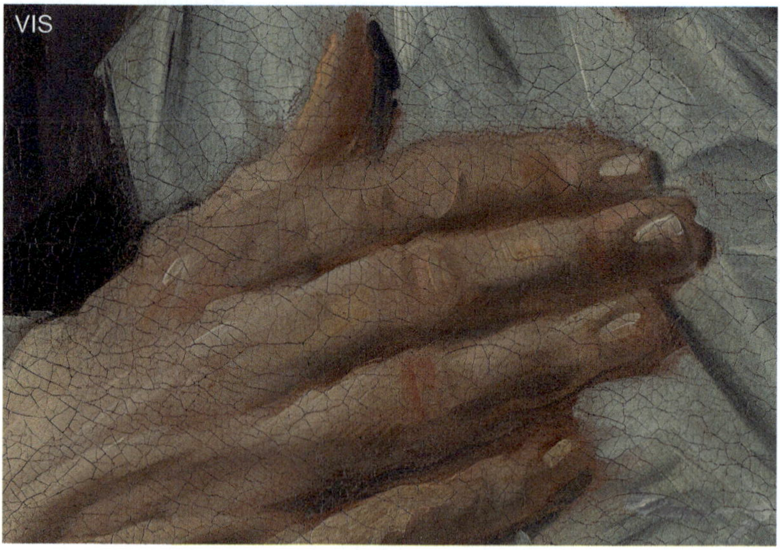

Detail of Fig. 6.5 (sn.pub/hkp533)

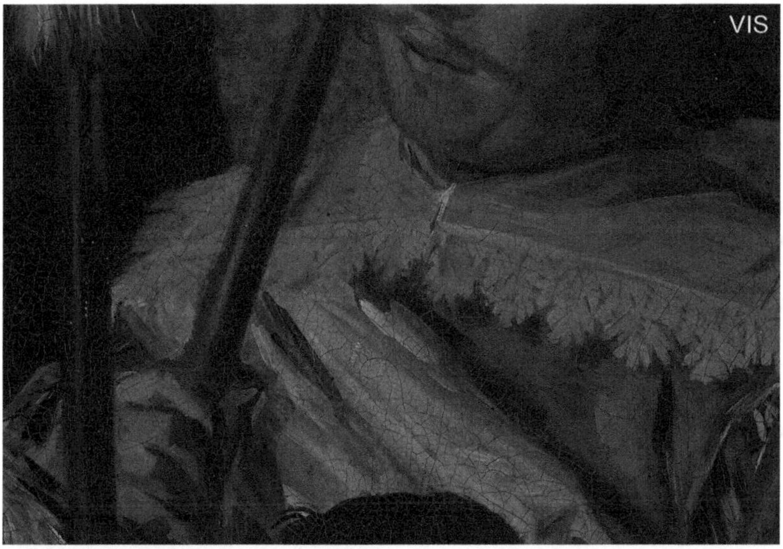

Detail of Fig. 6.5 (sn.pub/qmt0iw)

Detail of Fig. 6.5 (sn.pub/b88k2h)

A last striking feature concerns the summary indication of a background in between the two rightmost guardsmen, the standard-bearers Dirck Dicx and Pieter Schout, visible in the copper map. (Detail of Fig. 6.5: visible light image and copper (Cu-K) map, in 'sync mode'). Its bright signal and location (at the height of the trees in the background) suggest that it must be a copper containing green colour; it was later overpainted with concrete trees. Possibly, Hals simply had some green paint left over and used it to already roughly indicate the background here so as to ease the integration of foreground and background at a later stage in the painting process. As we saw above, he used a similar green to indicate the position of captain Damast's collar. Previously, in his 1633 civic guard portrait, he also indicated the greenish background around a face –although much more summarily there with just a thick contour- presumably also the facilitate the later integration of foreground and background: Detail of Fig. 6.4: visible light image and copper (Cu-K) map, in 'sync mode'.

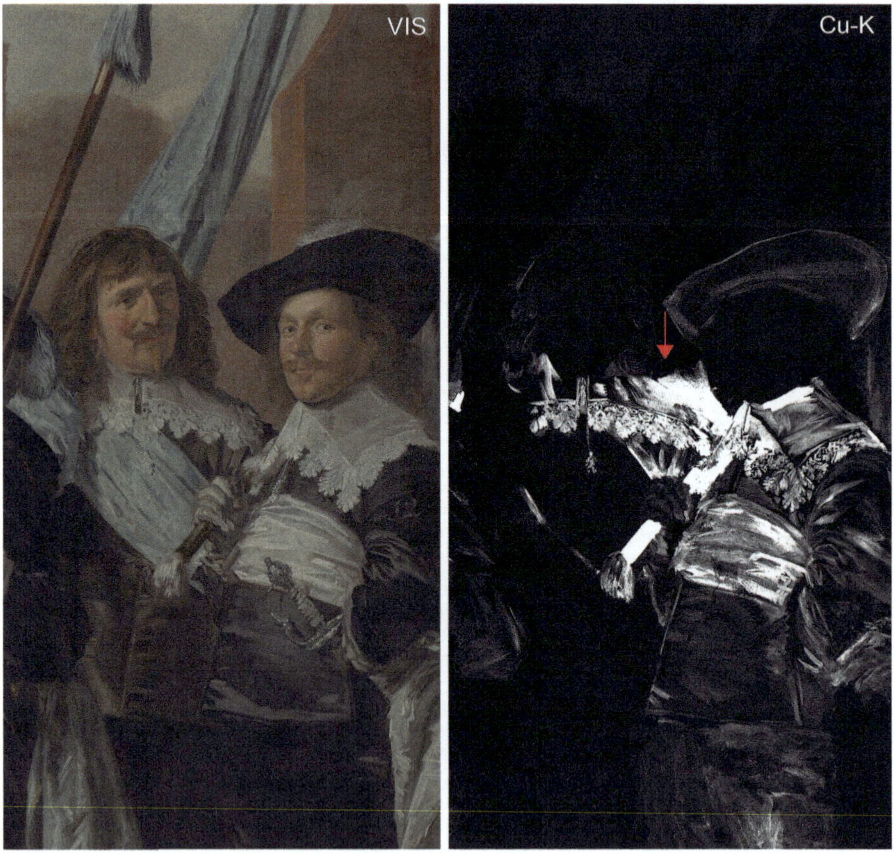

Detail of Fig. 6.5: visible light image and copper (Cu-K) map, in 'sync mode' (sn.pub/x4kylt)

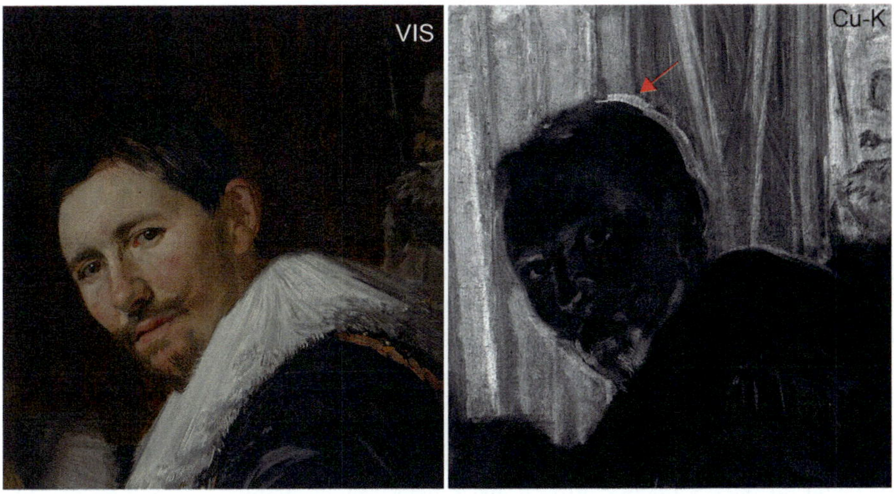

Detail of Fig. 6.4: visible light image and copper (Cu-K) map, in 'sync mode' (sn.pub/0uz1ix)

In short, the 1639 civic guard portrait shows an unusual number of variations and exceptions in the use of pigments and the level and quality of finish, as well as some clear evidence of workshop assistance. Even though the work contains one of the rare self portraits by the artist (the second figure from the left in the top row, Fig. 6.5, no. 19), its design and execution are not entirely on the same level as his earlier civic guard pieces. As it is close in date to *The Meagre Company*, it provides a important point of reference for the attribution of the latter.

6.8 New Light on the Attribution of *The Meagre Company*

Although many stylistic analyses of *The Meagre Company* have been done before, leading to contradictory results, thus far seventeenth-century art-critical terms have hardly been used in the process. These are nevertheless revealing when analyzing the characteristic features and the attribution of this group portrait. In particular, terms like *houding* (the balancing of colours and tones), *haspeling* (an unintentional chaotic effect), *koppeling* (the joining of adjacent shapes and directions) and *sprong* (spatial leaps), *eigentickheyt* ('real' in the sense of both convincingly natural and befitting the topic depicted) and *oogenblikkige beweeging* (instantaneous movement) help to clarify some of Hals's painterly goals and achievements and to distinguish his work from Codde's contribution. Specifically for this research project, new high-resolution photographs of *The Meagre Company* were taken, new technical analyses were carried out including MA-XRF scans and infrared reflectograms

(IRR), and digital tools were developed to help to read and interpret all the different data. We compared the results with both the new insights into Hals's painting process and use of materials discussed above and a new analysis of a reference work by Codde, *Plundering soldiers in a barn* of circa 1635 (Fig. 6.7), and related the findings to previous insights from the literature. The combination of new technical insights and an in-depth analysis of the paintings using seventeenth-century art-critical terms sheds new light on the attribution of *The Meagre Company*.

As mentioned above, Hals's teacher Karel van Mander used a lot of different terms to describe the importance of design (*voorbeworp, compositie, ordonnantie*) and explained how experienced painters went about inventing their compositions. Indeed, in early modern art theory, design was considered a crucial component of the art of painting. It was also the first step in the creation of an actual painting and it therefore makes sense to start our analysis of *The Meagre Company* by focusing on its design, from the initial sketch to the final result.

Based on what we have seen in his Haarlem civic guard portraits, Hals must have sketched his design for the entire portrait on the canvas before working up the different parts in oil paint. As per his habit, he must have done so *alla prima*, directly on the prepared canvas, presumably first in chalk and then in a dark, wet material, and subsequently proceeded to work up the different parts in his typical sketchy manner. Hals's manner of sketching is unique in that it is very sparse and free in its application: a kind of shorthand notation for himself. Consequently, his sketch lines did not provide an assistant –or a colleague like Codde- with much guidance when working up the different passages, which could explain some of the rather peculiar postures of the figures at the right in *The Meagre Company*. Many officers almost seem to fall backwards, as Grimm observed, such as the four most prominent guardsmen on the right: the officer in yellow, the one with the orange sash, the one in black with his hand on his hip and the one on the far right.[81] Before we zoom in on the final result, however, we should first focus on reconstructing the initial design for *The Meagre Company*.

From the legal documents we know that Hals received the commission to paint *The Meagre Company* in 1633; captain Reynier Reaal and lieutenant Cornelis Michielsen Blaeuw stated on 19 March 1636 that they had given him the commission already three years earlier (*al drye jaeren geleden*) (Slive (ed.) 1989–1990, Hals Doc. 74). In that same year Hals completed his fourth Haarlem civic guard portrait and as Slive pointed out, the overall design for both pictures bears a striking resemblance (sn.pub/ue7ze5). Apart from the fact that Hals depicted the Amsterdam guardsmen in full length –as was customary in the capital- the composition is indeed similar.[82] The group is split into several parts. On the left, a prominent standard-bearer leads the viewer into the painting (Details of Fig. 6.4 and Fig. 6.6, in 'gallery mode'). Several diagonal lines –i.e. the banners and weapons- create a rhythm and

[81] On Grimm's analysis, see the beginning of this chapter.

[82] As we have seen, Hals also adjusted his Haarlem designs to the specific setting of the paintings in the militia halls. Obviously, all the flanking civic guard portraits in Amsterdam also had full-length figures.

direct the viewer's attention to the highest ranking officer in the portrait, seated conspicuously beside the standard-bearer, surrounded by a group of officers of a somewhat lower status. The group is flanked by a standing officer in an elegant, contorted pose, wearing a striking yellow coat. Other structuring elements are also comparable, albeit with small variations: three prominent hats create a visual rhythm in the composition, and on the right a central figure in black is placed before the others to add some variety. In order to better understand Hals's initial design for *The Meagre Company*, we need to have a closer look at the painting and at various clues provided by different types of technical analyses.

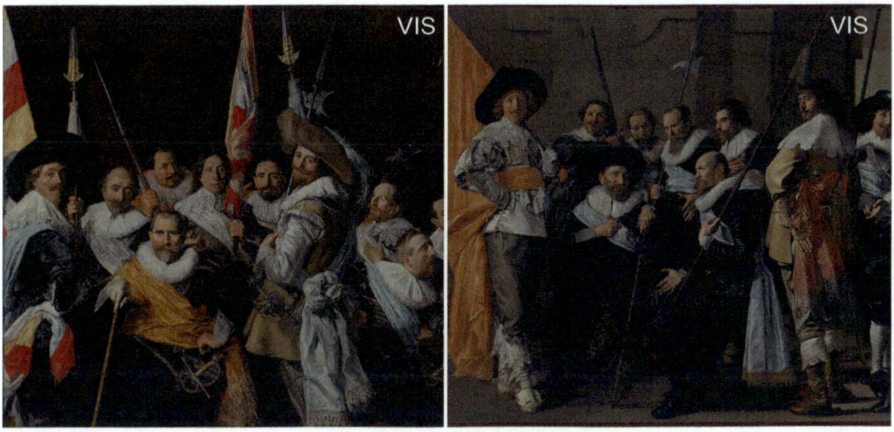

Details of Fig. 6.4 and Fig. 6.6, in 'gallery mode' (sn.pub/uzkm2g)

Although Hals may have first designed his figures in white chalk, as we saw in his 1627 Haarlem group portrait, such sketch lines are not visible in *The Meagre Company*. A few painted white strokes containing chalk can be detected in the MA-XRF calcium (Ca) map, and are partially visible at the surface. Below the eighth officer from the left, a few fluid strokes outline parts of the yellow coat, including the lower contours of his left sleeve as well as the side of his torso (Detail of Fig. 6.6: visible light image and calcium (Ca-K) map, in 'sync mode'). However, these overlap broad painted sketch lines and must thus have been applied at a later stage (Detail of Fig. 6.6: visible light image, calcium (Ca-K) map and x-radiograph, in 'sync mode') We did find evidence of Hals's usual underdrawing in a dark, carbon containing, wet material in the infrared reflectograms (IRR), which presumably outlines the entire composition. It is mostly visible in areas where the final design differs from the initial sketch, as we will see. Furthermore, Hals's characteristic sketchy way of subsequently setting up his paintings can also be seen in *The Meagre Company*, as will be discussed below.

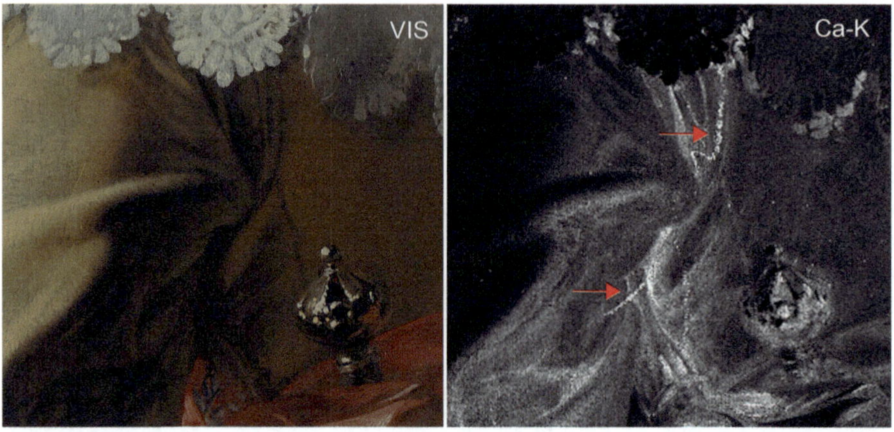

Detail of Fig. 6.6: visible light image and calcium (Ca-K) map, in 'sync mode' (sn.pub/zwcb14)

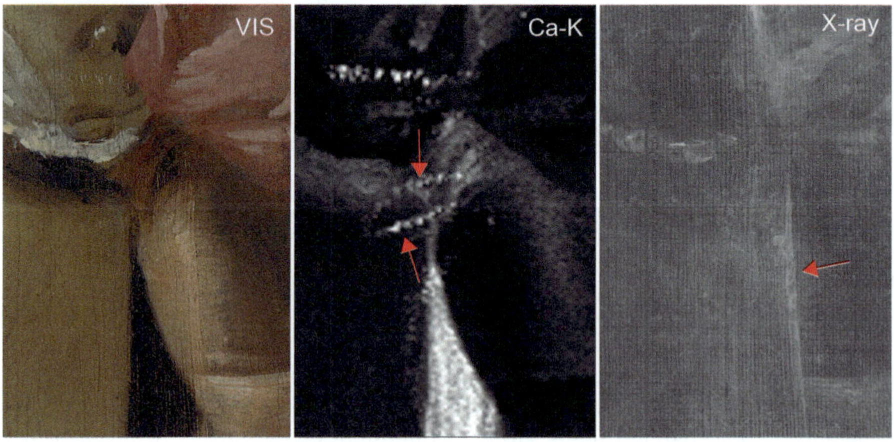

Detail of Fig. 6.6: visible light image, calcium (Ca-K) map and x-radiograph, in 'sync mode' (sn.pub/nuf081)

Hals originally intended to place the scene outdoors. The contours of a building behind the thirteenth figure from the left shine through somewhat at the surface, and were spotted in the radiographs made prior to the large Hals survey exhibition in 1989–1990. The contour lines can be seen even more clearly in the infrared reflectogram (IRR) and the lead-white element map (Pb-L) (Detail of Fig. 6.6: visible light image, IRR and lead (Pb-L) map, in 'sync mode'). Do note however, that the thick diagonal shadow line from the upper left to the lower right was added on top. The infrared reflectogram reveals the initial building most clearly, and also shows a summary sketch line indicating foliage (to the left of the head of officer 13).

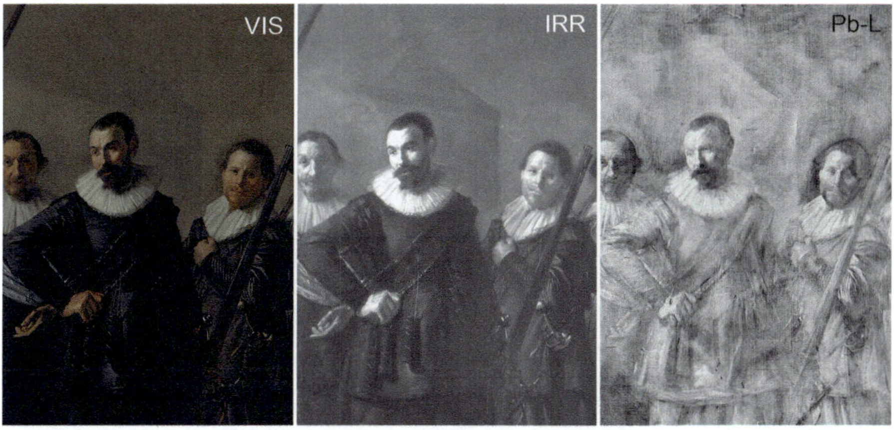

Detail of Fig. 6.6: visible light image, IRR and lead (Pb-L) map, in 'sync mode' (sn.pub/9uj40t)

The deeper green layer that Bijl saw shining through around the head of officer 14 can be seen very clearly in the copper map (Detail of Fig. 6.6: visible light image, copper (Cu-K) map and IRR, in 'sync mode'). It continues around an earlier outline of the hat of officer 15, which can also be seen in the infrared reflectogram, and is accompanied by a bold diagonal line above this officer's head—presumably indicating the direction of a staff weapon that was never executed. Hals also indicated the position of officer 14's shoe with one loose copper containing line.

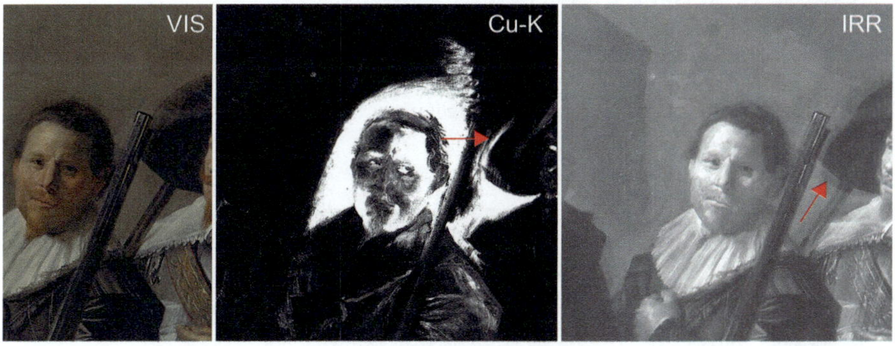

Detail of Fig. 6.6: visible light image, copper (Cu-K) map and IRR, in 'sync mode' (sn.pub/qs23oc)

Several other patches of copper containing paint can be seen around other figures on the right (Detail of Fig. 6.6: copper (Cu-K) map, visible light image and IRR, in 'fade mode'). A close comparison with the final painting reveals that the copper patches follow earlier outlines of some of these heads and collars. Note, for example, the different contour of the top of the head, the ear and collar of number 13 (moving the cursor from left to right allows one to fade from the high resolution photographic image to the copper element map). It strengthens the hypothesis that Hals (rather than Codde) must have painted these parts, just like he briefly outlined a part of the background behind several heads in his 1633 and 1639 Haarlem civic guard portraits.[83] A further clue comes from a loose copper containing sketch line in the vicinity, roughly outlining some foliage, just above the earlier indication of foliage visible in the infrared reflectogram (Detail of Fig. 6.6: copper (Cu-K) map, visible light image and IRR, in 'sync mode').[84]

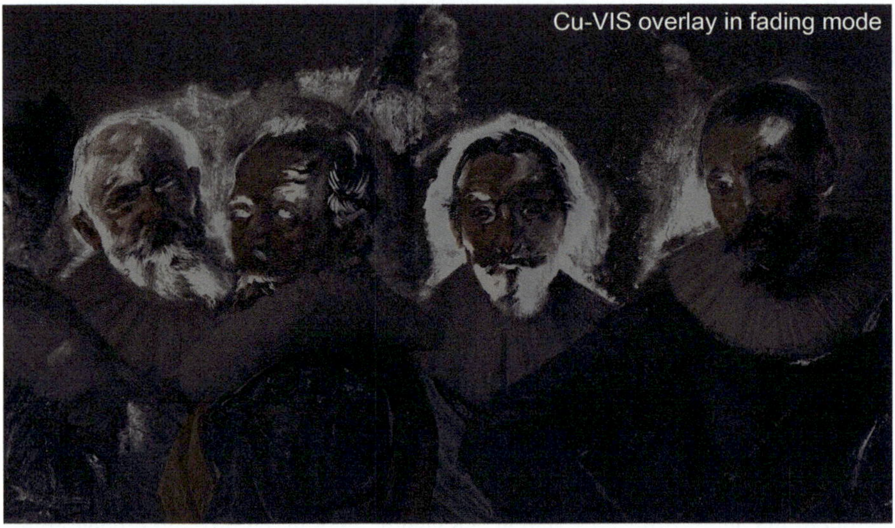

Detail of Fig. 6.6: copper (Cu-K) map, visible light image and IRR, in 'fade mode' (sn.pub/ph19so)

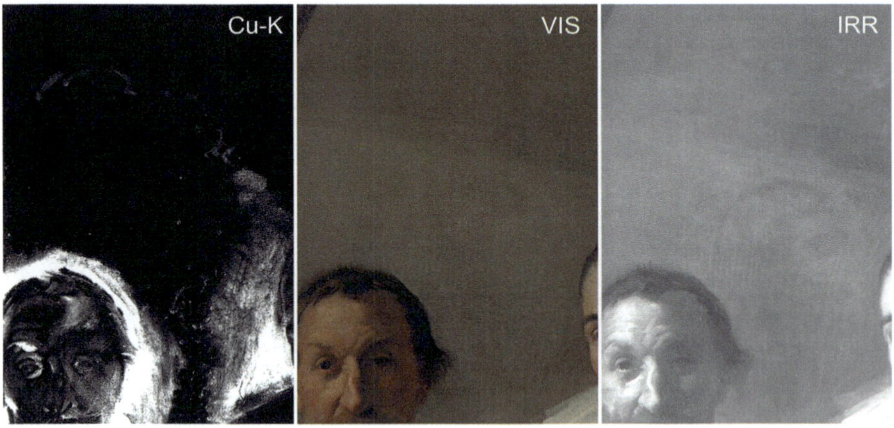

Detail of Fig. 6.6: copper (Cu-K) map, visible light image and IRR, in 'sync mode' (sn.pub/twmm20)

[83] See above: Revisions, inconsistencies, workshop practice and assistance. On the practice of painting partial backgrounds or 'halos' around heads and other main elements during the painting process, see (Derks et al. 2022).

[84] The copper containing paint in the faces will be discussed below.

The grey sketch lines below some of the figures that were discovered in 1989 can be seen most clearly in the radiographs (see Bijl 1989–1990), and to a lesser extent also in the MA-XRF lead (Pb) map: (Detail of Fig. 6.6: x-radiograph, visible light image and lead (Pb-L) map, in 'sync mode'). These indeed roughly indicate an earlier pose and outlines of the coat and arms, as described by Bijl (see above: a Contested Attribution). Apart from these lead containing strokes, MA-XRF also revealed very bold sketch lines containing umber (manganese and iron), some of which are partially visible at the surface and appear black or very dark.[85] (Detail of Fig. 6.6: visible light image and manganese (Mn-K) map, in 'sync mode'). The sketch lines can be seen particularly clearly below the officer in the yellow coat and the one in black with his hand on his hip, where the final design deviates somewhat from the first sketch. Similar umber-rich sketch lines have been found below dark passages in Hals's Haarlem civic guard portraits, as we have seen above.

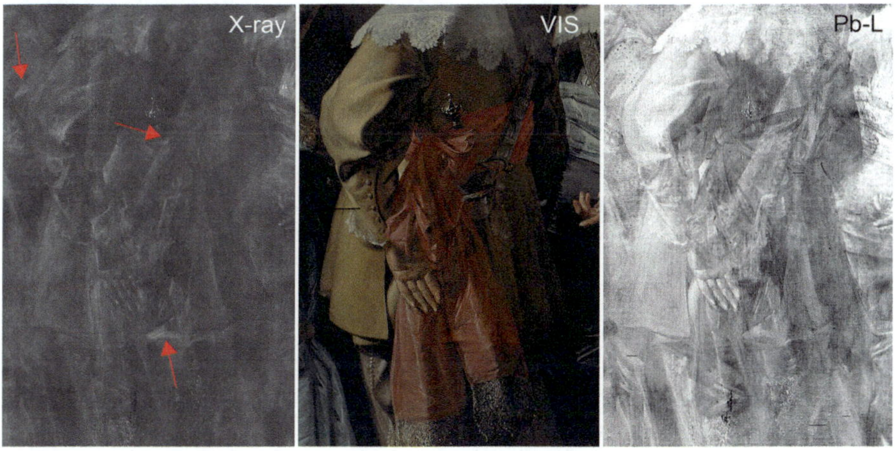

Detail of Fig. 6.6: x-radiograph, visible light image and lead (Pb-L) map, in 'sync mode' (sn.pub/3x51ft)

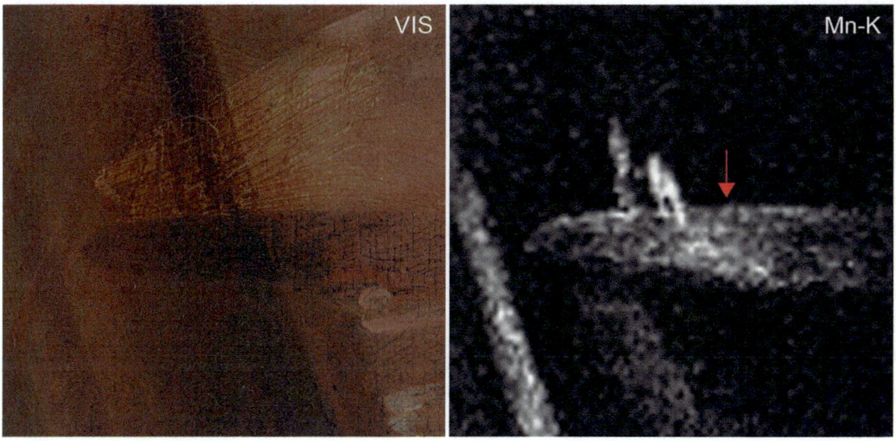

Detail of Fig. 6.6: visible light image and manganese (Mn-K) map, in 'sync mode' (sn.pub/aa6nut)

[85]As discussed above, art theorists recommend using umber as a siccative in certain types of black paint, and Hals indeed mixed it in with certain blacks.

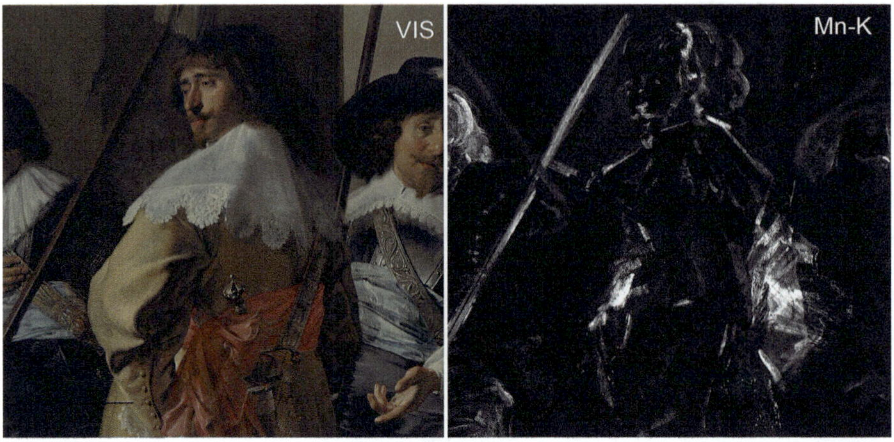

Detail of Fig. 6.6: visible light image and manganese (Mn-K) map, in 'sync mode' (sn.pub/qf40i6)

The most striking umber-rich sketch lines in *The Meagre Company* are visible below the officer in the yellow coat (the eight figure from the left). The sketch lines are very summary, such as the rough indication of the hair, hairline, face, beard and collar with curvy edges (Detail of Fig. 6.6: visible light image and manganese (Mn-K) map, in 'sync mode'). Brief strokes mark the beginning and end of a sleeve, the height of the sash, the end of the yellow jacket and the height of the boots, while zigzag lines denote a bent elbow (sticking out further than the final one (do note, however, that the signal is partially blocked here on the right by the lead white in the gorget, and a small amount of umber in the shadow of the sleeve in the top layer confuses the image a bit). A long diagonal line sketches a rapier carried from the hip (like the one belonging to the captain in the yellow coat in Hals's 1633 painting) and cuts through the officer's left hand in the top layer, as Bijl suspected (see the beginning of this chapter). Originally, the hand appears to have been placed more to the right (in the middle of his back). The manganese map also shows an early *pentimento* or change by Hals that has thus far gone unnoticed: with his right hand the officer originally held a weapon in front of him, possibly a rifle, as can also be seen in the infrared reflectogram, and vaguely with the naked eye: (Detail of Fig. 6.6: manganese (Mn-K) map, visible light image and IRR, in 'sync mode'). Comparing the initial design for the officer in the yellow coat with the upper layers here, it is possible to reconstruct its genesis with surprising precision. As discussed above, the attribution of this figure has led to a considerable amount of debate in the literature. Many experts stated that this figure was entirely or almost entirely painted by Codde, while others maintain that he was painted largely by Hals. Grimm for example, believed that the face was almost entirely by Hals, while Bijl suggested that Codde might have been working from either a fairly worked up lower layer or a rather incomplete sketch. The present research yielded not only a crucial new clue to solve the riddle but also a new digital tool to compare the distribution of the different elements with unprecedented precision, allowing us to reconstruct the different stages in much greater detail.

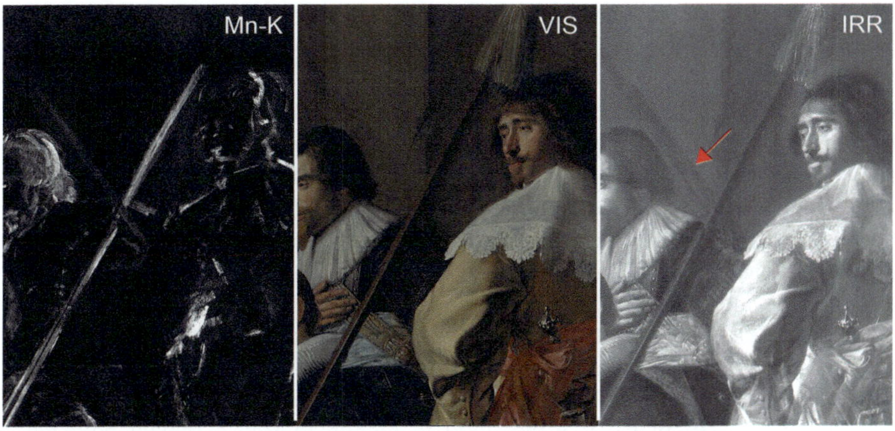

Detail of Fig. 6.6: manganese (Mn-K) map, visible light image and IRR, in 'sync mode' (sn.pub/6wearh)

When comparing the costume of officer 8 in the (manganese-containing) sketch lines precisely with its final appearance in the top layer, one gets the impression that a painter other than Hals took over and followed the -sometimes rather puzzling- sketch lines as best he could. The fairly odd placement of the collar behind the officer's neck and upper back (creating a kind of bump) in the final painting actually corresponds exactly with the sketch lines indicating its top (Detail of Fig. 6.6: manganese (Mn-K) map, calcium (Ca-K) map and visible light image, in 'sync mode'). The sketch for the lower part of the collar, however, seems to have been too summary for the second painter to make sense of it and neither does he seem to have fully understood the placement of the arms. He appears to have resolved the sometimes puzzling clues with minor adjustments. He painted a longer collar with curvy lace edges, adjusted the length of the coat a little, altered the position of the rapier hanging from his side somewhat, placed his left hand more to the left and made his right elbow stick out a bit less.

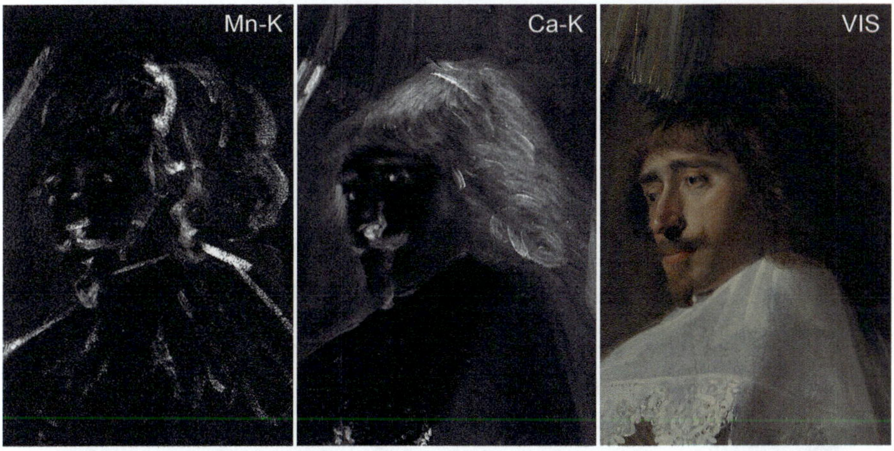

Detail of Fig. 6.6: manganese (Mn-K) map, calcium (Ca-K) map and visible light image, in 'sync mode' (sn.pub/f9ij5r)

The execution of the face is more complex to interpret at first. The final position of the officer's beard, nose, eyes, eyebrows and hairdo also follow Hals's indications, visible in the manganese map, rather faithfully in as far as these can be seen. However, given the confusion caused by the cursory indication of the collar, which led to a rather unconvincing suggestion of the underlying form, the face seems, in comparison, too successful in its shape and plasticity for both parts to be by the same hand. The top layer adds to the mystery in that it lacks Hals's characteristic bold brushwork and loose accents. Instead, the face consists for the most part of carefully blended brushstrokes, and is framed by strangely dull, repetitive hair on the forehead and back of the head. What are we looking at here?

The MA-XRF calcium map provides a crucial clue to the interpretation. On top of his umber-containing sketch, Hals apparently worked up the hair and face with a bone black containing paint (cleary visible in the calcium (Ca) map and vaguely in the phosphorus (P) map). It is not hard to recognize Hals's typical bold paint strokes here, with their characteristic raised edges, often on both sides, the so-called ribbon-touches. Given this more elaborate underpainting, the second painter thus had more guidance in this part, which helped to create a stronger end result. All in all, the deliberate variations by the second painter seem rather modest; he merely worked out parts that were not very clearly defined.

In fact, the loose indication of the hair resembles what we see elsewhere in the copper map. Hals appears to have used a copper containing greenish grey to summarily sketch shadows in the skin tones and a rough indication of the hair of many of the men on the right. Especially below officer 11 we can see a very free sketch of hair and shadows in the skin tones and eyes, partially overpainted in the top layer (Detail of Fig. 6.6: visible light image and copper (Cu-K) map, in 'sync mode' , p. 225).

Fig. 6.22 Copper map of Fig. 6.7

In short, the copper map also seems to reveal some of the underpainting by Hals. As we have seen, Hals used azurite regularly for his skin tones in the late 1620s and 1630 (Chaps. 3 and 5). Codde, on the other hand, did not use copper at all in the skin tones in his small guard scene of circa 1635 (Fig. 6.22).

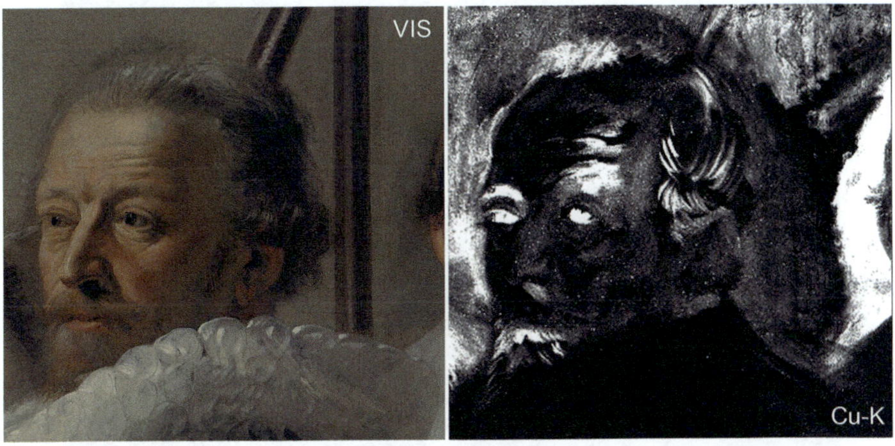

Detail of Fig. 6.6: visible light image and copper (Cu-K) map, in 'sync mode' (sn.pub/xw3vdp)

Below the officer in black (the thirteenth figure from the left), standing somewhat in front of his fellow officers at the right, the umber-containing sketch lines can also clearly be seen. Like the sketch below the officer in yellow, the lines are cursory and free (Detail of Fig. 6.6: visible light image and manganese (Mn-K) map, in 'fade mode'). Brief touches indicate the top of the shoulder, the direction of the arms, torso, sash and the end of the sleeves and pants. A decoration at the top of his right stocking is summarily indicated, but was only partially executed. Also, his jacket was originally longer. Furthermore, one of his accessories was misunderstood and/or changed significantly. The long object hanging at his side must have originally been intended as a forquet, witness its u-shaped ending (compare Figs. 6.23 and 6.24). It was a logical attribute for an officer with a musquet and a burning wick in his hand (yet another indication that the scene was originally planned outdoors). Nevertheless, the object misses its typical tip in the final version and was turned into a staff weapon instead.

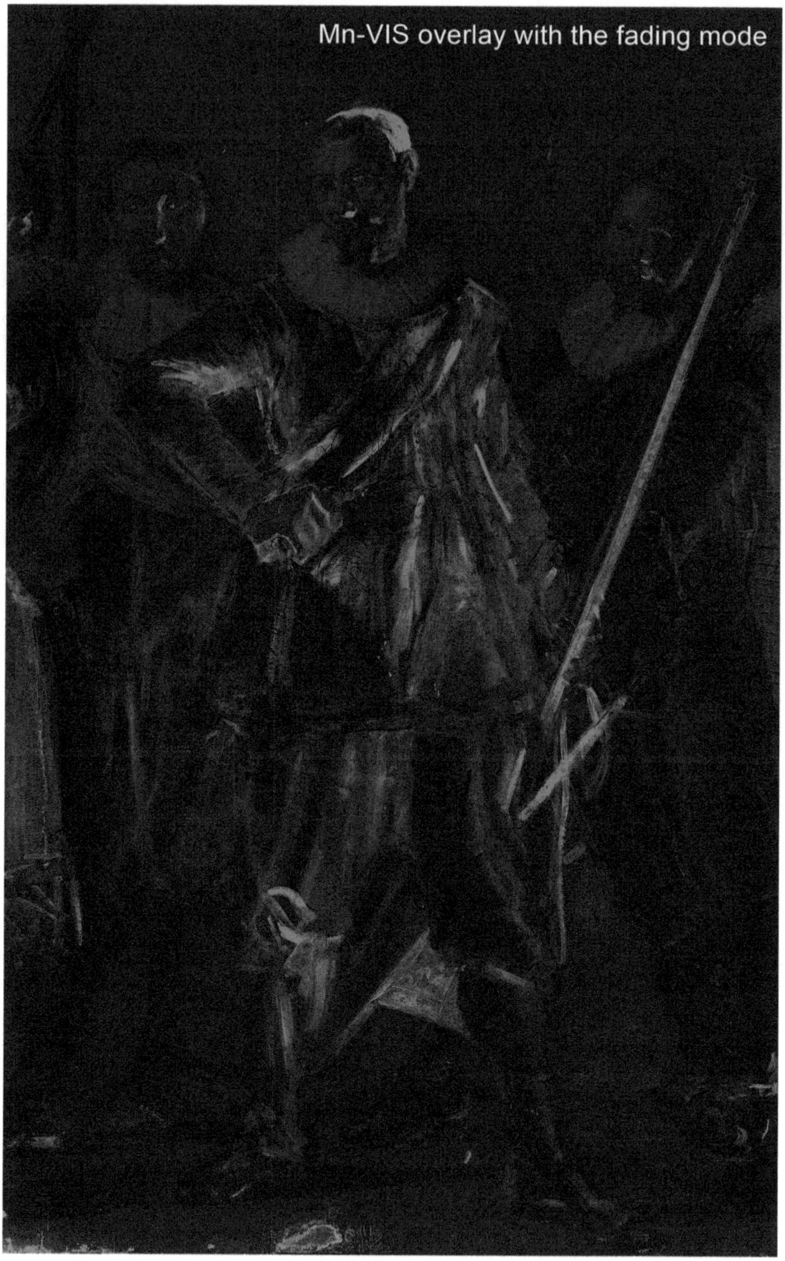

Detail of Fig. 6.6: visible light image and manganese (Mn-K) map, in 'fade mode' (sn.pub/l0z8f4)

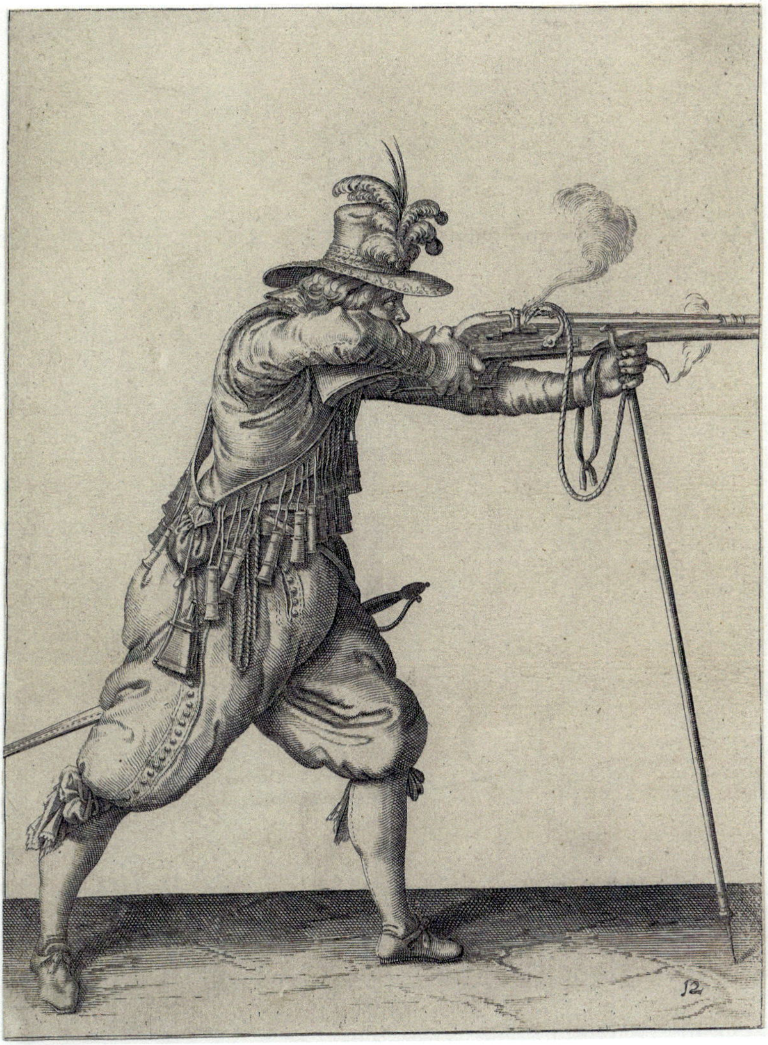

Fig. 6.23 Jacques Gheyn II, Illustrations 1 and 12 from (De Gheyn 1607)

As to Hals's initial design, a few more deviations from it can be detected in the infrared reflectogram (Detail of Fig. 6.6: visible light image and IRR, in 'sync mode'). In an early stage he outlined what appears to be a window in the background at the left and gave the banner a higher position. Neither of these elements was worked up in paint, but two others were: the black hats of both the standard-bearer Nicolaes van Bambeeck and captain Reynier Reael were originally larger, as

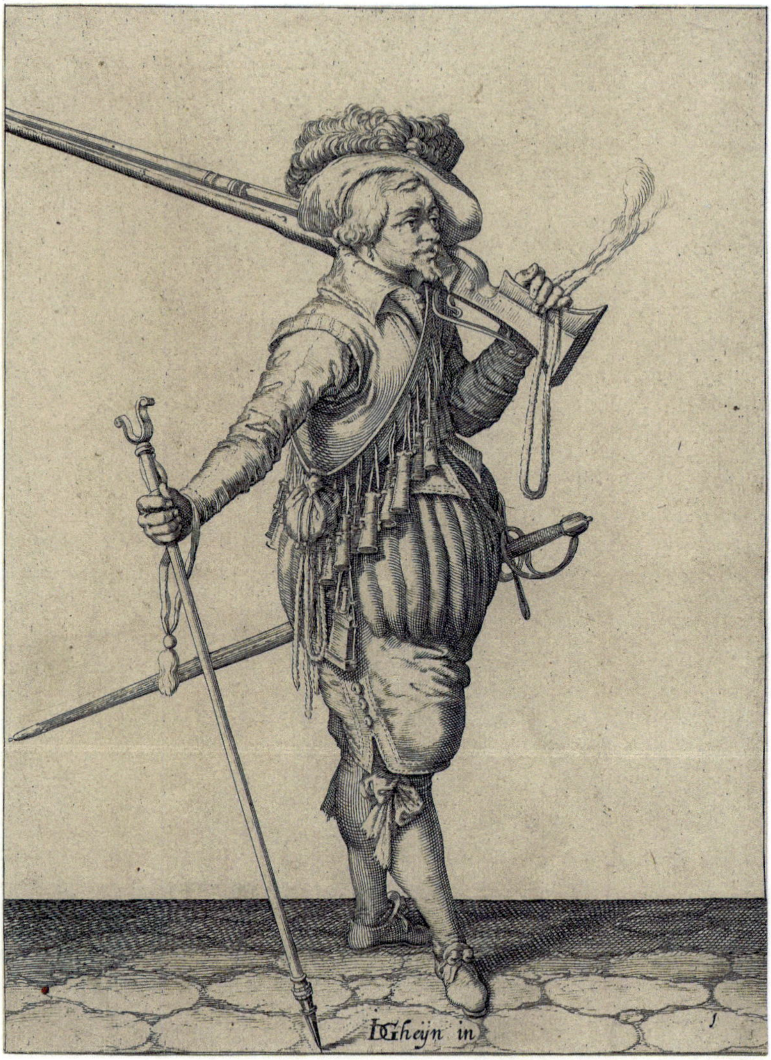

Fig. 6.24 Jacques Gheyn II, Illustrations 1 and 12 from (De Gheyn 1607)

can be seen in the Cu map (Detail of Fig. 6.6: visible light image, IRR and copper (Cu-K) map, in 'sync mode'). The overpainted parts can fairly easily be distinguished with the naked eye here, as these aged differently from the surrounding areas. The brushwork is also different here, suggesting that these changes were not made by Hals (Detail of Fig. 6.6: visible light image and IRR, in 'sync mode').[86]

[86] For a more extensive analysis of Hals's and Codde's brushwork, see below.

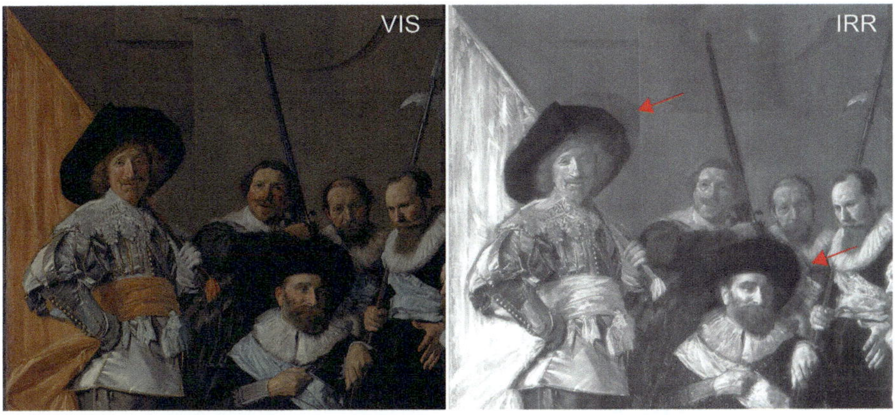

Detail of Fig. 6.6: visible light image and IRR, in 'sync mode' (sn.pub/o4pwft)

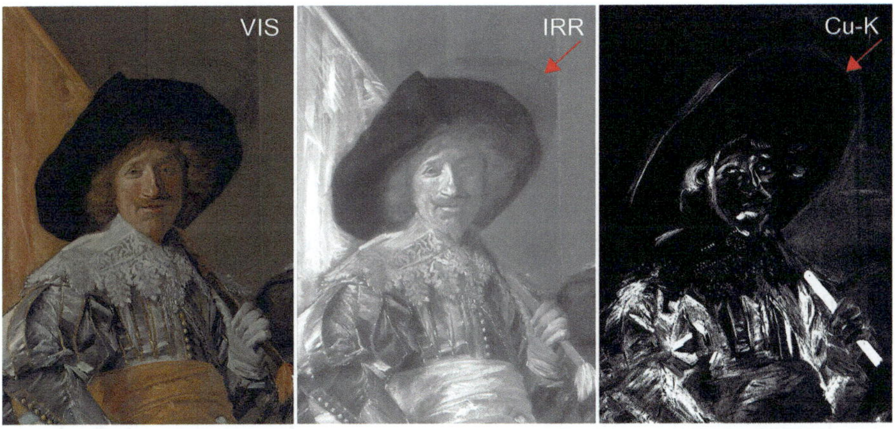

Detail of Fig. 6.6: visible light image, IRR and copper (Cu-K) map, in 'sync mode' (sn.pub/w9x9q2)

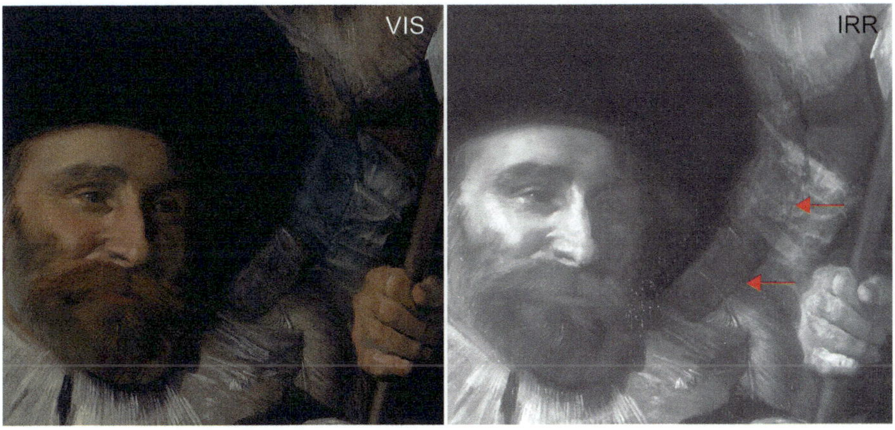

Detail of Fig. 6.6: visible light image and IRR, in 'sync mode' (sn.pub/vrx2nk)

The above reconstruction of Hals's original design, based on the new non-invasive imaging data and digital tools, allows us to make more precise comparisons to the final result than before, as in the case of the sketch lines below officer 8. Using some more specific seventeenth-century art critical terms helps to further evaluate to what extent the design is characteristic for Hals or deviates from his standard. It also helps to compare his inventions more thoroughly to works by Codde, and thus to gain more insight into Hals's and Codde's contribution to the painting as well as into its level of completion when Codde took over.

As briefly mentioned above, many of the figures on the right appear to lean backwards. A closer look at the various sketch lines visible in the middle and right parts of the painting and the comparison to the upper layers here showed that Hals left these figures in various states of completion. The costumes and postures of the men must have been indicated too summarily for Codde to work these up exactly as Hals had originally intended. As we have seen, Hals sketched the costumes of the figures very summarily using different paints. By contrast, their faces, which constitute the most important part for a portrait painter, could have been worked up further (as will be discussed in more detail below). Apart from their centre of gravity, other aspects of the postures of the different figures in the painting are also telling.

When designing a painting, it was important to create something original yet lifelike, according to seventeenth-century art theorists. In a speech to the Leiden painters guild in 1641, the painter and theorist Philips Angel explained that in his view the challenge was to invent 'something special yet natural' (*iets bisonders doch naturlicks*). A term he uses a lot in connection to this goal is '*eygentlick*', which meant 'real' in the sense of convincingly natural and befitting that character of what was depicted (Angel 1642; Tummers 2011b, p. 244).

In the left part of the painting it is striking how subtle aspects of the men's postures lend the sitters a certain ease and naturalness—a part of the invention that could both be called 'original yet natural' and '*eigentlick*'. Note, for example, how the captain does not place his feet neatly side by side but assumes a more relaxed pose, leaning casually on the sides of his shoes, as can be seen particularly clearly in the calcium map. (Detail of Fig. 6.6: visible light image and calcium (Ca-K) map, in 'sync mode')

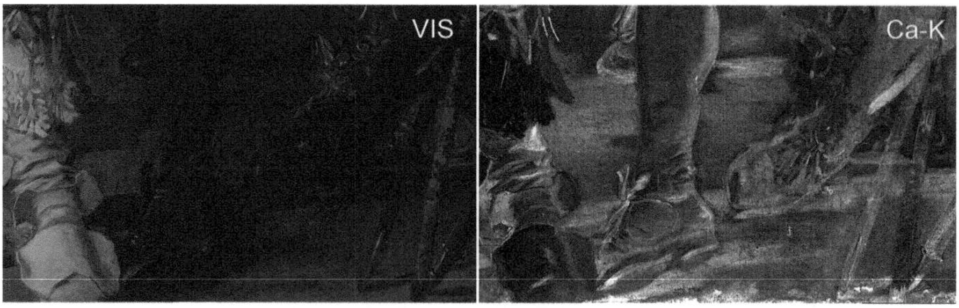

Detail of Fig. 6.6: visible light image and calcium (Ca-K) map, in 'sync mode' (sn.pub/686ysz)

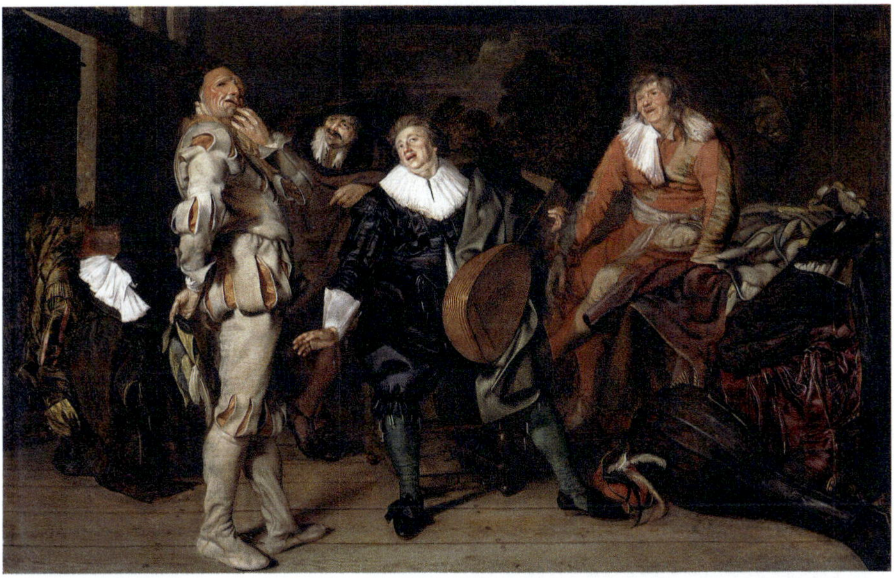

Fig. 6.25 Pieter Codde, *Actor's wardrobe*, c. 1630–1640, oil on panel, 33 × 52 cm. Gemäldegalerie, Staatliche Museen zu Berlin. (Credits: Staatliche Museen zu Berlin,/Jörg P. Anders)

At the same time, he also exudes authority and dignity, as he gazes calmly to his left and holds one hand on his heart (a sign of truthfulness). Likewise, the lieutenant sitting beside him casually wraps one lower leg around the other. The posture of the standard-bearer Nicolaes van Bambeeck is also remarkable: both confident and relaxed. He looks straight out at the viewer, one his hand firmly in his side, the other one loosely holding the banner, while he places one foot elegantly before the other.

By comparison, the postures of the men on the right do not display a similar ease. Although their feet are fairly wide apart, they seem nevertheless a bit unstable –due in part to their lopsided centre of gravity. Their hand gestures appear a bit contrived, and even the shape of their shoes is different, less pronounced—in fact these are typical Codde shoes both in their shape and in their application of colour, as we will discuss below (Fig. 6.25).

Indeed, these differences in shapes and postures agree with what we know of both artists. Codde tended to place his figures' legs fairly wide apart, and gave them often twisted wrists and somewhat contrived hand gestures (Figs. 6.7 and 6.25). Hals, by contrast, was praised in his time especially for his convincingly natural yet lively gestures, as we have seen. He even famously depicted a dignified gentleman casually leaning backwards on a chair, raising two of its legs off the ground (Frans Hals, *Willem van Heythuysen seated on a chair*, c. 1638, Private Collection) (see Slive 2014, Plate 125, p. 224).[87]

[87] The ease Hals introduced into his portraits seems to be a kind of visual equivalent of the ideal of the cultivated gentleman that had been popular since the renaissance. Such an ideal gentleman was not only dignified, wise and confident, but also able to relax; he demonstrated 'sprezzatura' (ease) as Baldasar Castiglione explained in his popular etiquette book *The Ideal Courtier*, which was translated into Dutch by the painter and art theorist Samuel van Hoogstraten.

The lifelike and somewhat casual poses of the figures reinforce the illusion that we encounter the gentlemen in Hals's paintings at a sudden moment in time, as if we walked into their space and some of the men just turned to us. This momentary effect was also described by seventeenth-century art theorists. Samuel van Hoogstraten stated in 1678 that painters of history scenes 'whether they depicted one figure or many together, should pay heed to show only one instantaneous movement' (*oogenblikkige beweeging*). His advice seems directly inspired by the immediacy of the paintings by his teacher Rembrandt, whose mise-en-scene could be so momentary that in a painting like *The Resurrection of Christ* (c. 1635–1639, Bayerische Staagemäldesammlungen, Alte Pinakathek, Munich) one can see a sword in mid-air before it hits the ground.[88] The so-called 'snapshot effect' often used in the secondary literature to describe Hals's paintings thus had a seventeenth-century equivalent. Certain details in Hals's pictures reveal that he must have very deliberately sought to create it, a decade before Rembrandt did so. For example, in Hals's 1627 Calivermen civic guard portrait, he depicted falling water drops, a phenomenon that, like Rembrandt's sword, can only be seen in a split second (Detail of Fig. 6.3: visible light image and copper (Cu-K) map, in 'sync mode').

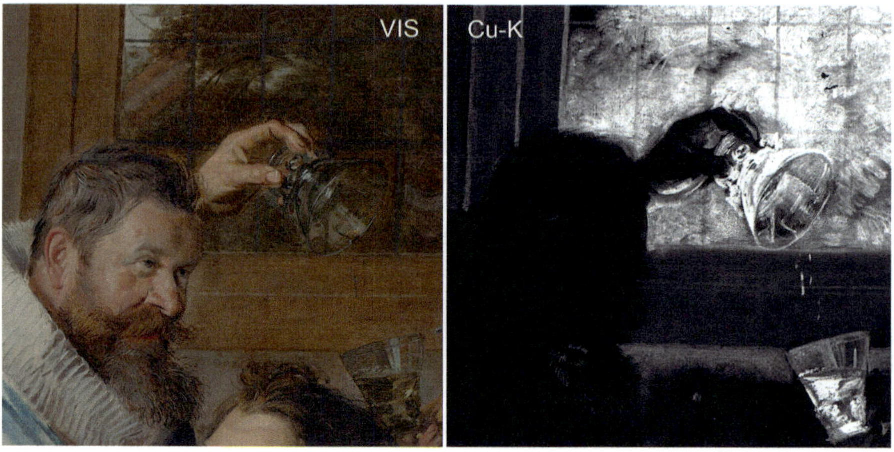

Detail of Fig. 6.3: visible light image and copper (Cu-K) map, in 'sync mode' (sn.pub/7kny9z)

A last aspect of the design worth noticing before we will look further into the colours, light and brushwork in *The Meagre Company*, concerns the way in which shapes influenced one another. This effect could be used to create unity though the

[88] On this quality, see also (Brenninkmeyer-de Rooy 1984, Fig. 70 and p. 65; Tummers 2011b, p. 225; Weststeijn 2005, pp. 117–119)

successful joining of related shapes and directions and alternating these with spatial leaps (*koppeling en sprong*) or, instead, cause confusion (see Tummers 2011b, p. 210; Van Hoogstraten 1678, p. 193). Seventeenth-century art theorists were very aware of the negative effects neighboring shapes could have. As we have seen, Karel van Mander explained in his *Schilder-boeck* of 1604 that it was crucial to avoid confusion of figures and limbs ('*haspeling*'); painters should make sure that the viewer could easily distinguish which limb belonged to which person (see above, p. 193ff; Van Mander 1604, fol. 18v; see also Tummers 2011b, p. 207 and note 101, p. 292). Frans de Grebber elaborated on the rule in 1649 by adding that figures should therefore not overlap too much (De Grebber 1649). At the end of the century, Gerard de Lairesse (1707, vol. 1, pp. 236 and 263–264; vol. 2, p. 256) even specified that neighboring shapes could have an involuntary chaotic effect which he called 'sprawl' (*sparteling*), if these had not been carefully balanced (see also Tummers 2011b, p. 210 and note 114, p. 292; De Vries 2004).

In the Haarlem civic guard pieces we saw that Hals made several adjustments to enhance the visual unity in these complex designs, and occasionally moved or removed hand gestures, presumably for this reason. He must have been very aware of which elements could strengthen or weaken the general design and visual coherence. After all, by the time he received the commission for *The Meagre Company* he was experienced in setting up such large-scale group portraits. Codde, on the other hand, had only worked on small-scale portraits and genre scenes before he took over Hals's commission, and must have been less aware of the pitfalls when designing large-scale, complex group portraits.

The left group of officers in *The Meagre Company* is a perfect example of a convincing '*koppeling en sprong*': connections and leaps in the spatial organization. All the shapes are connected in a coherent spatial whole, caused by many related diagonal shapes and thrusts, which lead the viewer's eye into the distance and back—Hals's famous diagonal compositional lines, as these were called in the twentieth century (see also above: Hals's design process). On the right-hand side of the painting, such a balanced coherence is largely missing. Moreover, there is a combination of hands that are put so close together that these actually cause confusion. It is not immediately clear to whom the different hands belong (Detail of Fig. 6.6).

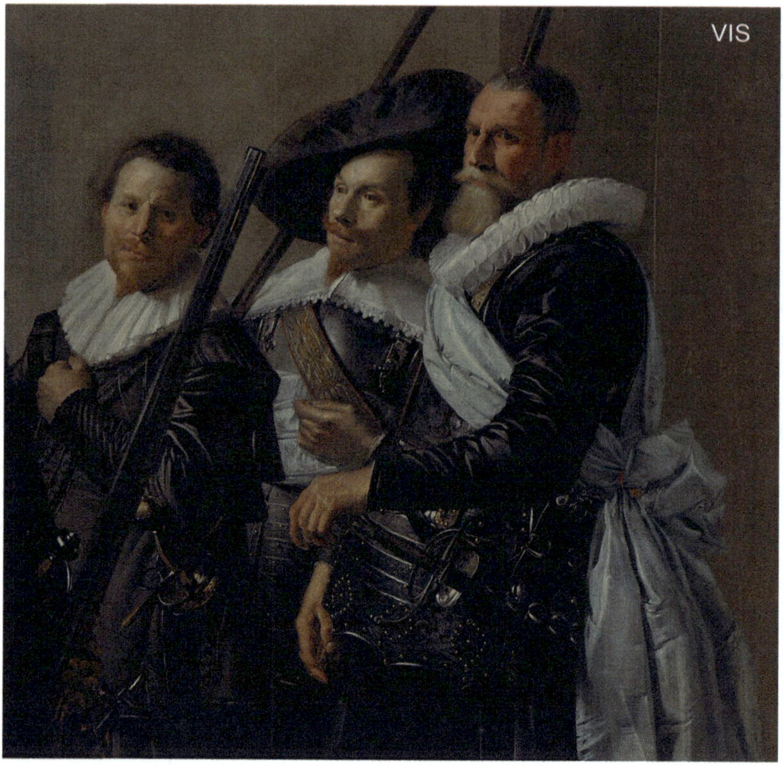

Detail of Fig. 6.6 (sn.pub/yhhk53)

The shapes also compete with one another since there is little space in between them, creating a chaotic effect. In short, these cause both '*haspeling*' and '*sparteling*'—betraying an inexperienced mind at work in this complex design. Indeed, the MA-XRF manganese map reveals that the top hand was not initially planned as it clearly overlaps an earlier design (Detail of Fig. 6.6: visible light image and manganese (Mn-K) map, in 'sync mode'). It must have been Codde's addition, which indicates that like the placement of the feet, the positions of most hands in the right part of the picture were not clearly defined when Codde took over (an exception being officer 13's hand, firmly holding the powder charge). In fact, even the placement of some of the hands on the left seems unusual for Hals—notably the ones of the sitting lieutenant, especially his left one with its strongly bent wrist in a rather unnatural position. Although the modeling of light and shade is successful in the hand, the transition of the wrist into the lower arm is not, suggesting that this arm was not fully worked up when Codde took over. The lieutenant's other hand had presumably only been sketched at the time. The copper and manganese maps reveal a bold initial indication of the position of his right hand, while the dull finishing and unconvincing modelling at the surface are not at all characteristic of Hals (Detail of Fig. 6.6: visible light image, copper (Cu-K map) and manganese (Mn-K) map, in 'sync mode').

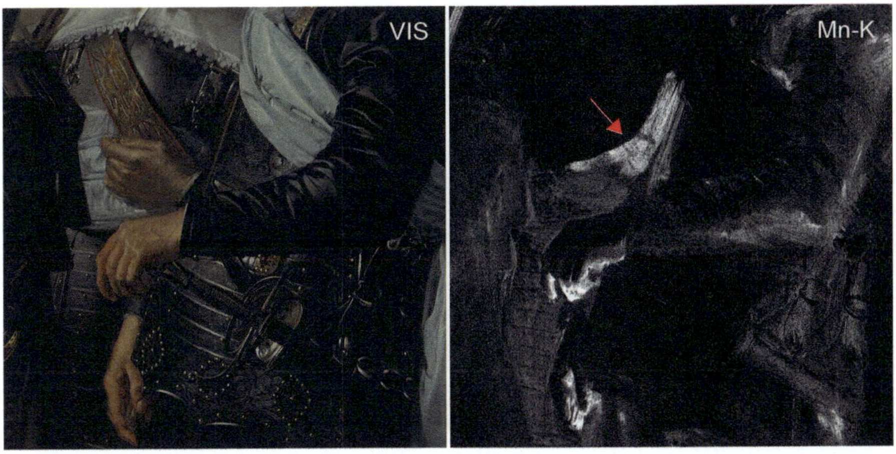

Detail of Fig. 6.6: visible light image and manganese (Mn-K) map, in 'sync mode' (sn.pub/s7serq)

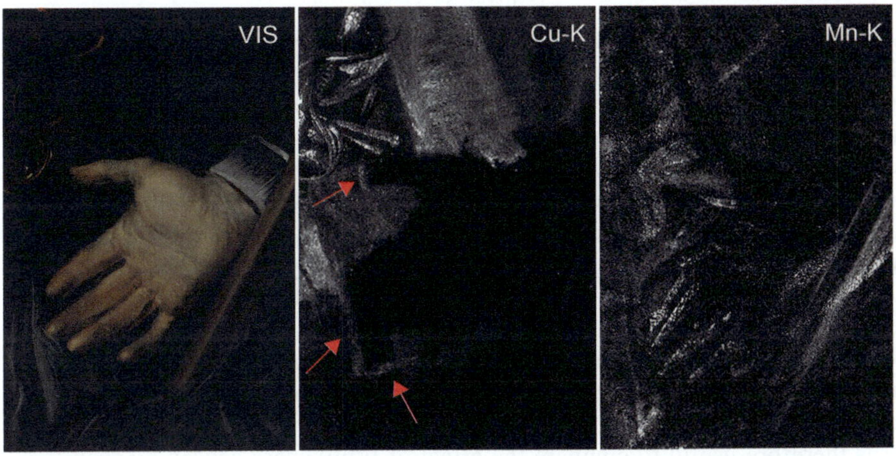

Detail of Fig. 6.6: visible light image, copper (Cu-K map) and manganese (Mn-K) map, in 'sync mode' (sn.pub/3arbsx)

The use of colour, light and brushwork in *The Meagre Company* is also revealing. The momentary effect discussed above did not, of course, depend on design alone. According to seventeenth-century art theorists, especially the application of colour and the brushwork were crucial for the effect of lifelikeness and, as we have seen, Hals was greatly admired for precisely this effect: 'as if life were moved into the paint'.[89] Especially his loose touches strongly contributed to the

[89] This citation is taken from a poem by Govert Bidloo witten between 1683 and 1685, see note 58 above: '*Als was het leeven, juist, het leeven afgekeken, / En in de verf verplaatst*' (Bidloo 1719, pp. 173–185).

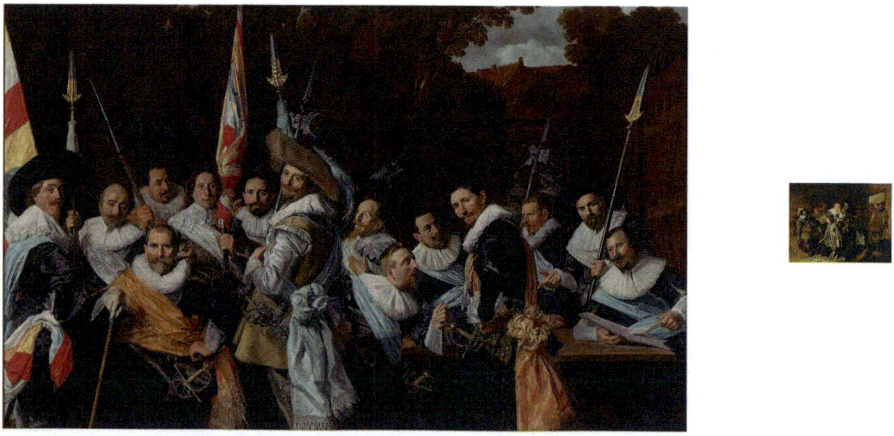

Fig. 6.26 Scale difference of Figs. 6.4 and 6.7

suggestion of movement, the impression that we are witnessing a sudden glance in a split second. The loose brushwork triggers the viewer's imagination to automatically fill in the details and complete the effect, as Samuel van Hoogstraten (1678, p. 27) explained. It was an art to place such loose touches confidently and accurately, as many art theorists, beginning with Karel van Mander (1604, fol. 47r-48v), stressed throughout the century (see also Atkins 2012, ch. 1; Tummers 2011b, pp. 219–221). It required not just talent but also a significant amount of experience. Moreover, especially the finishing of a painting was crucial. According to Van Hoogstraten, the danger when finishing a picture was that the edges of brushstrokes, which had been placed swiftly and with a sure touch, would blur and become unrecognisable something that could only lead to stiffness. Carrying on painting for too long, correcting too much and working out details too precisely had the same effect (Van Hoogstraten 1678, p. 27 and 264). It is therefore useful to compare the effect of lifelikeness caused by such loose touches, bearing in mind that although Codde must have tried to mimic Hals's brushwork as best he could, his lack of experience in painting on a large scale with extremely loose brushwork may well betray him. Just for reference (Fig. 6.26), note the scale difference between Hals's 1633 civic guard portrait (207 × 337 cm) and the guard scene by Codde (35 × 44 cm), as well as the difference in paint handling (a more painterly versus a more linear approach) visible in the infrared reflectogram of two details depicting guardsmen (Fig. 6.27).

In fact, looking at the effect of lifelikeness caused by loose brushstrokes goes a long way towards distinguishing Hals' contribution. The standard-bearer Nicolaes van Bambeeck at the left is a good example of Hals's abilities in this respect, and if we compare his face, for example, to the second figure from the right, the difference in liveliness and expression is obvious (Details of Fig. 6.6, in 'gallery mode'). Indeed, as many art theorists had indicated, the art of portraiture was not so much

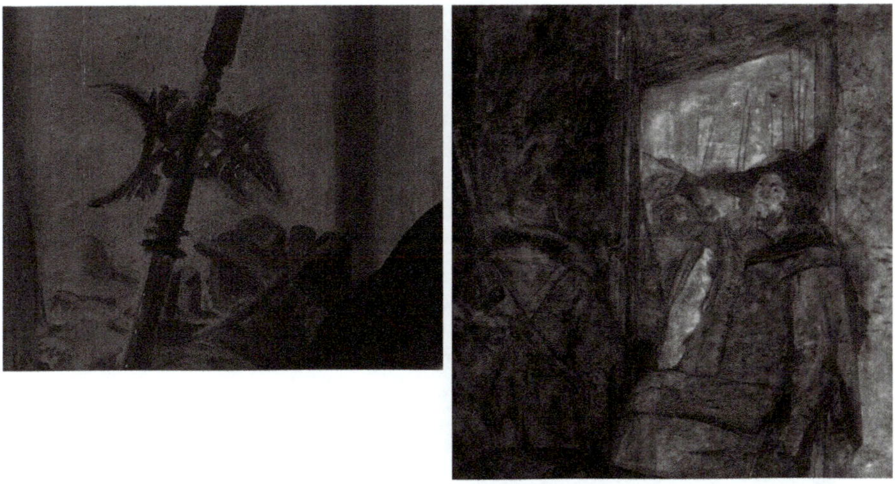

Fig. 6.27 Infrared reflectograms (IRR) of details of Figs. 6.3 and 6.7

about the accurate representation of form as it was about infusing the portrait with a sense of inner life. The fifth and seventh figure from the right, on the other hand, are much more animated in their expressions; it looks as if they were barely touched by Codde. These faces must therefore have been as good as finished when Codde took over (Detail of Fig. 6.6).

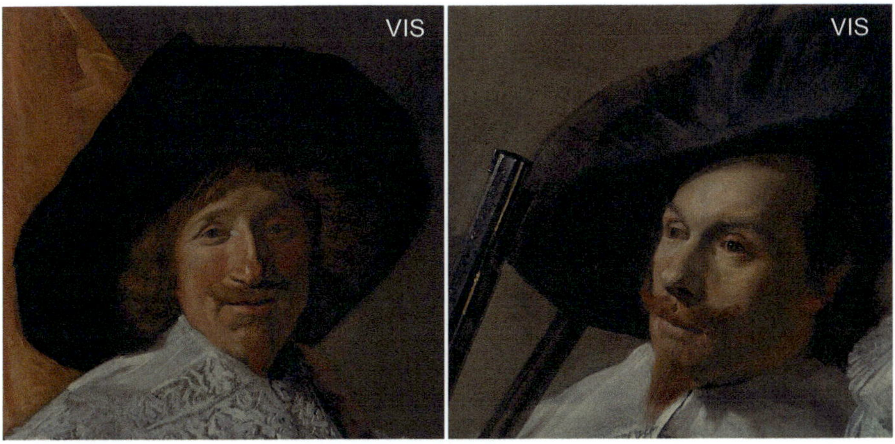

Details of Fig. 6.6, in 'gallery mode' (sn.pub/po1p9h)

Detail of Fig. 6.6 (sn.pub/me3n3v)

This brings to mind the early reception of Hals's paintings. Already in 1628 Schrevelius had remarked that it was Hals's 'exceptional manner of painting' that made the people in his portraits 'seem to breathe and live' (see above, note 9). Furthermore, around 1632 Van Dyck reputedly remarked that once Hals 'had applied the underpainting of a portrait, he could give the characteristic features, highlights and shadows their proper place with one brushstroke, without tempering or change', as Arnold Houbraken noted down. Van Dyck was thus apparently impressed with the exceptional ease and accuracy of Hals's loose brushwork in the finishing. According to Houbraken, Hals even had a habit of calling these final touches 'the recognizable features of the master' (*het kennelyke van de meester*).[90]

[90] 'Men zegt dat hy [Hals] voor een gewoonte had, zyn Pourtretten vet, en zachtsmeltende aan te leggen, en naderhand de penseeltoetsen daarin te brengen, zeggende: *Nu moet 'er het kennelyke van den meester noch in*' (Houbraken 1718–1721, vol. 1, p. 92; see also above, note 17).

In *The Meagre Company*, quite a few other faces, though convincing in their plasticity, fall short in their finishing and thus in their liveliness and expression. The eighth and ninth figure from the left have a somewhat troubled look caused by the way in which their eyebrows are raised in the middle, which does not really seem fitting for the occasion. Moreover, like officers 11, 13, 15 and 16, they appear somewhat wooden in their expression, betraying the hand of Codde in their execution.

The overall use of colour and tonal contrast in the different parts of *The Meagre Company* is also very revealing as to its attribution. In seventeenth-century Dutch art theory the use of colour and tonal contrast was considered crucial to the overall effect of lifelikeness, one of the most important goals of the entire art of painting. A key term art theorists used in this respect was '*houding*', which could be translated as the 'balancing of colours and tones' (like *verhouding* in modern Dutch), that is: the successful use of colours and tones to create a convincing suggestion of three-dimensionality and to highlight the key elements (Taylor 1992; Tummers 2011b, p. 214 ff; Van de Wetering 1991; Van de Wetering 1997, pp. 149–152). It was well known at the time that intense, undiluted colours and strong tonal contrasts attracted attention and seemed to come forward, whereas 'broken' or blended colours and soft tonal contrasts made figures and other elements seem to recede into the distance. Of all the colours, red and yellow were known to be the most powerful (*krachtig*) in their purest form. Painters used this knowledge to create a convincing suggestion of depth in their paintings and to unify their designs by highlighting the most important elements. The successful use of colour and tone in a painting was therefore not simply a matter of matching lifelike colours and giving a given object a convincing sense of three-dimensionality, but a much more complex balancing act. According to Gerard de Lairesse (1707, vol. 1, pp. 12–13), in order to achieve a successful balance of colours and tones (*houding*), one should therefore start by carefully adjusting the tonal values in the dead colouring stage of a painting (see also Van Eikema Hommes 2004, pp. 13–14).

Of course, some painters were better at balancing the colours and tones in their pictures than others. Rembrandt, in particular, was highly praised for the *houding* in his paintings. Hals also had a knack for it. Notice how, in his 1627 portrait of the Caliverman, he reserved the starkest highlight for the figure most in the front, the standard-bearer Adriaen Matham, and the softest tonal tonal transitions for the hall ward, Willem Ruychaver, who is furthest away from the viewer, while drawing attention to the highest ranking officer, colonel Willem Claesz Vooght, by fully illuminating his face (he faces the light) and showing a large part of his bright orange sash (Frans Hals, *Banquet of the Officers of the Calivermen Civic Guard*, 1627, oil on canvas, 183 × 266.5 cm. (Frans Hals Museum, Haarlem)).

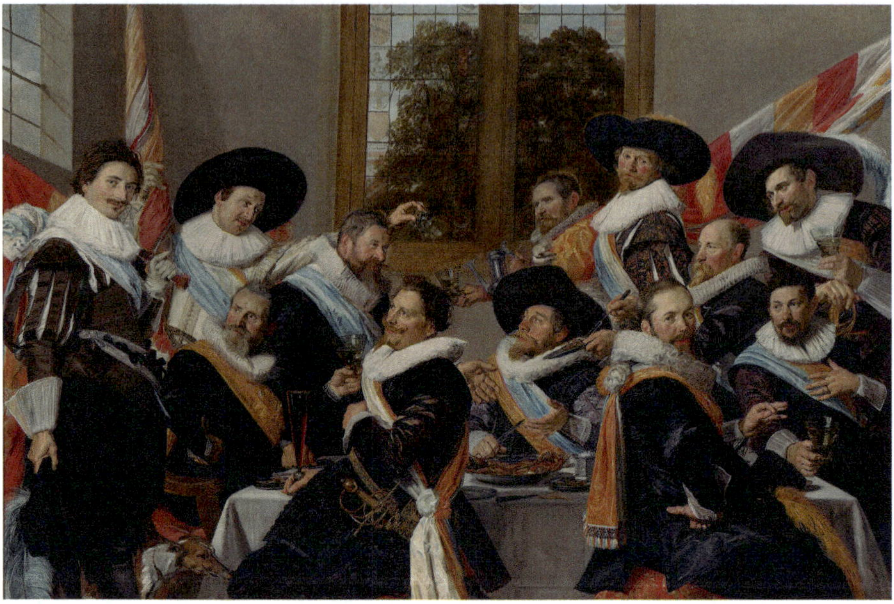

Frans Hals, *Banquet of the Officers of the Calivermen Civic Guard*, 1627, oil on canvas, 183 × 266.5 cm. (Frans Hals Museum, Haarlem) (sn.pub/gg2ack)

In *The Meagre Company* we can see a sharp contrast between the balance of colours and tones (*houding*) on the left of the painting and in the rest of the picture (Frans Hals and Pieter Codde, *Militia Company of District XI under the Command of Captain Reynier Reael, Known as 'The Meagre Company'*, 1637, oil on canvas, 209 × 429 cm. (Rijksmuseum, Amsterdam)). On the left the tonal values have been carefully adjusted so as to make the eye focus naturally on the most important parts: the faces and gestures of the men, and hence their interaction. The colours of the costumes complement the design of the composition. The captain has the starkest colour contrast in his costume, the deepest blacks juxtaposed with a bright white collar; while the costumes of the men around him are slightly more subdued. The bright orange banner and sash of the standard-bearer make him stand out, emphasising his key role in the composition, drawing the viewer into the painting.

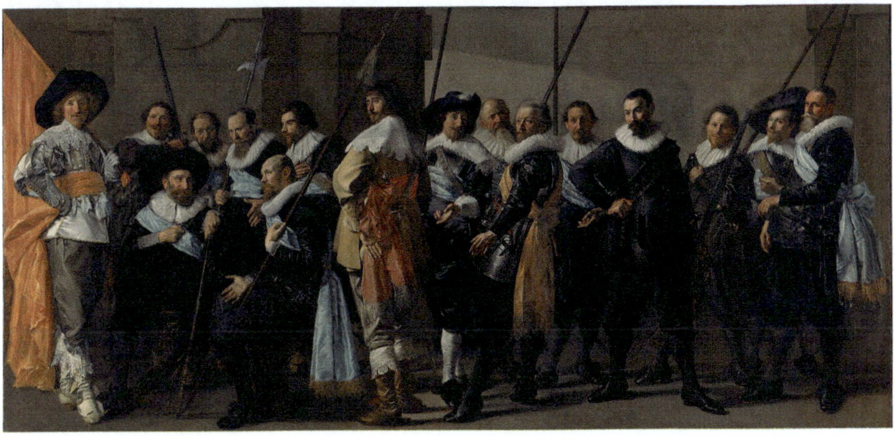

Frans Hals and Pieter Codde, *Militia Company of District XI under the Command of Captain Reynier Reael, Known as 'The Meagre Company'*, 1637, oil on canvas, 209 × 429 cm. (Rijksmuseum, Amsterdam) (sn.pub/izlv0w)

Ideally, the orange of the banner and sash on the left should be balanced out by a strong red in the middle (as in Hals's 1639 civic guard portrait see Fig. 6.5) and by a powerful orange at the right. However, these colours lack the appropriate intensity. The orange sash on the right does not contain the same pigments as the one on the left; it lacks the rather expensive and powerful red pigment vermillion, which may have also caused it to age differently. The calcium map points to the use of a lake pigment, which is sensitive to degradation, in the sash on the right. The sash in the middle does contain vermillion but mixed in such a way that it has turned into a rather dull pinkish red. These discolorations -possibly related to the application of thin glazes- disrupt the '*houding*' here pointing to Codde's hand, for such contrasts do not occur in Hals's other civic guard portraits. Moreover, the paint handling confirms that the sashes in the middle and on the right are not by Hals. The folds and highlights are indicated with soft transitions and rather dull, repetitive touches instead of Hals's characteristic bold, zigzag strokes (Details of Fig. 6.6, Fig. 6.4 and Fig. 6.5, in 'gallery mode').

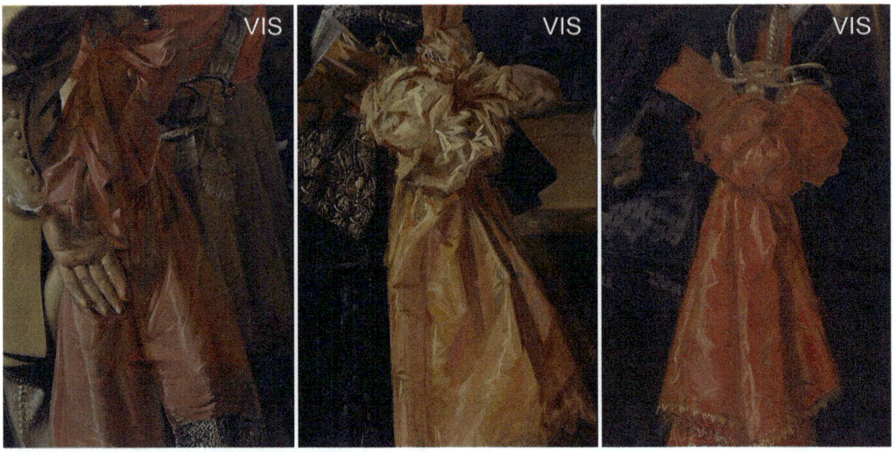

Details of Fig. 6.6, Fig. 6.4 and Fig. 6.5, in 'gallery mode' (sn.pub/2grtsp)

By contrast, the indigo sashes in *The Meagre Company* are in harmony with the overall balance of colours and tones in the painting, suggesting that Hals set these up. The difference here is in the finishing; when compared carefully to the sashes in the Haarlem civic guard portraits, most of the ones in *The Meagre Company* lack the swift and spontaneous brushwork characteristic of Hals's depiction of drapery, as Van Eikema Hommes has already noted (see above: A contested attribution).

Furthermore, in the centre and on the right of *The Meagre Company*, there are many sharp light-dark contrasts that do not highlight crucial parts. The shoes have intense highlights that make them pop out, as does the shiny armour and even the collars have been given very strong contrasts (Detail of Fig. 6.6). Admittedly, these do give the separate objects a convincing suggestion of three-dimensionality, but on the composition as a whole these have a rather scattering effect, attracting the viewer's attention in many different directions. It is a clear sign that this part was not worked up by the master himself, but by Codde instead, confirming our previous observations about the design.

Detail of Fig. 6.6 (sn.pub/4e440m)

In conclusion, the above analysis has yielded unique new insights into the genesis of *The Meagre Company*, from Hals's initial thoughts and an early *pentimento* (change) to the way he built up a face from a cursory sketch to the 'dead colouring' or underpainting. It also gave a lot of concrete insight into the different levels of completion the various parts of the painting had when Codde took over. In general, the faces turned out to be more worked up than other parts and the poses and costumes on the right were far less finished than the ones on the left, although the two key figures for the composition were sketched up a little further. Moreover, the use of seventeenth-century art theory in combination with the in-depth analysis of the use of materials and the new visualisations of the underlying sketch layers gave a number of new insights into both Hals's achievements and his characteristic use of materials as well as the differences with Codde. In particular, Hals's original yet natural (*eigentlick*) postures and his careful balancing of forms, colours and tones stood out (avoiding *haspeling* and increasing *koppeling*, *sprong* and *houding*), while Codde's use of tonal contrasts to lend particular objects such as shoes, armour and collars a convincing three-dimensionality gave him away. As for their use of pigments and paint application, Hals's use of umber and azurite blue were especially revealing, while Codde occasionally betrayed himself by creating a superficially similar colour with a rather different layer build-up that disrupted the overall *houding* or colour balance (as in the case of the orange banners). All in all, the new approach and rich data set provided an important stepping stone for a deeper understanding of this fascinating painting, and can yield many further insights in the future when specific elements will be studied in more detail. Moreover, the comparison with the five Haarlem civic guard portraits proved very fruitful in distinguishing between Hals's and Codde's contributions to the contested *The Meagre Company*, thus resolving earlier contradictions in its interpretation.

Appendixes

Appendix 1

1636, 19 March—FORMAL NOTIFICATION AND SUMMONS FROM REYNIER REAEL DEMANDING THAT FRANS HALS COMPLETE THE CIVIC GUARD PORTRAIT KNOWN AS *'THE MEAGRE COMPANY'*

Op huyden den 19en martii anno 1636 compareerden voor mij Frans Bruijningh etc. ter presentie etc., d'edele Reynier Reael, out schepen ende capiteyn van een vaendel burgers binnen deeser stede, ende Cornelis Michielsen Blaeuw, sijn luytenant, voor henselven ende vervangende d'andere officiren van 't voorsz. vaendel. Ende verclaerden hoe dat sij comparanten al drye jaeren geleden Frans Halsch, schilder tot Haerlem woonachtich, hadden aenbesteet seecker stuck schilderije, te weeten de conterfeytsels van alle de officiren van't voorsz. vaendel. Ende hoewel de voors. Frans Halsch behoort hadde't voors. stuck schilderije al over langh volcoomentlijck affgeschildert ende affgedaen te hebben, gelijck hij sulcx op Sint Jan lestleeden aennam ende belooffde te doen ende te presteren, soo is hij van sulcx te doen tot noch toe wel onbehoorlijck in gebreecke gebleeven, nietteegenstaende verscheyden interpellatien, soo mondelingh als schriftelijck daeromme aen hem gedaen, hebbende't voors. stuck schilderije maer voor een gedeelte gedaen, streckende't selve tot schaede, intresse ende naedeel van de comparanten, die daeromme alsnoch mits deesen versoecken, dat de voors. Frans Halsch hem binnen veerthien daegen nae de insinuatie deses sal hebben te vervoegen binnen deeser stede omme't voors. stuck ende't gene daeraen noch ontbreect ende resteert voorts aff te schilderijen, in behoorlijcke forme, off bij faute van dien, dat de comparanten verstaen sullen alsdan niet langer in den voors. Frans Halsch gehouden te sijn, maer't selve stuck alhier door een ander goet meester sullen laeten voltrecken, gereserveert de comparanten actie nopende de penningen die op't voors. stuck alreede betaelt sijn, daervan expresselijck protesterende mits deesen, alsmede van allen costen, schaeden ende intressen geleeden ende te lijden tottet uyteynde van de saecke toe.

Authoriserende voorts alle notarissen ende publycque persoonen, ende bijsonder dengeenen hiertoe versocht sijnde, omme't gunt voors. is den voors. Frans Halsch te insinueren, ende hen comparanten daervan te leeveren acten, een oft meer, in debita forma. Gedaen binnen der voors. stede van Amsterdamme ter presentie van Jacob Bruijningh, mede notaris publycq, ende Anthony Meerhout, getuygen hiertoe versocht.

[Signed] *Quod attestor ego F. Bruijningh, Nots. publ.1636*
Municipal Archive Amsterdam, NA 833 (notaris Frans Bruijningh), f°s 110v°-111.

Appendix 2

1636, 20 March—ANSWER FROM FRANS HALS TO THE SUMMONS FROM REYNIER REAEL REGARDING THE PROGRESS OF HIS WORK ON THE COMPLETION OF THE CIVIC GUARD PORTRAIT KNOWN AS *'THE MEAGRE COMPANY'*

Op huyden den 20en martii 1600 zes ende dertich, soo hebbe ick Egbert van Bosvelt, secretaris van de weescamer der stadt Haerlem ende Notaris publycq bij den Hove van Hollant geadmitteert, mij metten ondergeschreven getuygen ten versoucke van den edele Dirck Willemsz Abbas van wegen den edele heeren comparanten in de aengehechte acte genoemt, gevonden aen den persoon van Mr. Frans Hals, schilder (die sieckelyck aen een quaet been te bedde lach) ende hebbe hem de voors. acte van de 19en deser voorgelesen, ende aen hem antwoort versocht. Die mij notaris ten antwoorde gaff : dat hij't stuck wel aengenomen hadde te maecken, nyet dan tot Amstelredamme maer tot Haerlem; dan naederhant, alhoewel nochtans ongehouden, bewillicht dat hij de troinges tot Aemstelredamme soude beginnen ende tot Haerlem voorts opmaecken, gelijck hij oock al hadde begonnen te doen ende oock volbracht soude hebben, indyen hij de persoonen bijeen hadde connen crijgen, ende daertoe nyet connen geraecken, soodat hij daeromme in zijn huys veel versuympt ende tot Amstelredamme in de herberge veel verteert heeft in plaetse dat geseyt was dat men hem soude defroyeren.

Nyettemin, omme te presteren dat hij aengenomen ende belooft heeft, es als noch tevreden, indyen de persoonen tot Haerlem believen te comen, dat hij't werk datelyck bij der hant nemen ende sonder vertouven affmaecken sal ende zijn eere daerin betrachten, ende dat het met meerder lust tot Haerlem als tot Amsterdam sal gedaen werden, overmits hij dan binnenshuys ende bij zijn volck zijnde't oge oock daer op mach hebben. Aldus gedaen t'zijnen huyse op't Groote Heyligelant in presentie van Adriaen van Bosvelt, deurwaerder van den voors. Hove ende Jacob Maertensz, beyde wonende binnen Haerlem als getuygen van gelove tot kennisse van desen gerequireert.

[signed] *d'oirconde A. van Bosvelt, Jacob Maertens, E. van Bosvelt Nots. publ.*
Municipal Archive Haarlem, NA 63 (notaris Egbert van Bosvelt), f° 8or°.

Appendix 3

1636, 29 April—REYNIER REAEL'S REBUTTAL OF FRANS HALS'S EXPLANATION FOR THE DELAY IN COMPLETING THE CIVIC GUARD PORTRAIT KNOWN AS *'THE MEAGRE COMPANY'*

Op huyden den XXIX aprilis anno 1636 compareeden voor mij Frans Bruijningh etc., ter presentie etc. d'edele Reynier Reael, out schepen ende capiteyn van een vaendel burgers binnen deeser stede, ende Cornelis Michielsen Blaeuw, sijn luytenant, voor henselven ende vervangende d'andere officiren van't voors. vaendel, ende verclaerden, dat sij gesien hebbende de frivole ende onwaerachtige andtwoorde bij mr. Frans Halsch, schilder tot Haerlem, gedaen, op haerluyder verclaeringe, insinuatie ende proteste, seggen bij replycq, dat sij eerst metten voorn. Frans Halsch geaccordeert ende verdraegen sijn, dat hij de troniges alhier ter stede soude beginnen ende tot Haerlem voorts opmaecken, daervooren hij soude hebben ende genieten tsestich guldens van yeder personagie, dan daernae metten voorn. Frans Halsch naeder accorderende soude van yeder personagie oft conterfeytsel hebben ses guldens meerder, te weeten sesentsestich guldens, mits dat hij daertegens weeder alhier ter stede, ende niet tot Haerlem, de personagien, soowel van lichaemen als troniges, ende sulcx als behoort soude schilderen ende volcoomentlijck opmaecken, gelijck hij oock alreede eenige personagien hier ter stede alsoo heeft beginnen te doen; soo ist, dat sij comparanten alsnoch versoecken dat den voorn. Frans Halsch hem binnen den tijdt van thien daegen nae de insinuatie deeses sal hebben te vervoegen binnen deeser stede, omme't voors. stuck schilderije voorts te vervolgen ende naer behooren op te maecken ende voltrecken. Op welck versoeck den voornoemde Frans Halsch sijn ronde verclaeringe van jae oft neen, sal hebben te doen, opdat men mach weeten, waernae men sich sal hebben te reguleren, ende dit alles onvermindert haerluyder voirige gedaene verclaeringe ende proteste, daerbij de comparanten alsnoch sijn persisterende. Authoriserende voorts alle notarissen ende publycque persoonen, ende bijsonder dengeenen hiertoe versocht sijnde, omme 't gunt voors. is den voorn. Frans Halsch te insinueeren, ende hen comparanten daervan te leeveren acten, een oft meer, in debita forma. Gedaen binnen der voors. stede van Amsterdamme ter presentie van Jacob Bruijningh, mede notaris publycq, ende Anthony Meerhout, getuygen hiertoe versocht.

[signed] *Quod attestor ego F. Bruijningh, Nots. publ. 1636*
Municipal Archive Amsteram, NA 833 (notaris Frans Bruijningh), f°s 143 v°-144 v°.

Appendix 4

1636, 26 juli—SECOND ANSWER FROM FRANS HALS REGARDING THE SUMMONS FROM REYNIER REAEL CONCERNING THE PROGRESS OF HIS WORK ON THE CIVIC GUARD PORTRAIT KNOWN AS *'THE MEAGRE COMPANY'*

Op huyden den 26 julii 1636 soo hebbe ick Jacob van Bosvelt openbaer notaris bij den Hove van Hollant geadmitteert binnen de stadt Haerlem residerende mij metten ondergeschreven getuygen ten versoucke van den edele Pieter Pietersz provoost van de burgerije tot Amsterdam, van wegen de edele heeren comparanten in de geannexeerde acte genomineert gevonden ten huyse en aen den persoon van Mr. Frans Hals schilder alhier ter stede, ende hebbe hem d'voorsz. acte wesende gedateert den 29 aprilis 1636 lestleden voorgelesen ende aen hem sijne ronde verclaringe van jae ofte neen daerop versocht. Die mij notaris ten antwoorde gaff dat hij persisteerden bij de antwoorde dien hij Egbert van Bosvelt notaris binnen Haerlem den 20 martii voorleden achtervolgende sijne insinuatie aen hem geinsinueerde vanwegen de comparanten geexpresseert in den aengehechte acte gedaen gegeven heeft. Voegende dien onvermindert daerbij dat hij tevreden es het stuck schilderije in denselve acte verhaelt datelijck van Amsterdam te haelen ende tot sijnen huyse te brengen, omme bij hem de ongeschilderde clederen aldaer eerst opgemaeckt te werden. Ende d'selve gedaen sijnde, d'troinges van de goetwillige personen, van hier tot Haerlem te comen tot sijnen huyse door dien niemant daerbij om den corten tijt en es geinteresseert ofte oock niet vertrout dat daer ymant tegens es, te schilderen ; ende oock soo het mochte gebeuren dat daer onder ses ofte seven onwillige, ofte die het haer niet gelegen en soude mogen comen mochten sijn dat hij om alle voldoens wille d'selve ses ofte seven personen, het voors. stuck soo nae voltoyt hebbende, tot Amsterdam schilderen ende 't selve stuck werxs dan naer behooren voltrecken sall, waervan de voorsz. Pieter Pietersz vanwegen als boven versocht heeft dese acte die gepasseert es binnen der voorsz. stadt Haerlem in presentie van Barent Deteringh ende Vechter Hasewindius inwoonders deser stadt als getuygen hiertoe versocht.

[signed] *V. Hasewindius, B. Deteringh, 23/1636 d'oirconde Jacob van Bosvelt, Nots. publ. 1636*

Municipal Archive Haarlem, NA 165, (notaris Jacob van Bosvelt), f° 265 r°.

Based on the transciptions by Levy-van Halm in (Slive (ed.) 1989-90), Hals Docs. 73, 74, 75 & 78.

References

Abrahamse, L.: Zo goed als nieuw. TOON. Magazine. **4**, 45–46 (2018)

Alfeld, M., Pedroso, J.V., Van Eikema Hommes, M., Van Der Snickt, G., Tauber, G., Blaas, J., Haschke, M., Erler, K., Dik, J., Janssens, K.: A mobile instrument for in situ scanning macro-XRF investigation of historical paintings. J. Anal. At. Spectrom. **28**(5), 760–767 (2013). https://doi.org/10.1039/c3ja30341a

Ampzing, S.: Beschryvinge ende lof der Stad Haerlem in Holland. A. Rooman, Haarlem (1628)

Angel, P.: Lof der schilderkonst. Willem Christaens, Leiden (1642) (facsimile edn. Davaco Publishers, Utrecht/Soest 1969)

Atkins, C.D.M.: Frans Hals's virtuoso brushwork. Ned. Kunsthist. Jaarb. **54**, 280–307 (2003)

Atkins, C.D.M.: The Signature Style of Frans Hals: Painting, Subjectivity, and the Market in Early Modernity. Amsterdam University Press, Amsterdam (2012)

Atkins, C.D.M.: Frans Hals in Amsterdam and his impact on Rembrandt. In: Tummers, A. (ed.) Frans Hals: Eye to Eye with Rembrandt, Rubens and Titian, exh. cat., pp. 55–70. Frans Hals Museum/NAI010 Publishers, Haarlem/Rotterdam (2013)

Beurs, W.: De groote waereld in't kleen geschildert, of schilderagtig tafereel van 's weerelds schilderyen, kortelijk vervat in ses boeken. Verklarende de hooftverwen, haare verscheide mengelinge in oly, en der zelver gebruik.... Johannes en Gillis Janssonius van Waesberge, Amsterdam (1692)

Bidloo, G.: Mengel Poëzy, met konstplaaten versiert. Joh. Arnold Langerak, Leiden (1719)

Bijl, M.: The meagre company and Frans Hals's working method. In: Slive, S. (ed.) Frans Hals, exh. cat., pp. 103–108. National Gallery of Art/Royal Academy of Arts/Frans Hals Museum, Washington/London/Haarlem (1989–1990)

Bikker, J. (ed.): Dutch Paintings of the Seventeenth Century in the Rijksmuseum Amsterdam, vol. 1: Artists born between 1570 and 1600. Nieuw Amsterdam, Amsterdam (2007)

Borchert, T.-H., Ainsworth, M.W., Van der Glas, M., Van Calster, P.: De portretten van Memling, exh. cat. Ludion, Gent (2005)

Bredius, A.: Die Meisterwerke des Rijksmuseum zu Amsterdam. Hanfstaengl, Munich (1890)

Bredius, A.: De Geschiedenis van een Schutterstuk. Oud. Holl. **31**(1), 81–84 (1913). https://doi.org/10.1163/187501713X00115

Brenninkmeyer-de Rooy, B.: Theorie van de kunst. In: Haak, B. (ed.) Hollandse schilders in Gouden Eeuw, pp. 60–70. Meulenhoff/Landshoff, Amsterdam (1984)

Carasso-Kok, M., Knevel, P., Levy-Van Halm, K., Van den Berg-Hoogterp, L.E., Ekkart, R.E.O., Tümpel, C., Kist, J.B.: Schutters in Holland: Kracht en zenuwen van de stad, exh. cat. Frans Hals Museum, Haarlem (1988)

De Bie, C.: Het gvlden cabinet vande edel vry schilder const: inhovdende den lof vande vermarste schilders, architectē, beldthōwers, ende plaetsnyders van dese eevw. Ian Meyssens, Antwerp (1661)

De Clippel, K., Vermeylen, F.: Frans Hals and Flemish artists: dialogue or one-way street? In: Tummers, A. (ed.) Frans Hals: Eye to Eye with Rembrandt, Rubens and Titian, exh. cat., pp. 41–54. Frans Hals/NAI010 Publishers, Haarlem/Rotterdam (2013)

De Gheyn, J.: Wapenhandelinghe van Roers, Musquetten ende Spiessen. Robbert de Baudous, The Hague (1607)

De Grebber, P.: Regulen welke by een goet schilder en teykenaer geobserveert en achtervolght moeten werden: tesamen ghestelt tot lust van de leergierighe discipelen. Pieter Casteleyn, Haarlem (1649)

De Lairesse, G.: Het Groot Schilderboek, 2 vols. Erfgenamen van Willem de Coup, Amsterdam (1707)

De Monconys, B.: Journal Des Voyages de Monsieur de Monconys. Horace Boissat and George Remeus, Lyon (1665–1666)

De Vries, L.: Gerard de Lairesse: the critical vocabulary of an art theorist. Oud. Holl. **117**(1/2), 79–98 (2004). https://doi.org/10.1163/187501704X00296

Derks, K., Van Der Snickt, G., Legrand, S., Van Der Stighelen, K., Janssens, K.: The dark halo technique in the oeuvre of Michael Sweerts and other Flemish and Dutch baroque painters. A 17th c. empirical solution to mitigate the optical 'simultaneous contrast' effect? Herit. Sci. **10**(5), 1–16 (2022). https://doi.org/10.1186/s40494-021-00634-w

Erftemeijer, A., Fuhri-Snethlage, H., Kohler, N.: Hoogtepunten. Frans Hals Museum/Die Keure, Haarlem/Bruges (2014)

Gibson, W.S.: Pleasant Places. The Rustic Landscape from Brueghel to Ruisdael. University of California Press, Berkeley (2000)

Gifford, M., Glinsman, L.D.: Collective style and personal manner: materials and techniques of high life genre painting. In: Waiboer, A.E. (ed.) Vermeer and the Masters of Genre Paintings: Inspiration and Rivalry, pp. 65–83. Yale University Press, New Haven/London (2017)

Grimm, C.: Frans Hals: The Complete Work (1989). Translated from German by Riehle, J. H.N. Abrams, New York (1990)

Groen, K., Hendriks, E.: Frans Hals: a technical examination. In: Slive, S. (ed.) Frans Hals, exh. cat, pp. 109–127. National Gallery of Art/Royal Academy of Arts/Frans Hals Museum, Washington/London/Haarlem (1989–1990)

Hendriks, E.: Johannes Cornelisz. Verspronck: the technique of a seventeenth century Haarlem portraitist. In: Hermens, E., Ouwerkerk, A., Costaras., N. (eds.) Looking through Paintings: the Study of Painting Techniques and Materials in Support of Art Historical Research. Leids Kunsthistorisch Jaarboek, vol. 11, pp. 227–267. Prom, Baarn (1998)

Hendriks, E., Levy-Van Halm, K., in collaboration with Van Asperen De Boer, J.R.J.: Report Concerning a Preliminary Technical Investigation of Paintings exhibited during the Frans Hals Exhibition, held from May 11 to July 22 1990 in the Frans Hals Museum Haarlem. (1991)

Hendriks, E., Van Eikema Hommes, M., Levy-Van Halm, K.: Indigo used in the Haarlem Civic Guard group portraits by Frans Hals. Stud. Conserv. **43**(1), 166–170 (1998). https://doi.org/10.1179/sic.1998.43.Supplement-1.166
Henny, X.: Hoe kwamen de Rotterdamse Schilders aan hun verf. In: Schadee, N. (ed.) Rotterdamse meesters uit de Gouden eeuw, exh. cat., pp. 43–53. Historisch Museum/Waanders, Rotterdam/Zwolle (1994)
Honig, E.: The beholder as work of art: a study in the location of value in seventeenth-century Flemish painting. Ned. Kunsthist. Jaarb. **46**(1), 252–297 (1995). https://doi.org/10.1163/22145966-90000133
Houbraken, A.: De groote schouburgh der Nederlantsche konstschilders en schilderessen, 3 vols. B.M. Israël, Amsterdam (1718–1721)
Kirby-Talley, M.: Portrait Painting in England: Studies in the Technical Literature Before 1700. Paul Mellon Centre for Studies in British art, London (1981)
Knevel, P.: Burgers in het geweer: de schutterijen in Holland, 1550–1700. Historische Vereniging Holland/Verloren, Hilversum (1994)
Köhler, N., Levy- Van Halm, K.: Frans Hals: Schuttersstukken. Schwartz/SDU, Maarssen/The Hague (1990)
Krekel, C., Burmester, A.: Pharmacy price-lists as a new type of documentary source for research into historical artist's materials: the Münchner Taxenproject. In: Kirby, J. (ed.) Dyes in History and Archeology 19: Papers Presented at the 19th Meeting, Edinburgh 2000, pp. 261–434. Archetype Publications, London (2003)
Krekeler, A., Smeenk-Metz, E., Benders, Z., Van Der Laar, M.: Consistent choices: a technical study of Johannes Cornelisz Verspronck's portraits in the Rijksmuseum. Rijksmus. Bull. **62**(1), 2–23 (2014). https://doi.org/10.52476/trb.9839
Kurtz, G.H.: Het Proveniershuis te Haarlem. Vereniging Haerlem, Haarlem (1979)
Languri, G. M.: Molecular studies of Asphalt, Mummy and Kassel earth pigments: their characterisation, identification and effect on the drying of traditional oil paint. Dissertation, University of Amsterdam (2004)
Levy-Van Halm, K.: De Haarlemse Schutterstukken. In: Carasso Kok, M., Knevel, P., Levy Van Halm, K., Van den Berg Hoogterp, L.E., van den Ekkart, R.E.O., Tümpel, C., Kist, J.B. (eds.) Schutters in Holland: Kracht en zenuwen van de stad, exh. cat., pp. 104–123. Frans Hals Museum, Haarlem (1988)
Liedtke, W.: Frans Hals: style and substance. Metrop. Mus. Art. Bull. **69**(1), 1–48 (2011)
Middelkoop, N.E.: Schutters, gildebroeders, regenten en regentessen: Het Amsterdamse corporatiestuk 1525–1850. Dissertation, University of Amsterdam (2019)
Miedema, H.: Over kwaliteitsvoorschriften in het St. Lucasgilde: over "doodverf". Oud Holl. **101**(3), 141–147 (1987). https://doi.org/10.1163/187501787X00411
Miedema, H.: De ontwikkeling van de kunsttheorie in de Hollandse Gouden Eeuw II. Karel van Mander. Oud. Holl. **126**(1), 17–30 (2013). https://doi.org/10.1163/18750176-90000028
Newman, A.D., Nijkamp, L. (eds.): Many Antwerp Hands: Collaborations in Netherlandish Art. Harvey Miller Publishers, London/Turnhout (2021)
Nicolaus, K.: Het schilderij: materiaal, techniek, behoud, 2nd edn. Cantecleer, De Bilt (1990)
Niessen, J.: Cornelis van Haarlem, 1562–1638, exh. cat. Frans Hals Museum/Nai010 Uitgevers, Haarlem/Rotterdam (2012–2013)
Packer, L., Roy, A.: The making of the laughing cavalier. In: Packer, L. (ed.) Frans Hals: the Male Portrait, exh. cat., pp. 81–91. Philip Wilson Publishers, London (2021–2022)
Rosen, J.: Pieter Codde (1599–1678): Catalogue Raisonnè. Cambridge Scholars, Newcastle upon Tyne (2020)
Rosenberg, A. (ed.): Rubensbriefe. E.A. Seemann, Leipzig (1881)
Rowlands, J.K.: Holbein: The Paintings of Hans Holbein the Younger, complete edn. Phaidon, Oxford (1985)
Schaep Pietersz, G.: Begin, vervolg ende verandering vande schutterien, vaendelen der burgeren ende wijcken.... (1630 1653)

Schama, S.: 2001 [1987], Overvloed en onbehagen: De Nederlandse cultuur in de Gouden Eeuw (1987). Translated from the English edition. Contact, Amsterdam (2001)

Scheltema, P.: De schilderijen in de drie doelens, te Amsterdam, beschreven door G. Schaep 1653. In: Scheltema, P. (ed.) Aemstel's Oudheid of gedenkwaardigheden van Amsterdam, vol. 7, pp. 121–141. Ten Brink & De Vries, Amsterdam (1885)

Schrevelius, T.H.: ofte om beter te seggen, de eerste stichtinghe der stadt Haerlem…. Thomas Fonteyn, Haarlem (1648)

Six, J.: Opmerkingen omtrent eenige meesterwerken in's Rijks Museum. Oud. Holl. **11**, 96–104 (1893)

Slive, S.: Frans Hals, 3 vols. Phaidon, London (1970–1974)

Slive, S. (ed.) Frans Hals, exh. cat., National Gallery of Art/Royal Academy of Arts/Frans Hals Museum, Washington/London/Haarlem (1989–1990)

Slive, S.: Frans Hals, revised edn. Phaidon, London (2014)

Spring, M.: Pigments and color change in the paintings of Aelbert Cuyp. In: Wheelock Jr., A.K. (ed.) Aelbert Cuyp, exh. cat., pp. 64–71. Thames & Hudson/National Gallery of Art, New York/London/Washington DC (2001–2002)

Tamis, T.M.: Van twee handen geschildert: werkverdeling tussen schilders in de Nederlanden in de zestiende en zeventiende eeuw. Dissertation, University of Amsterdam (2016)

Taylor, P.: The concept of Houding in Dutch art theory. J. Warbg. Court. Inst. **55**, 210–232 (1992). https://doi.org/10.2307/751425

Tummers, A. (ed.): Celebrating in the Golden Age, exh. cat. Frans Hals Museum/NAI010 Publishers, Haarlem/Rotterdam (2011a)

Tummers, A.: The Eye of the Connoisseur: Authenticating Paintings by Rembrandt and his Contemporaries. Amsterdam University Press/Getty Publications, Amsterdam/Los Angeles (2011b)

Tummers, A.: Frans Hals: Frans Hals: rightly admired by the greatest painters. In: Tummers, A. (ed.) Frans Hals: Eye to Eye with Rembrandt, Rubens and Titian, exh. cat., pp. 13–40. Frans Hals Museum/NAI010 Publishers, Haarlem/Rotterdam (2013a)

Tummers, A. (ed.): Frans Hals: Eye to Eye with Rembrandt, Rubens and Titian, exh. cat Frans Hals Museum/NAI010 Publishers, Haarlem/Rotterdam (2013b)

Tummers, A., Gration, J.: Eye to eye with Frans Hals: the location and reception of some of his masterpieces. In: Tummers, A. (ed.) Frans Hals: Eye to Eye with Rembrandt, Rubens and Titian, exh. cat., pp. 71–93. Frans Hals/NAI010 Publishers, Haarlem/Rotterdam (2013)

Tummers, A., Wallert, A., Hartwieg, B., Kleinert, K., Laurenze-Landsberg, C., Dik, J., Groves, R., Anisimov, A., Papadakis, V., Erdmann, R.: Supplementing the eye the technical analysis of Frans Hals's paintings—I. Burlingt. Mag. **161**, 934–941 (2019)

Van Asch van Wijck, A.M.C.: De schut- of schuttengilden in Nederland. Ber. Hist. Gez. Utr. 1(2), 92–202 (1848) (part 1), 3(2), 1–171 (1851) (part 2)

Van Bueren, T.: "de beste Schilders van het gantsche Nederlandt" Karel van Mander en het Haarlemse cultuurbeleid 1603-1606. Oud. Holl. **105**(4), 291–305 (1991). https://doi.org/10.1163/187501791X00164

Van Bueren, T.: Tot lof van Haarlem. Het beleid van de stad Haarlem ten aanzien van de kunstwerken uit de geconfisqueerde geestelijke instellingen. Verloren, Hilversum (1993)

Van Dantzig, M.M.: Het "Corporaalschap van kapitein Reynier Reaal". Phoenix. **I**(10), 5–9 (1946)

Van de Graaf, J.A.: Betekenis en toepassing van 'loodwit' en 'schelpwit' in de 17e eeuwse Nederlandse schilderkunst. Bull. L'Institut R. Patrim. Artist. **4**, 198–201 (1961)

Van de Wetering, E.: Rembrandt: the Painter at Work. Amsterdam University Press, Amsterdam (1997)

Van Dyck, J.: Kunst en historiekundige beschryving en aanmerkingen over alle de schilderyen op het stadhuis te Amsteldam. J. De Jongh, Amsterdam (1758)

Van Eeghen, I.H.: Pieter Codde en Frans Hals. Amstelodamum. **61**, 137–141 (1974)

Van Eikema Hommes, M.: Changing Pictures: Discoloration in 15th–17th-Century Oil Paintings. Archetype Publications, London (2004)

Van Gogh, V.: Verzamelde brieven van Vincent van Gogh, 4 vols. Wereldbibliotheek, Amsterdam/Antwerp (1953)
Van Hoogstraten, S.: Inleyding tot de hooge schoole der schilderkonst, anders de zichtbaere werelt: verdeelt in negen leerwinkels, yder bestiert door eene der zanggodinnen. Francois van Hoogstraeten, Rotterdam (1678)
Van Hout, N.: On Dead Colour. KMSKA, Antwerp (2008)
Van Loon, A., Speleers, L.: The use of blue and green verditer in green colours in the mid-seventeenth-century paintings of the Oranjezaal, The Hague. In: Spring, M. (ed.) Studying Old Master Paintings, Technology and Practice, pp. 260–286. Archetype in association with The National Gallery, London (2011)
Van Mander, K.: Het schilder-boeck waer in voor eerst de leer-lustighe ieught den grondt der edel vry Schilder-Const in verscheyden deelen wort voorghedraghen. Paschier van Wesbuch boeck-vercooper, Haarlem (1604)
Van Thiel-Stroman, I.: Frans Franchoisz Hals. In: Köhler, N., Levy Van Halm, K., Runia, E., Van Thiel, P., Van Thiel-Stroman, I. (eds.) Painting in Haarlem, 1500–1850: the Collection of the Frans Hals Museum, pp. 178–184. Ludion, Gent (2006)
Van de Wetering: E.: Rembrandt's method—technique in the service of illusion. In: Brown, C., Kelch, J., Pieter, J.J. (eds.) Rembrandt: the Master and his Workshop, pp. 12–39. Yale University Press, New Haven (1991)
Ward, G.W.R. (ed.): The Grove Encyclopedia of Materials and Techniques in Art. Oxford University Press, Oxford/New York (2008)
Weststeijn, T.: The concepts of affectus, Enargeia and Ornatus in Samuel van Hoogstraten's judgement of his master. In: Van den Doel, M., Van de Wetering, E. (eds.) The Learned Eye: Regarding Art, Theory and the Artist's Reputation: Essays for Ernst van de Wetering, pp. 111–130. Amsterdam University Press, Amsterdam (2005)

Part III
The Digitally Enhanced Eye

Chapter 7
Connoisseurship and Smart Tools

Robert G. Erdmann, Anna Tummers, and Marie-Noëlle Grison

Abstract As the Frans Hals & Co case study (see Chap. 6) comprises six huge paintings and the new techniques used to study these yield enormous amounts of data, it soon became clear that data science and computer tools would make a substantial difference in facilitating and enhancing the analysis. It was the reason to start the Seed Money Project *21st Century Connoisseurship: Smart Tools for the Analysis of Seventeenth-Century Paintings* (2018–2022), funded by the Netherlands Institute for Conservation, Art and Science (NICAS). Building on very high resolution photographs and advanced technical analyses, online computer tools are causing a turning point in the early twenty-first century. For the first time in history, it has become easier to study paintings online than offline. This chapter focuses on the different computer tools that have been developed and adjusted to the study the large amount of data generated in the context of the *Frans Hals or not Frans Hals* project and discusses their respective applications.

Parts of this chapter were first published in Tummers, A. and Erdmann, R.: The Eye Versus Chemistry? From Twentieth to Twenty-First Century Connoisseurship. In: Colombini, M.P., Degano, I., Nevin, A. (eds). Analytical Chemistry for the Study of Paintings and the Detection of Forgeries, pp. 3–45. Springer International Publishing, Cham (2022).

R. G. Erdmann (✉)
University of Amsterdam, Amsterdam, The Netherlands
e-mail: r.g.erdmann@uva.nl

A. Tummers (✉)
Ghent University, Ghent, Belgium
e-mail: Anna.Tummers@UGent.be

M.-N. Grison (✉)
KU Leuven, Leuven, Belgium
e-mail: mn.grison@gmail.com

A. Tummers, R. G. Erdmann, *Frans Hals or not Frans Hals*, Cultural Heritage Science, https://doi.org/10.1007/978-3-031-59489-2_7

7.1 The Data Deluge and the Need for Smart Tools

While only a small fraction of the paintings of well-known masters have been subject to in-depth technical research, the data gathered is nevertheless substantial and increasing rapidly. Especially, the data collected with advanced photography and scanning methods such as MA-XRF and HI/RIS is considerable. We have entered a digital age that comes with entirely new challenges and potential. How to manage and properly interpret vast amounts of data? In what way and to what extent can digital techniques facilitate art authentication?

In the early twenty-first century, various computer scientists have been developing computer programs with the intention of facilitating the attribution process, focusing in particular on brushstroke analysis. For example, at the end of 2004, a team from Dartmouth College in Hannover, New Hampshire, developed a method to analyse pen lines and brushstrokes, based on an algorithm that proved useful in court for the identification of manipulated photographs (Lyu et al. 2004). With the aid of so-called 'wavelets', Siwei Lyu, Daniel Rockmore and Hany Farid were able to isolate pen- and brushstrokes and analyse their direction, scale and relation to surrounding strokes. They assumed that every artist had a unique way of applying ink and paint to the surface, which results in a kind of virtual signature that can be analysed by a computer regardless of the subject matter. Since subject matter is however likely to affect the variability of the strokes, they only compared works with similar subjects.

The program they developed successfully confirmed existing attributions in a training set, and so did several other programs (Johnson et al. 2008; Hendriks and Hughes 2009; Li et al. 2012; Van Noord et al. 2015; Ji et al. 2021). However, thus far such programs have not been used to make new attributions or to confirm or exclude attributions in court—which is related to the complexity of such decisions. Complicating factors include, among other things, the condition of the painting studied (for example, old master paintings virtually always include restorations and later retouches which 'muddy' the data). Also, as we have seen, artists sometimes deliberately varied their styles and techniques, and/or used assistants in the execution of their works, which further complicates the analysis. In short, computer programs that can make decisions in the expert's place remain hitherto elusive. As of yet, it is unclear if these can be realised in the near future; the need for contextual knowledge may prove too formidable a hurdle.

However, digital techniques have shown great potential in facilitating in-depth comparisons. Based on Erdmann's work in this area over the past 10 years, and the NICAS project *21st Century Connoisseurship*, we will discuss some of our main conclusions regarding the merits, challenges and potential of a number of digital techniques aimed to facilitate comparison below. These pioneering tools are currently mostly in use at the Rijksmuseum; the aim, however, is to open these up for wider use in the future.

7.1.1 Behind-the-Scenes: Infrastructure

Given the situation in which we now find ourselves, with an overabundance of data coming from a variety of sources, with an emphasis on imaging data, there is a strong need to process the data in a way that promotes honest comparisons. These comparisons rely heavily on accuracy and consistency, since fundamentally, we must ensure that when the viewer sees differences between images of artworks it is because they are in fact different, and not due to differences arising from the capture or subsequent processing of the imaging data. Furthermore, given the importance of materiality for judging both the condition and the attribution of paintings, imaging should strive to capture as much of the materiality of the artworks as practically possible. The following principles, while not comprehensive, serve as prescriptive guidelines for enhancing the utility of digital tools for authentication purposes:

Consistent High-Resolution Imaging Modern digital cameras and lens systems are capable of capturing images with extremely high spatial sampling resolutions, so that even small artworks can be captured with many overlapping tiles of high-resolution photographs. The value of such photographs is immense since they reveal microstructural details that are hidden at normal resolutions: paint pigment particles, cracks, retouches, areas of abrasion, nuances of brushstrokes or pen lines, subtle or small-scale *pentimenti* (corrections) and other details of artistic technique, details of the support, and many other aspects of the materiality of the artwork. In the absence of an ability to physically inspect all relevant artworks side-by-side, consistently sharp and well-lighted digital photos are the best available option.

Consistent Colour Management and Processing Images collected by different photographers, with different light sources, or at different institutions will invariably display differences due to imaging technique. To the degree possible, these differences in equipment and technique should be minimised. Even so, differences will remain, but these variations can be attenuated by performing careful colour management during the processing of the raw photos. The variations in lighting and wavelength-dependent pixel sensitivity are lessened by always photographing a colorimetric standard (e.g., an XRite ColorChecker SG colour card) as part of a standard workflow, so that a colour profile can be made which will adjust the as-captured colours to their correct values. Imaging guidelines such as the Metamorfoze Guidelines[1] or the FADGI Guidelines[2] have proven useful in defining best practices and criteria for acceptable image quality.

[1] https://www.metamorfoze.nl/sites/default/files/documents/Metamorfoze_Preservation_Imaging_Guidelines_1.0.pdf, accessed 21 February 2024.

[2] http://www.digitizationguidelines.gov/guidelines/FADGI%20Federal%20%20Agencies%20Digital%20Guidelines%20Initiative-2016%20Final_rev1.pdf, accessed 21 February 2024.

Careful State-of-the-Art Processing The use of high-resolution imaging generally leads to a collection of images spanning the artwork, and the use of additional technical imaging modalities such as radiography, infrared reflectography, or reflectance imaging spectroscopy further leads to multiple images describing any given point on the object. Among the desirable features of such a system are the following: (a) it should respect the details of human colour perception since naive averaging of RGB pixel values does not result in perceptually averaged colours; (b) it should avoid any tears or duplications when assembling the component tiles in a single whole-artwork image; (c) it should use high-order interpolation kernels to avoid introducing blurring or ringing artefacts when performing the inevitable resampling of the images during the stitching; and (d) it should ensure subpixel precision when performing registration among the different imaging modalities, so that data from one imaging modality is fused with the data from another modality at the same physical location on the artwork. Erdmann (2016b) developed such a system as part of a comprehensive campaign of imaging and study of the oeuvre of Hieronymus Bosch, and this system is now in permanent use at the Rijksmuseum.

7.2 Curtain Viewer

Meaningful comparisons between artworks or between different areas of an artwork are essential to the expert's judgement. Even with a collection of consistent colour-managed high-resolution images, traditional image-editing tools such as Photoshop are ill-suited to making frictionless comparisons among many works or among different imaging modalities of a single work. The problem is exacerbated when the images themselves are very large; 20 μm/pixel resolution (1270 ppi) 16-bit colour imaging consumes 15 GB/m^2, so large-format paintings such as Hals' militia company portraits or Rembrandt's *Night Watch* consume hundreds of gigabytes each. Side-by-side comparisons of such artworks may then be practically impossible using standard image-editing software due to memory limitations. Furthermore, such an approach makes it very difficult to save a comparison for later review, and collaborative inspections are impractical.

In response to these difficulties, in 2012 Erdmann developed an internet-based viewer for very high-resolution images with an explicit design goal of enabling smooth and seamless comparisons between images. As with other web-based image viewers, the viewer, which he named the Curtain Viewer, utilises image pyramids so that a cascade of image resolutions is pre-computed and stored on the server as small tiles to enable immediate on-demand zooming and panning without the need to pre-load large images.

The Curtain Viewer enables comparisons using a variety of modes: A 'gallery mode' (Fig. 7.1) allows the user to juxtapose an unlimited number of viewing panes, each of which is unconstrained. Constraints can be placed among the panes in a 'sync mode' (Fig. 7.2), so that they zoom or pan the same way simultaneously. The way an artist depicts a detail is strongly dependent on the scale of the depiction, so

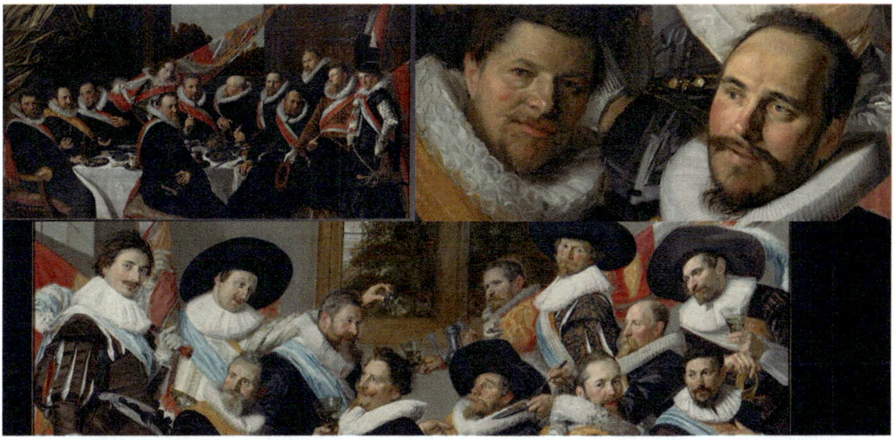

Fig. 7.1 Clockwise from top left: three militia company portraits from the Frans Hals Museum collection, *Banquet of the officers of the St. George's Civic Guard* (1616), *Banquet of the officers of the St. George's Civic Guard* (1627), and *Banquet of the officers of the Kloveniers militia* (1627), displayed using the 'gallery mode' of the Curtain Viewer

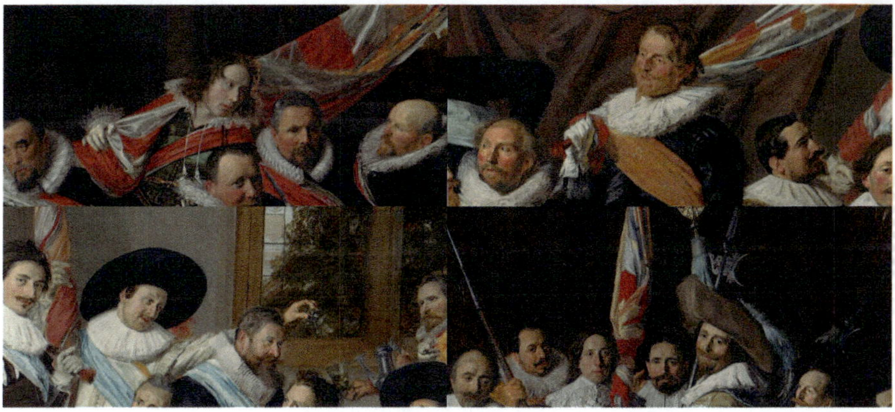

Fig. 7.2 Clockwise from top left: four militia company portraits from the Frans Hals Museum collection, *Banquet of the officers of the St. George's Civic Guard* (1616), *Banquet of the officers of the St. George's Civic Guard* (1627), *Meeting of officers and non-commissioned officers of the Kloveniers militia* (1633), and *Banquet of the officers of the Kloveniers militia* (1627), displayed using the 'sync mode' of the Curtain Viewer

this feature enables a user to zoom in to a pair of details in different artworks while ensuring that each is presented at the correct scale, and by synchronising their scales they can zoom out to compare their contexts or zoom in to compare their details without fear of being deceived by scale differences. The sync mode is also useful in comparing different image modalities, since it is often difficult, for example, to associate a feature in a radiograph with the same feature in a visible-light photograph.

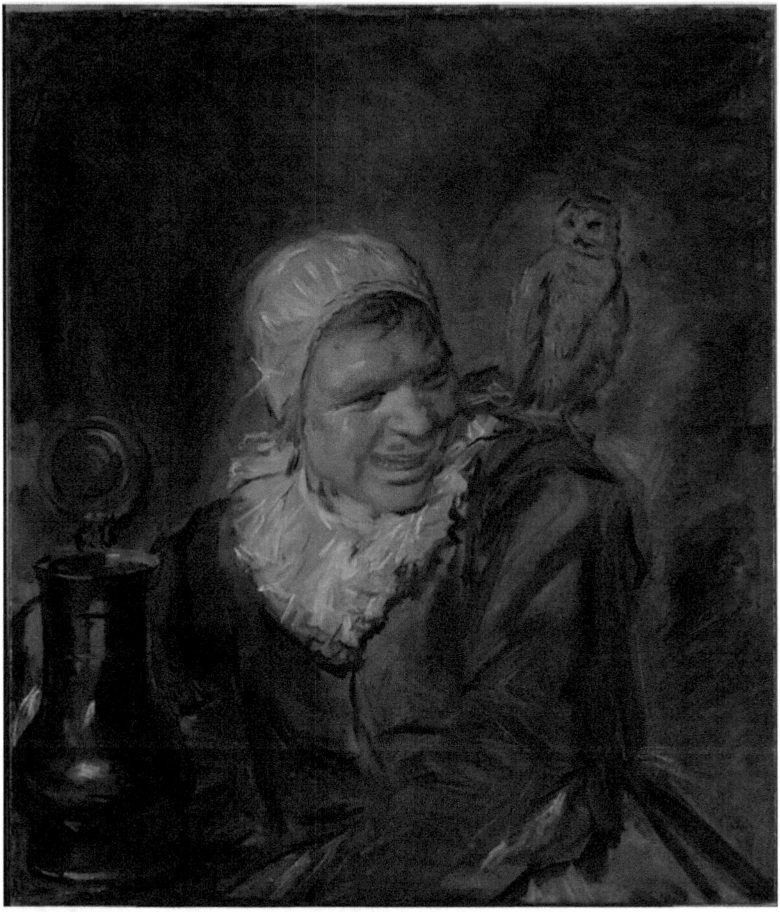

Fig. 7.3 A hybrid image of the *Malle Babbe* (1632–1635, Staatliche Museen zu Berlin, Gemäldegalerie) mixing the visible light photography and infrared reflectography in equal amounts with the 'fade mode' of the Curtain Viewer

A 'fade mode' enables an overlay of two or more panes, with relative opacities computed based on the position of the mouse. The fade mode can be used to create hybrid images such as those mixing visible-light photography and infrared reflectography (Fig. 7.3), and has also been used to simulate a moving light source attached to the mouse by fading among a collection of raking-light images collected with the light source at different positions.

The Curtain Viewer is named after the 'curtain mode', in which a single pane is split into multiple regions at the location of the mouse cursor, with a different imaging modality or artwork shown in each region. The movement of the mouse pointer thus gives an impression of 'pulling back the curtain' or of 'lifting the curtains' as it reveals additional imaging modalities (Fig. 7.4). The system is designed to enable the user to focus carefully on an area of interest on the artwork and to repeatedly

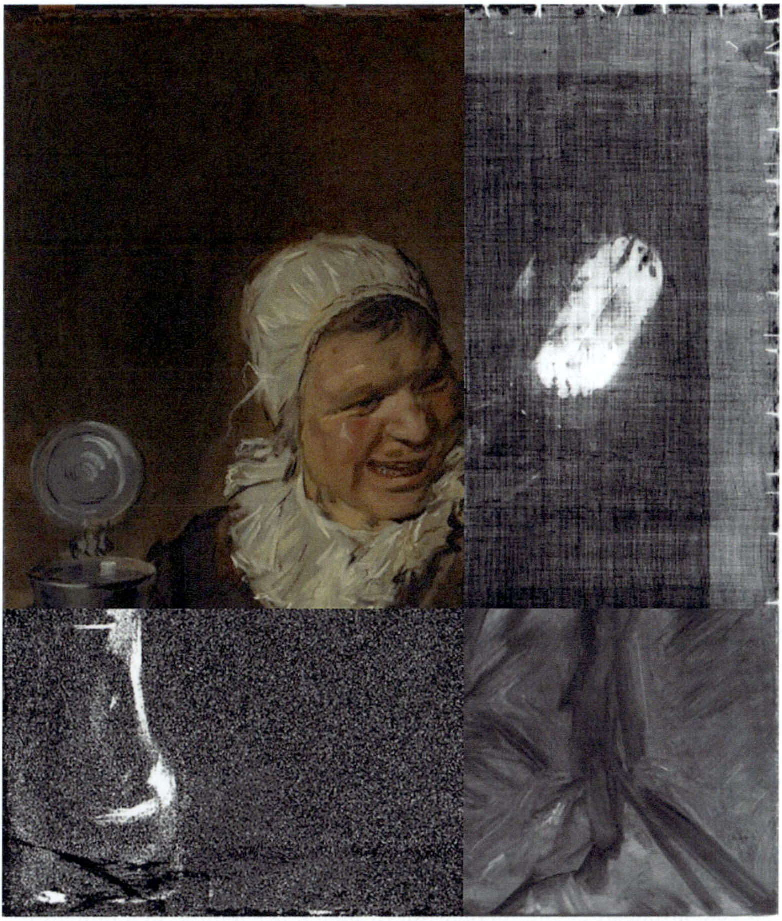

Fig. 7.4 Four different imaging modalities of the *Malle Babbe* (1632–1635, Staatliche Museen zu Berlin, Gemäldegalerie) are viewed simultaneously with the 'curtain mode' of the Curtain Viewer. Clockwise from top left: visible light photography, x-radiography, infrared reflectography, Cu-K MA-XRF map

brush over it to show the exact relation among the features revealed by different imaging modalities. A traditional side-by-side view has been used to make these kinds of comparisons, but it induces a kind of 'visual context switching' in which the user is forced to change focus from one location to another, diminishing the effectiveness of the comparison due to our limited visual memory. In contrast, the 'curtain mode' enables the user to remain fixed on the same location in an artwork while comparing the different modalities.

The avoidance of visual context switching is another one of the major design goals of the Curtain Viewer system. Unlike other image viewers, its interface is not cluttered with visually distracting overlays such as logos, viewer control buttons, magnification sliders, and the like. The user is thus able to focus exclusively on studying the art without the visual noise of user interface elements. Furthermore, switching between modes is always done in a smooth way, with image panes animating their positions and opacities gradually to promote object constancy; the user is never forced to break their focus to reestablish their bearings in a new viewing configuration.

The Curtain Viewer also features a system where every aspect of the view is encoded in the URL, enabling easy bookmarking of an exact configuration for later study or for sharing and collaboration. As a demonstration of the technology, every figure from the Bosch Catalogue Raisonné (Ilsink et al. 2016; Erdmann 2016a), is also presented online (Erdmann 2016b) using the Curtain Viewer, enabling readers to understand the exact context and details of every featured detail.

As the book figures show, the Curtain Viewer's URL scheme allows for the design of displays that rely heavily on what Tufte (1990) calls 'small multiples':

> At the heart of quantitative reasoning is a single question: Compared to what? Small multiple designs, multivariate and data bountiful, answer directly by visually enforcing comparisons of changes, of the differences among objects, of the scope of alternatives. For a wide range of problems in data presentation, small multiples are the best design solution.

These small multiple visualisations are crucial for connoisseurship since they ease and promote the fundamental act of comparison across scales, imaging modalities, and artworks.

7.3 Draper

The Curtain Viewer affords great freedom to configure complex views involving multiple artworks, multiple imaging modalities, and a variety of viewer modes in terms of synchronisation and display, all of which is captured in a human-readable URL. However, this flexibility comes at the price of complexity. The solution is an additional tool to help precisely configure a desired Curtain Viewer display, dubbed as the 'Draper' (one who makes curtains) by Erdmann. The tool consults a database of technical images for a specified artwork and finds those which are co-registered with each other, presenting the user with a menu of different technical images. The resulting images can be arranged into a desired configuration via a drag-and-drop interface, and additional Curtain Viewer options can be specified. A live viewer preview at the bottom allows immediate exploration and tweaking of the viewer parameters.

Thus it becomes basically frictionless for the expert to quickly answer complex questions about image collections that can easily occupy several terabytes on disk. For example, did Hals consistently use the rather costly red pigment vermillion for all the faces and hands of the 68 men he portrayed in his five prestigious large-scale civic guard group portraits? It also dramatically eases the process of documenting a judgement with supporting evidence.

7.4 Morelli's Vision

Facilitating fast and meaningful comparisons among small details from one or more artworks is the design goal of Erdmann's 'Morelli's Vision' technique. It derives its name in honour of Giovanni Morelli, an art historian who advocated for the careful study of small habitually-painted details to discern the characteristic 'handwriting' of an artist.[3] It is driven by a system of user- or computer-generated rectangular selections on artworks which are given a semantic tag such as 'ear' or 'hand'. The model hinges on a recent breakthrough in computer vision and machine learning: Contrastive Language–Image Pre-training (CLIP) (Radford et al. 2021). This new approach makes it possible to jointly embed images and text within a high-dimensional semantic space in order to map out their degree of similarity relative to one another. This ability of the CLIP model, which we use without further specialised training on our images, to perform this task arises from the training procedure, in which it learns how to pair images with their original captions from an enormous set of image-caption pairs taken from the internet. To succeed at this task, the network must simultaneously 'understand' both how to read images as well as English text. While the details of the process are beyond the scope of this chapter, the key point is that the way the model does this is by learning how to compute an appropriate location in a high-dimensional embedding space for both images and captions. During training, the network is rewarded when, in this embedding space, the closest image to a given caption is the one it was originally paired with. Similarly, the network is rewarded if the closest caption to a given image is the one it was originally paired with. This implies that the network learns how to organise images (and captions) semantically within the space. Thus, the original objective of the CLIP model of comparing captions to images indirectly induces a means of comparing images with each other. In other words, images that are nearby each other in the embedding space would be well-described by the same set of captions.

This high-level ability to compare images is the key functionality that we utilise in our CLIP-based Morelli tool. It takes as input square snippets from Hals' paintings that are tagged by the user according to what they depict, for example hands, noses, and lace. When these image tiles are fed into the CLIP model, an embedding is calculated for each: every tile is assigned a 768-dimensional coordinate within this space. The pairwise similarity between every tile and every other one is calculated using a cosine metric, so that for each we can calculate an ordering from most to least similar (according to the CLIP model). The display is web-based and interactive: when the user clicks on a particular tile, all the other individual tiles are dynamically sorted according to the pairwise similarity between the clicked image and all other images, placing the clicked tile first in the list with all the others sorted in decreasing order of their similarity to the clicked one.

The consequence for the expert is that they can see as many comparisons in a single field of view as possible (Fig. 7.5). Every detail can be easily compared with every other detail. The system also allows the selection of a subset of interesting

---

[3] On Morelli, see also Sect. 2.5.

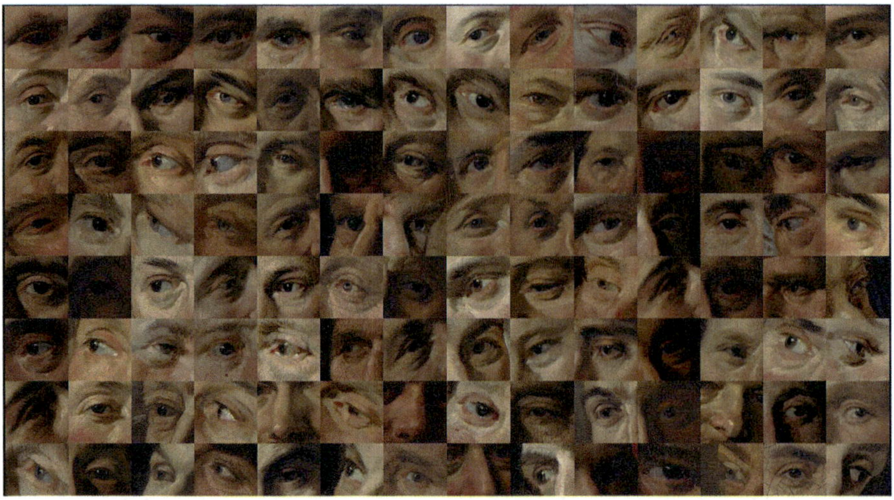

Fig. 7.5 All the tiles tagged as 'eyes' in Hals' militia company portraits

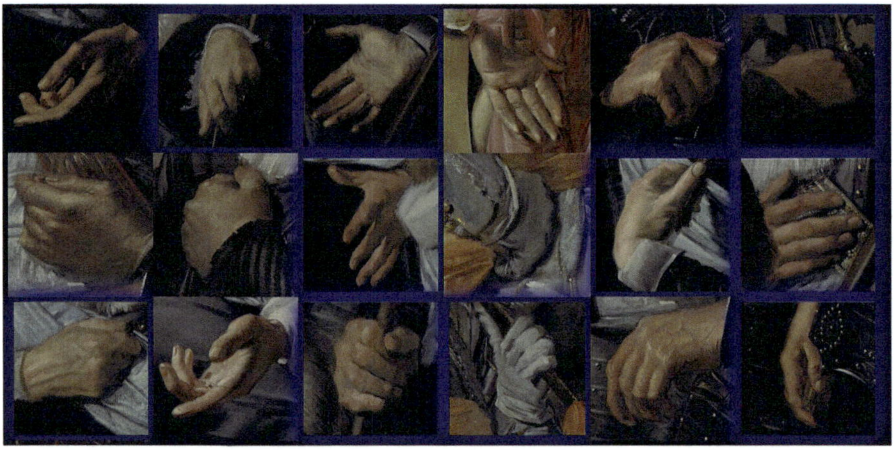

Fig. 7.6 From all the tiles tagged as 'hands' in *The Meagre Company*, the user selected a group for further study, outlined in blue

details and to instantly launch a Curtain Viewer in 'sync mode' to show the selected details side by side, zoomable and in high resolution. As an example of its use, consider its application to investigating attribution of *The Meagre Company* by studying the rendering of hands. The user has selected a set of distinctive hands (shown in Fig. 7.6, outlined in blue)—in this case, hands showing a bold brushwork that does not entirely resemble that of other hands—for further examination. They then launch this selection into a custom Curtain Viewer (shown in Fig. 7.7) that enables studying the details in context. This tool proves particularly helpful when comparing subtle features such as brushwork since scale is conserved across all studied regions when viewed in 'sync mode'.

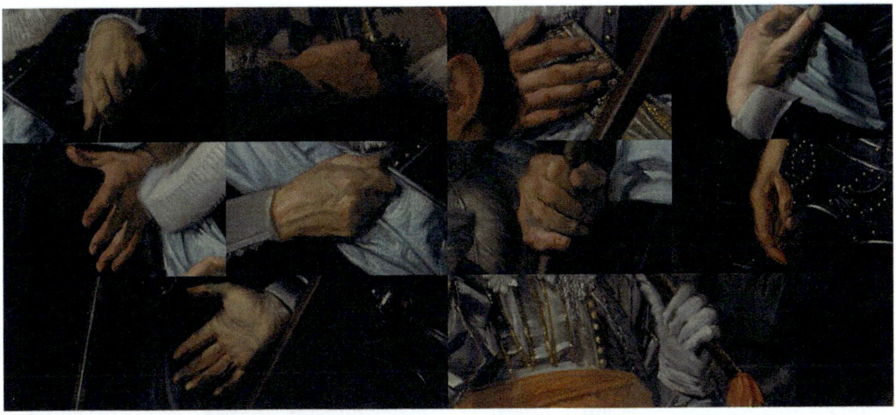

Fig. 7.7 Once the selection is completed, the user can launch it in a separate Curtain Viewer window in 'Sync' mode, thereby allowing seamless comparison between the elements, and dynamic zooming in and out of all views at once

7.5 PixelSwarm

The PixelSwarm tool is an online interactive tool to allow the user to draw insights from high-dimensional data arising from multimodal imaging of artworks. A visible-light photograph associates five numbers with each pixel: the three components of the colour (RGB, e.g.), and two coordinates of its location in the painting. In this framing, every pixel can be conceived of as occupying a five-dimensional space. A pair of pixels close to each other on the painting and with similar colours will be near each other in this five-dimensional space, and large collections of similar pixels form clusters and complex topological structures there. When additional co-registered images are added, such as the many element maps arising from MA-XRF scanning, the dimensionality of the space grows because then every pixel has many elemental compositions, each of which corresponds to a separate axis in the high-dimensional space. Clusters and other structures in these high-dimensional elemental composition spaces form from areas of similar layer buildup, so visualising them can help to make sense of the deluge of data.

Our solution to this problem of overwhelming amounts of data is allow the user to explore any combination of projections and colourings of the space interactively. In this approach, each pixel is drawn individually, and changes to the positions of the pixels are animated. Elemental compositions or colours can be used to position the pixels directly, or dimension-reduction techniques such as Principal Component Analysis (PCA), t-distributed Stochastic Neighbour Embedding (t-SNE) (Van der Maaten and Hinton 2008), or Uniform Manifold Approximation and Projection (UMAP) (McInnes and Healy 2018) can be used to cluster the pixels according to their composition or other criteria. In any configuration, the user is able to 'lasso' points and temporarily colour them so that they can be easily tracked from one view to another. The iterative projection-lasso-colouring-reprojection sequence thus allows

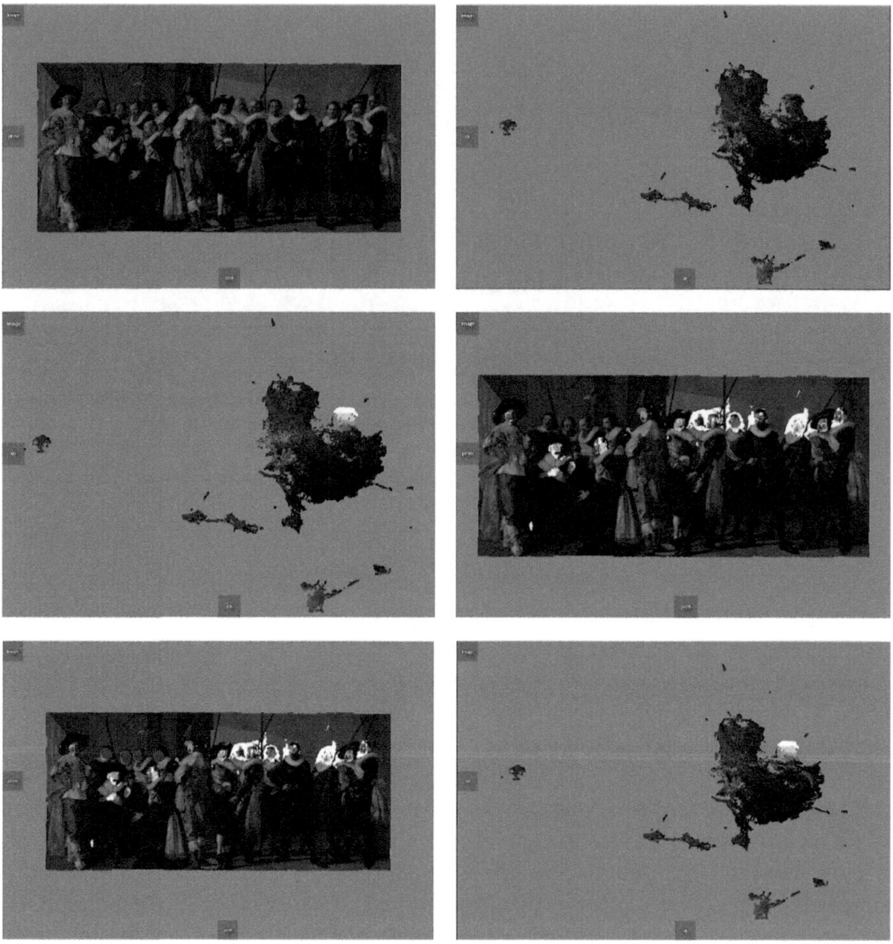

Fig. 7.8 The PixelSwarm tool in use on Hals' *The Meagre Company* (1633–1637), Rijksmuseum, Amsterdam). Steps a-f are described in the text and shown here from top to bottom, left to right

the user to gain an understanding of the layer buildup and painterly technique that cannot easily be obtained by inspecting the individual technical images of the painting.

Figure 7.8 shows a sequence of states from a typical use in an investigation of *The Meagre Company*. In (a), the initial display shows the pixels of the painting in their normal positions with normal colours. In (b) the display is animated to rearrange the pixels into an alternate 2D arrangement as calculated using the UMAP algorithm applied to the elemental compositions obtained from MA-XRF element-line maps. In this display, pixels associated with a similar chemical makeup will be clustered together regardless of their position on the painting. In our example, the user lassoes pixels in the upper centre lobe of the main cluster, and (c) tags them by colouring them in four shades from white to grey, in order to achieve a more finely grained spatial distribution when the swarm of pixels is animated back to their initial positions (d), maintaining their coloured tags from step (c). This reveals that the coloured pixels are distributed both in and/or around many painted faces—most prominently on the right side—and draws our attention to rather curious halo-shaped forms around some of the faces. A closer visual inspection of both the surface of the paintings and the MA-XRF element maps demonstrates that the element copper is present in these areas, both in a green underlayer and in certain skin tones (see Chap. 6). By contrast, none of the pixels in four faces in the left half of the painting—belonging to officers in the second row—got tagged at all, indicating a different chemical composition. To further explore this discrepancy, the user then lassoes these four faces and tags these pixels blue, as shown in (e). The display is then animated back into the UMAP projection, now showing the distribution of the blue-tagged pixels relative to the white-grey group (f). It becomes clear that these two groups exhibit very distinct characteristics in their chemical makeup, although these differences do not translate everywhere into visible differences in colours. Based on a powerful data exploration and visualisation tool such as the PixelSwarm, the user can therefore draw insights from complex datasets and form hypotheses about the pigments used, the layer buildup, and artists' techniques in general. In the specific case of *The Meagre Company*, it helped to enhance our understanding of the painting's very complex genesis and the consistencies and inconsistencies in Hals' and Codde's work process.

For the first time, these tools enable precise in-depth comparison, not just of the visible surface of very large paintings, but also of the chemical properties of deeper layers in a heartbeat, zooming in and out from an overall view of a large-size painting to microscopic observation, changing seamlessly from visible light to different wavelengths, x-radiography, IRR and various elemental maps, while juxtaposing comparable elements in shape and size and material composition. Smart digital tools enhance the expert's eye and help experts to make meaningful observations, guided by the art historical knowledge necessary to frame these observations within the broader context of an artist's hallmark styles, techniques and studio practices.

References

Erdmann, R.G.: Image processing for the Bosch research and conservation project. In: Hoogstede, L., Spronk, R., Erdmann, R.G., Klein Gotink, R., Ilsink, M., Koldeweij, J., Nap, H., Veldhuizen, D. (eds.) Hieronymus Bosch, Painter and Draughtsman, Technical Studies, pp. 31–51. Yale University Press, Brussels (2016a)

Erdmann, R.G.: Bosch project (2016b). http://boschproject.org/#/book/

Hendriks, E., Hughes, S.: Van Gogh's brushstrokes: marks of authenticity?. Art, conservation and authenticities: material, concept, context: proceedings of the international conference held at the University of Glasgow, 12–14 September 2007, 143–151 (2009)

Ilsink, M., Koldeweij, J., Spronk, R., Hoogstede, L., Erdmann, R.G., Klein Gotink, R., Nap, H., Veldhuizen, D.: Hieronymus Bosch, Painter and Draughtsman. Catalogue Raisonné. Mercatorfonds, Brussels (2016)

Ji, F., McMaster, M.S., Schwab, S., et al.: Discerning the painter's hand: machine learning on surface topography. Herit. Sci. **9**, 152 (2021). https://doi.org/10.1186/s40494-021-00618-w

Johnson, C.R., Hendriks, E., Berezhnoy, I., et al.: Image processing for artist identification. IEEE Signal Proces. Mag. **25**(4), 37–48 (2008)

Li, J., Yao, L., Hendriks, E., Wang, J.Z.: Rhythmic brushstrokes distinguish van Gogh from his contemporaries: findings via automated brushstroke extraction. IEEE Signal Proces. Mag. **34**, 1159–1176 (2012)

Lyu, S., Rockmore, D., Farid, H.: A digital technique for art authentication. Proc. Natl. Acad. Sci. **101**(49), 17006–17010 (2004)

McInnes, L., Healy, J.: UMAP: uniform manifold approximation and projection for dimension reduction. ArXiv. **1**, 1–63 (2018)

Radford, A., Kim, J.W., Hallacy, C., Ramesh, A., Goh, G., Agarwal, S., Sastry, G., Askell, A., Mishkin, P., Clark, J., Krueger, G., Sutskever, I.: Learning Transferable Visual Models from Natural Language Supervision. arXiv:2103.00020v1 [cs.CV] (2021). https://doi.org/10.48550/arXiv.2103.00020

Tufte, E.: Envisioning Information. Graphics Press, Cheshire (1990)

van der Maaten, L.J.P., Hinton, G.E.: Visualizing high-dimensional data using t-SNE. J. Mach. Learn. Res. **9**, 2579–2605 (2008)

van Noord, N., Hendriks, E., Postma, E.: Toward discovery of the artist's style: learning to recognize artists by their artworks. IEEE Trans. Pattern Anal. Mach. Intell. **32**(4), 46–54 (2015)

Part IV
Concluding Remarks

Chapter 8
Epilogue

Anna Tummers

Abstract The epilogue discusses some of the most revealing findings in this book, ranging from spectacular sketches found underneath some of his most famous paintings to a surprising new insight into seventeenth-century attribution practice (i.e. Hals monogrammed a painting that he did not technically paint himself). Building on the survey of the development of Hals connoisseurship and the extensive case studies, the epilogue reflects on how (relatively) new techniques, digital tools and a close reading of primary sources have altered and expanded our understanding of Frans Hals, and explores their potential for enhancing the eye in attribution matters.

Despite Hals's *virtuoso* loose painting style, which appears to be highly individual, works in his manner have repeatedly been at the centre of attribution controversies since the early twentieth century. The lack of clear criteria for attributing works to the master—due in part to the differing opinions among experts—have long left Hals prone to doubt and misattribution. Nevertheless, Hals connoisseurship has evolved tremendously over the past century. From the implicit knowledge or intuitive insight of a single expert it gradually became a predominantly evidence-based practice building on the specialized insights of multiple experts. Although implicit knowledge still plays an important role in the authentication process, it is no longer accepted at face value; instead, the rational arguments underlying authentication decisions have come to the fore.

Chemical evidence was first introduced in a Dutch court in 1925 to settle a forgery dispute. Since then, chemical analysis has become more integrated in the decision-making process—albeit relatively slowly. While in Rembrandt research scientific analysis has played a major part since 1968, it was only in 1990 that such analysis became an integral part of Hals attributions, when a technical investigation of Hals paintings was launched in response to a major attribution controversy. The resulting research report (1991), which examined 56 paintings, is still the largest

A. Tummers (✉)
Ghent University, Ghent, Belgium
e-mail: Anna.Tummers@UGent.be

A. Tummers, R. G. Erdmann, *Frans Hals or not Frans Hals*, Cultural Heritage Science, https://doi.org/10.1007/978-3-031-59489-2_8

technical study to date, and has served as a benchmark for all subsequent studies in the field. While this study provides useful reference material, in practice, attributions to Hals are often still based almost exclusively on a style analysis of the visible surface in the early twenty-first century.

Admittedly, technical evidence rarely provides conclusive answers in attributions matters. Only in the case of anachronistic materials can analytical chemistry provide conclusive evidence and expose forgeries or misattributions. In all other cases, the authentication process remains a matter of interpretation. However, technical analysis is a crucial component of the attribution process in order to gain an in-depth understanding not only of the final result and visible surface of a painting, but also of the entire creative process, from the initial sketch to the various stages of execution and the specific techniques and materials used. Significantly, no positive attribution (i.e. an attribution to a particular painter) can be made without visual analysis. This latter task, the visual analysis, has become both more challenging and easier in the early twenty-first century, due to a paradigm shift in the humanities, myriad new technical possibilities and the development of advanced digital tools. Within the humanities, the very notion of what constitutes 'authenticity' was redefined, particularly in the field of old master painting. As a result, art experts moved away from a simple binary perspective (either by the master or not), became more aware of the complex range of possibilities, and began to introduce more nuanced categories of thought (as in Grimm's forthcoming Hals oeuvre catalogue). In addition, relatively new techniques such as MA-XRF and HI/RIS scanning are currently transforming the field, and yielding a wealth of new information. Moreover, the amount of data produced by these techniques is often substantial and requires digital tools to aid in interpretation, to focus on the most relevant parts, and to perceive both larger patterns and subtle characteristics.

The NWO and NICAS projects *Frans Hals or not Frans Hals* and *21st Century Connoisseurship* explored the use of a range of analytical techniques and (newly developed) digital tools for attributing works to Frans Hals. These relatively new methods were combined with careful naked-eye and microscopic observations, as well as a close reading of seventeenth-century art theory and other relevant primary sources. The combination of approaches proved successful. The case studies (Chaps. 3, 4, 5 and 6) allowed us to significantly refine our understanding of Hals's characteristic style and achievements, his painting techniques, his use of materials, and his workshop practice. One case study even provided a surprising insight into his own attribution practice (see below). Moreover, the digital tools, especially the Draper, proved to be indispensable for the interpretation of the large group portraits, as they allowed for a quick and precise analysis of a very large amount of divergent data and image material.

One of the most spectacular discoveries of the research have been the various types of sketches found underneath Hals's paintings. Since there are no known drawings by Hals, these sketches are all the more revealing, providing unique insights into Hals's working process. For the first time, we found concrete evidence of a very early sketch in white material, probably dry chalk, on a coloured ground. Although it was widely known that artists often sketched their designs in white chalk on coloured grounds (as shown in Vermeer's famous *The Art of Painting*), no concrete evidence of such drawings had been found. The MA-XRF scan of Hals's

1627 *Portrait of the Saint George Civic Guard* not only revealed grainy lines containing calcium, we were also able to confirm the presence of fairly large white particles visible on the surface of the painting with a hand-held digital microscope. The free sketch lines describing various possible contours of a face confirm the idea that Hals sketched his designs directly on the prepared canvas. He did not need to trace previously prepared drawings. Instead, he worked freehand, quickly and effectively from start to finish.

In the same painting we found evidence of a second sketching phase in the infrared reflectogram (IRR): a very cursory sketch under two faces and collars in a wet material containing carbon. The sketch lines are extremely bold: just a rough indication of the position of the nose, mouth and beard, and a quick curly or straight line to indicate a collar. These sketch lines show just how economically Hals set up his paintings; he simply indicated the positions of the main elements without describing the forms in detail.

These initial sketch lines, which show exactly how Hals designed his paintings can only be seen in rare cases, such as when an intermediate layer is missing or part of the design was never executed. In most cases, these deepest layers are not visible with modern analytical techniques, because they are covered by other layers that block the signal or, because they are difficult to distinguish among many similar signals. For example, if the initial indication of the position of the eyes or the outline is accurate, it is hard to distinguish it from the top layer. The study of Hals's *Portrait of a Woman* in Berlin showed that phosphorus-containing sketch lines (presumably bone black) under the face and collar were visible in the neutron autoradiographs (NAR) but not in the MA-XRF scan and only partially in the infrared reflectogram (IRR). Because NAR analysis is very costly and therefore not widely used, such lines seldomly come to light. Nevertheless, the rare detailed glimpses of Hals's initial design for a face, collar or composition provided by the various case studies give us a clear insight into his efficient working method.

A third type of sketch line further illustrates how Hals set up his paintings. Once he had thought out the rudimentary design for a painting, he roughly indicated the postures of his figures and their clothing with bold, broad brushstrokes. Thus, he still sketched while he worked up his painting in oil paint. In the civic guard portraits that were finished under his supervision such lines are most easily seen under the black costumes and hats: broad manganese-containing strokes (probably a black mixed with umber) indicating the direction of a leg, a separation between different parts of a costume, or the position of a hat, for example. He also used lead-containing paints for similar lines in lighter areas; sometimes he even used a paint on hand that bore no particular relationship to the final colour, as in the case of the dark green, copper-containing indication of a white collar in his 1639 civic guard portrait.

In the contested *The Meagre Company* many different sketch lines have come to light which allowed us to reconstruct its genesis in great detail, and to resolve some earlier contradictions in its interpretation. The infrared reflectogram (IRR) revealed parts of the design for the painting that were never executed. In addition, it was possible to distinguish various stages of the painting process. Below the central guardsman in yellow, we found broad sketch lines in a manganese-containing material, roughly indicating his initial pose, as well as evidence of further elaboration of this figure's head: Hals roughly indicated dark passages with a calcium-containing paint (presumably bone black). In other areas of the painting, particularly in the MA-XRF manganese (Mn), lead (Pb) and copper (Cu) maps, similar sketch lines have appeared, in part due to the unfinished state of the painting and Codde's somewhat different working process.

In the case of this complex attribution, our integrated approach was key to our new understanding of the painting. Particularly when the insights gained with different methods—including stylistic analysis, a close reading of seventeenth century art theory and new material-technical observations facilitated by state-of-the-art digital tools—were brought together, the integrated approach led to illuminating new conclusions. For example, art theory helped to shed new light on Hals's careful balancing of colours and tones (*houding*), facilitating the identification of divergent passages, that also revealed differences in the brushwork, painting techniques and materials used, while a deeper material-technical analysis—made possible with great precision by MA-XRF scanning and the digital 'Draper'—allowed us to reconstruct the stages of completion of the various areas when Codde took over. In addition, seventeenth-century art-critical terms such as *koppeling* (joining) and *sprong* (leap) enabled us to recognise Hals's careful placement of his sitters and their attributes and his characteristic grouping of these, while in-depth technical analysis confirmed that a very obvious instance of *haspeling* (confusion) in *The Meagre Company* was not part of the original design.

The integrated approach also helped to significantly improve the criteria used in Hals attributions. Whereas the 1991 report stated that Hals used a limited range of pigments that remained constant throughout his career (Hendriks et al. 1991, p. 51), the extensive new reference material (comprising 69 individual portraits) allowed us to gain new insights into Hals's chronological development, the consistency of his choice of materials, his painting techniques and his workshop practice. The new analyses not only provided a deeper understanding of Hals's design process and the construction of his paintings as discussed above, but also revealed patterns and characteristics in his use of materials, techniques and workshop assistance. In terms of his use of materials, Hals's lead white showed a considerable consistency in its isotopes, which differed greatly from, for example, Van Meegeren's forgery in Hals's style. As for his choice of specific pigments, his use of bone black, manganese and copper (probably azurite) in the facial shadows is characteristic, as is his use of manganese and copper in certain blacks, and copper for bluish shadows in white clothing. While bone black and manganese are standard in Hals's mixture for skin tones, his use of copper has so far only been found in paintings dating from the late 1620s to the late1630s. It is also noteworthy that he used vermillion in all his large group portraits created for semi-public display, though not for every single face and hand. In his individual portrait of a woman in Berlin he reserved it for the face only, while creating a similar pink skin tone in the hands with cheaper, earth pigments. His use of this pigment thus varied somewhat, which may well be related to its price. Furthermore, as noted before, his use of indigo blue (mixed with pure, rather coarse lead white) is unique for his time.

The Malle Babbe case study showed that both Hals and one of his workshop assistants occasionally painted *alla prima:* they designed the painting directly on the canvas and worked it up completely wet-in-wet. Furthermore, a close reading of seventeenth-century art theory improved our understanding of what was meant by the term in Hals's time. While the 1991 study convincingly showed that by far the most of Hals's paintings are built up in distinct layers, the conclusion that he did not paint *alla prima* at all, was thus incorrect. He did use this technique, albeit sparingly and presumably mostly around 1645–1650, and so did at least one of his workshop assistants.

Moreover, this same assistant also applied lead white paint with a similar viscosity and forceful touch as the master's, resulting in so-called 'ribbon touches' with raised edges on both sides. This type of loosely applied accent is therefore not unique to Hals himself, as was previously thought. In this case, their placement betrayed another hand, as they lacked Hals's characteristic ease and accuracy. Apparently, individual traits such as Hals's ribbon touch and *alla prima* painting technique were in practice created not only by the master himself, but also by others in his studio. It reminds us of the different notions of authenticity that prevailed in the seventeenth century, which also explains a curious feature of this picture: the presence of Hals's monogram in the original paint layer. Even though Hals does not appear to have painted any part of the New York *Malle Babbe* himself, he must have approved of its overall quality and considered it worthy of bearing his name. Although we knew that this was possible according to seventeenth-century guild rules, this is the first time that an in-depth material analysis indicates that Hals actually did so.

The fact that Hals authenticated paintings that were not necessarily painted by him alone, but sometimes partially by assistants—as the close analysis of the 1639 civic guard portrait showed very clearly in Chap. 6—and occasionally not by him at all—as in the case of the New York *Malle Babbe*—is important to bear in mind when attributing paintings to the master. *Frans Hals or not Frans Hals?* is not always the right question, as a Hals was not necessarily *by* Hals. From a seventeenth-century perspective, the question is rather whether or not a painting came from Hals's studio and if Hals would have considered it worthy of bearing his name. Of course, not everything created in his studio would have been sold under his name; after all, the master's name was a guarantee of quality. To judge from the New York *Malle Babbe*, Hals sold a range of studio products as his own, and he may have made certain quality distinctions between these works and priced them accordingly, as his Antwerp colleague Rubens, for example, was wont to do.

Another distinctive feature of Hals's painting technique and further evidence of the speed and efficiency of his working process is the partial exposure of the ground. Hals used the colour of the ground effectively; hair and shadows in the skin tones owe much of their colour to the ground, which he covered with a thin translucent layer, as the hyperspectral imaging (HI/RIS) of the Berlin *Portrait of a Woman* has shown very clearly. In addition, a kind of halo effect around many of his contours reveals Hals's fast and very functional painting technique: he worked on adjacent areas while the paint was still wet, and prevented the different wet colours from smearing by keeping the areas apart.

In short, the case studies provided both new and improved standards for attributions to Hals and significantly sharpened our understanding of Hals's workshop practice and his notion of authenticity. Indeed, they proved very useful when we were consulted about two paintings that had caused confusion in the market (see the

Introduction and Chap. 5). While the reference material and technical analysis provide firmer standards for attribution to the master, the question of what exactly counts as a 'Hals' became more complex. The range of studio products that may have carried his name sheds new light on the attribution controversies. As we have seen, Hals experts had very different approaches to delimiting the master's oeuvre, ranging from a broad contextual approach, like Slive's, which included seventeenth-century attributions in reproductive prints, to a sharp focus on distinguishing the master's hand on the basis of visual characteristics, like Van Dantzig and Grimm. A broader notion of authenticity resolves some of the earlier contradictions by allowing a wider variety of workshop products to pass as an authentic Hals. It also opens up the possibility that original Halses came in a range of qualities and prices. Furthermore, the new technical data and digital tools allow for much more precise analysis and comparisons, revealing hidden layers and patterns, and thus creating a firmer technical basis for Hals attributions. Especially when combined with seventeenth-century art theoretical sources, the new data and tools have provided much deeper insights into Hals' *virtuoso* work process and his characteristic, artistic achievements. While this book discusses some of the most revealing and striking new findings, the very rich collection of data gathered for the above mentioned research projects can be used for many further explorations and analyses of the material. To facilitate further research, we have made the image material, the newly developed digital tools and a digital version of this book freely available online.

References

Hendriks, E., Levy-Van Halm, K., in collaboration with Van Asperen De Boer, J.R.J.: Report Concerning a Preliminary Technical Investigation of Paintings exhibited during the Frans Hals Exhibition, held from May 11 to July 22 1990 in the Frans Hals Museum Haarlem (1991)

Acknowledgments

This book could not have been written without the input of many specialized co-authors: Andrei Anisimov, Silvia Centeno, Joris Dik, Nouchka De Keyser, Marie-Noëlle Grison, Roger Groves, Babette Hartwieg, Erma Hermens, Katja Kleinert, Annelies van Loon, Dorothy Mahon, Claudia Laurenze-Landsberg, Vassilis Papadakis and last, but certainly not least, Arie Wallert. We are deeply grateful to all of them.

The online digital tools that were key to the interpretation of the largest and most complex case study (Chap. 6) and greatly facilitated the interpretation of the different paintings of Malle Babbe (Chap. 5) were created by Robert G. Erdmann and have been made freely available online at the occasion of this publication (https://images.erdmann.io/Draper/?collection=/NICAS/Frans_Hals; https://images.erdmann.io/Draper/?manifest=/NICAS/Frans_Hals/image_manifest_MB.json).

The research presented in this book was funded by the Netherlands Organisation for Scientific Research as part of the projects *Frans Hals or not Frans Hals: Defining the Oeuvre of the Painter Frans Hals (1582/83–1666)* (NWO Museumbeurs 2016–2018) and *21st Century Connoisseurship: Smart Tools for the Analysis of Seventeenth-Century Paintings* (NICAS Seed Money Grant 2018–2022); by the Prins Bernard Cultuurfonds Conservatoren Stipendium, 2018, by a private benefactor and by a grant from the Leiden University Centre for the Arts in Society (LUCAS). The authors wish to thank the following institutions and people for enabling and supporting the research: at the Frans Hals Museum (Haarlem): Lidewij de Koekkoek, Ann Demeester, Liesbeth Abraham, Mireille de Marvelde, Jessica Roeders, Marrigje Rikken, Norbert Middelkoop, Ulrich Roldanus, Herman Bessels, Wilfred Zwart and Judith Harren; at the Rijksmuseum (Amsterdam): Taco Dibbits, Robert van Langh, Pieter Roelofs, Gregor Weber, Friso Lammertse, Petria Noble, Anna Krekeler, Susan Smelt, Gwen Tauber, Gerrit Albertson, Nienke Woltman, Laura Raven, Jolanda van Iperen, Carola van Wijk, Rik Klein Gotink, Suzanne Plaum, Ton Nuijen, Ivo Rijkers, Gijs Smit, Tessa Hogeboom, Frederique van Reij and Pauline Kintz; at the Metropolitan Museum of Art (New York): Michael Gallagher, Adam Eaker and Scott Geffert; at the Gemäldegalerie, Staatliche Museen

A. Tummers, R. G. Erdmann, *Frans Hals or not Frans Hals*, Cultural Heritage Science, https://doi.org/10.1007/978-3-031-59489-2

(Berlin): Michael Eissenhauer, Nelleke de Vries; at the University of Amsterdam: Maarten van Bommel and Ella Hendriks; at Delft University of Technology: Joris Dik. Apart from the experts mentioned above, Eric Jan Sluijter, Carol Pottasch, Rosa Hoogenboom, Jasper Hillegers and Lotte Kokkedee also gave valuable feedback during five expert meetings that were organized in the context of the research projects.

The 20 micron pixel high resolution photography of *The Meagre Company* was done by Carola van Wijk and Rik Klein Gotink at the Rijksmuseum. The latter also photographed the five civic guard portraits at the Frans Hals Museum in the same resolution, as well as the *Portrait of a Woman* and *Malle Babbe* at the Gemäldegalerie in Berlin. Rik Klein Gotink also made the infrared reflectograms (IRR) at the Frans Hals Museum and at the Rijksmuseum, while Arie Wallert made the IRRs at the Metropolitan Museum of Art in New York and at the Gemäldegalerie in Berlin. The high resolution photography at the Metropolitan was done by Scott Geffert.

At the Metropolitan Museum of Art the MA-XRF scanning and processing was done by Louisa Smieska and Silvia Centeno. At the Rijksmuseum, the MA-XRF scanning of *The Meagre Company* was done by a large team consisting of: Annelies Van Loon, Nouchka De Keyser, Anna Krekeler, Susan Smelt, Gwen Tauber, Erma Hermens, Gerrit Albertson, Nienke Woltman, Laura Raven and Petria Noble. The MA-XRF scanning of the five civic guard paintings at the Frans Hals Museum was led by Nouchka De Keyser with assistance from Jennie Allred and Rosa Hoogenboom. At the Gemäldegalerie in Berlin, the MA-XRF scanning was done by Arie Wallert with assistance from Babette Hartwieg and Anna Tummers. All paintings were mapped with a step size of 700 micron and a dwell time of 70ms/step, using a Bruker M6 Jetstream, and the method described by (Alfeld et al. 2013), apart from the Metropolitan *Malle Babbe*, which was mapped with a step size of 1000 microns, and a dwell time 90 ms/step. The hyperspectral imaging or Reflectance Imaging Spectroscopy of the *Portrait of Woman* and *Malle Bobbe* at the Gemäldegalerie in Berlin was carried out by Andrei Anisimov and Tigran Mkhoyan using the instruments and methods described in Groves et al. (2018).

The paintings were studied in situ by the main authors of the different chapters, while the observations through the microscope and microscopic photography were led by Babette Hartwieg at the Gemäldegalerie in Berlin and Dorothy Mahon at the Metropolitan in New York. At the Rijksmuseum and at the Frans Hals Museum a digital microscope was used in the galleries and the microscopic photographs were taken by Nouchka De Keyser, Anna Tummers, Arie Wallert, Robert G. Erdmann, Jung Roubos, Erma Hermens and Annelies van Loon. The neutron autoradiographs (NAR) of the *Portrait of a Woman* and *Malle Babbe* at the Gemäldegalerie were made at the Helmholtz Zentrum Berlin für Materialien und Energie, the former Hahn-Meitner Institut (NAR) in collaboration with Claudia Laurenze-Landsberg from the Gemäldegalerie. All lead isotope analyses were done by Gareth Davies and Paolo d'Imporzano of the Geochemical Laboratory for Ultra-Low Isotopic Analyses in the Faculty of Science, Vrije Universiteit, Amsterdam. The GC-MS analyses were done by Henk van Keulen and Saskia Smulders at the Cultural Heritage Agency of the Netherlands (RCE).

Two introductory videos were made about the *Frans Hals or not Frans Hals* and *21st Century Connoisseurship* research projects by Marcel van der Velde and Krista Ariens. The results can be seen online: https://www.youtube.com/watch?v=G5VDHIztv4U; https://www.youtube.com/watch?v=gwHRH0YPPqI.

The texts for this book were first edited by Carlijn Vos, who made all the references consistent with the MathPhys Style and also compiled the bibliography. Furthermore, the authors wish to thank the following editors at Springer Nature: Annelies Kersbergen and Barbara Amorese, as well as the editors of the Cultural Heritage Science series: Klaas Jan van de Berg, Aviva Burnstock, Koen Janssens, Robert van Langh, Jennifer Mass, Austin Nevin, Bertrand Lavedrine, Bronwyn Ornsby and Marija Strilc, and production editor Rathika Ramkumar, production supervisor Divya Rajasekaran, the Straive Springer Art Team, and production coordinator Cynthia Kroonen.

Lastly, we would like to thank our loved ones at home for seeing this multi-year project slowly come to fruition and for their patient support: Joa Chen, Jung Roubos, Maarten Roubos and David Roubos (AT) and Marie-Noëlle Grison (RE).

References

Alfeld, M., Pedroso, J.V., Van Eikema Hommes, M., Van Der Snickt, G., Tauber, G., Blaas, J., Haschke, M., Erler, K., Dik, J., Janssens, K.: A mobile instrument for in situ scanning macro-XRF investigation of historical paintings. J. Anal. At. Spectrom. **28**(5), 760–767 (2013). https://doi.org/10.1039/c3ja30341a

Glossary of Technical Methods

Arie Wallert

Methods in the scientific examination of paintings are often classified as 'non-destructive' and 'destructive', where 'destructive' refers to the requirement of taking a small sample to carry out the examination. In our study we tried to rely as much as possible on methods of non-destructive examination. In many cases the analysis of paint samples provides us with more precise and reliable information. But because the art works that we study are unique and irreplaceable, we tried to limit our number of samples (even though invisible to the unaided eye), as much as possible. We did that also because it is always difficult to extrapolate the results of those 'destructive' point examinations to areas larger than the immediate surroundings of the sampling point.

Recent advances in analytical instruments have now expanded the range of possibilities to perform non-contact examinations of paintings. Some of these methods are tried-and-tested, like X-radiography (XR), observation with the stereo-microscope, and photography with normal-, ultraviolet-, and infrared light.

Others have been developed in the second half of the twentieth century, such as infrared reflectography (IRR) and the still rather exotic, neutron activation imaging, or neutron autoradiography (NAR). As the costs of equipment have come down in recent years, IRR has quickly become more accessible. It is now an easily handled, standard tool for every researcher of paintings. The opposite is true for neutron activation techniques (NAR) that require access to one of the - fairly scarce - nuclear reactor research facilities in the world. Those techniques also require highly specialised researchers, with a good understanding of both paintings-, and radiation issues. Those people are not plentiful.

A. Wallert
University of Amsterdam, Amsterdam, The Netherlands

Rijksmuseum, Amsterdam, The Netherlands
e-mail: arie.wallert@gmail.com

A. Tummers, R. G. Erdmann, *Frans Hals or not Frans Hals*, Cultural Heritage Science, https://doi.org/10.1007/978-3-031-59489-2

But from the beginning of this century, other, very much more accessible methods have entered the field of painting examination. Non-destructive 'point-examinations' can now also be done on site with handheld digital microscopy (DM). Because of their ease of use and low cost, digital microscopes have now often replaced the use of conventional stereo microscopes in the examination of paintings.

Previously an exotic beast, energy dispersive X-ray fluorescence spectrometry (XRF), is now leaving its dedicated laboratory in the form of portable hand-held instruments that can be brought to the museum for non-invasive analysis of the art objects in situ. Similarly, reflectance spectrometry (RS) has been equipped with fibre optics for light transmission into a portable system that allows for non-contact examination of coloured objects: fibre optic reflectance spectrometry (FORS).

Macro X-ray fluorescence spectrometry (MA-XRF) and reflectance imaging spectroscopy (RIS) respectively, provide essentially the same information that the point examinations from XRF and RS give us. But both techniques, MA-XRF and RIS alike, give their types of information in a two-dimensional manner, rather than from individual points.

DM

(Hand held) digital microscopy

Digital microscopes come as extremely advanced instruments (Hirox, Keyence) with impressive features that allow for very precise 2D/3D measurements and can capture images up to 10,000 times magnification. Especially at larger magnification it is difficult to get everything within the field of view in focus. High-end digital microscopes, therefore, are provided with digital 'stacking' features that make it possible to capture images with extended depth of field.

In the on-site examination of paintings, however, much cheaper hand-held microscopes are used with magnifications with digital zoom capability adjustable from c. 10x to 50x. These microscopes have built-in LED illumination, some even in the UV (375 nm)-, and NIR (815 nm) ranges. Some of these hand-held microscopes even have high dynamic range (HDR) capability. In that case, the camera captures multiple frames of the same scene but with different exposures. These frames are then digitally combined into one. The combined image has a much higher dynamic range than each individually captured frame would have. For the examination of very reflective varnished surfaces of paintings, a set of polarisation filters is used. With such microscopes, the surfaces features of the painting are observed through a set of small lenses that are arranged to focus at close distances. Images are acquired with small sensors, usually not exceeding 5-megapixel resolution, and recorded via USB connection onto a laptop computer.

XR

X-radiography

The x-ray tube placed at some distance from the back side of the painting to be examined sends out X-rays that penetrate through the painting. Some features in the painting may obstruct the passage of the x-rays. These features then cast "shadows" upon a sensitive film, placed in close contact with the other side of the painting. In those areas the film is not, or just less blackened by the transmitted radiation. The amount of blackening, and hence the amount of transmittance, is largely determined by the position in the periodic system of the obstructing features. X-ray absorption rises steadily with a material's atomic number. The inorganic pigments of paintings are produced from a variety of elements. Pigments made of heavier elements such as lead or mercury tend to obstruct the x-rays more than pigments based on low-Z elements like carbon black, organic lake pigments on alumina substrate, or chalk. Those lighter elements will allow most of the x-rays to pass through, resulting in a stronger blackening of the radiographic film.

IRR (IRP)
Infrared reflectography (infrared photography)
Infrared techniques are non-destructive methods to penetrate below the surface of paint and visualize compositional paint changes and preparatory sketches. They are used to 'see through' paint layers that are otherwise impenetrable to the human eye.

With conventional photography objects are recorded in visible light. That is the portion of the light spectrum that has a wavelength ranging from c. 400 to just above 700 nm. Anything outside of that range is invisible to the naked eye.

Infrared waves are longer than those in the visible range. Waves ranging from c. 750 to 1000 nm, commonly described as the near-infrared (NIR), can be recorded with IR photography. Conventional digital cameras have sensors that are also sensitive in the NIR range. Those cameras can be adapted for IR photography by removing the special filter over the sensor that cuts out infrared light.

Recording longer wavelength ranges requires special cameras with dedicated sensors. Shortwave infrared images (SWIR) are made in the region of 1000–2500 nm. Those images are usually called infrared reflectograms, to distinguish them from infrared photographs taken in the near infrared range. It has become convention that any imaging above c 1000nm is called IR-reflectography. Collecting infrared images in the lower wavelength ranges is called IR-photography. However, both techniques are based on the same principle. Infrared light passes through the paint layers to the white ground and is then reflected back to the camera (hence reflectogram). On its way through the paint layers the light can be absorbed by certain pigments. Especially carbon black pigments—which were often used for the underdrawing of paintings—are good infrared absorbers. When the light is absorbed by those pigments it can no longer be reflected to the camera. The differences between absorption and reflection are captured by the infrared camera. The captured light is converted into a black-and-white image. If an artists has begun a painting by drawing the design in carbon black on a white reflective ground the image can make this visible. Therefore, the infrared images may provide information on underdrawings, the evolution of the composition, pentimenti (changes), construction methods, signatures and dates.

NAR
Neutron activation autoradiography
Examination with neutron autoradiography (NAR) is based on a very moderate, non-destructive radioactivation of the painting with thermal neutrons. In this process free neutrons are captured by the nuclei of atoms in the sample, i.e. pigments

on the painting. The neutron capture creates new radioactive nuclei, which will decay in time to a stable ground state. Thus, a small number of the atoms of the pigments in the paint are temporarily transformed into a radioactive species. A proportion of the energy is released as radiation. In the layers of paints about a dozen isotopes, emitting β- (electrons) and γ-radiation, are created. This radioactivity, the β-radiation, is sufficient to blacken a highly sensitive film in direct contact with the painting and reveal the spatial distribution of pigments. As the radioactivity of different elements decays at different rates, a series of films can be made (at different half-life times) that show the different elements, allowing us to represent the distribution of a number of pigments over the painting in series of separate films. The method does not equally work for all elements. The maximum of energy for aluminium (^{28}Al) for instance already occurs at 2.3 min. Too short, and weak, for a proper exposure of a film. The first film exposure from a painting is usually for manganese. The activation product ^{56}Mn has an ideal half-life of 2.578 h and shows quite strongly on the films. Manganese oxides are together with iron oxides important components in the reddish and brownish earth pigments ochre and umber that were often employed for the initial sketches in painting. The next film, for copper (half-life for ^{64}Cu is 12.8 h) would be ready after one to three days of exposure. That film would show us the distribution of copper-containing pigments like the blue azurite or the green verdigris over the picture. Films for arsenic would show us the distribution of blue smalt or yellow orpiment. Somewhat later a film for phosphorus would show passages painted with bone black. The last film, taken after a month would show the mercury in the orangy-red vermillion.

RS/FORS

Reflectance spectroscopy / Fibre optic Reflectance spectroscopy

Reflectance spectroscopy (RS) is a type of spectroscopy where light which is reflected from the surface of the object, i.e. the painting, is measured. The painting is illuminated with a source projecting light, usually in the ultraviolet, visible and near-infrared ranges (c. 250–1000 nm).

In point examinations the projected light from the source as well as the reflected light from the painting is guided through a fibre-optic probe to the instrument (FORS). The light that scatters back from the painting is collected on a detector that produces a characteristic reflectance spectrum. If the light interval investigated corresponds to the wavelengths perceived by the human eye (c. 400–700 nm), the measurement results in the objective characterisation of the colour of an object. This spectrum contains information on the electronic transitions of the pigments in the sample. Thus, dependent on the specificity of the spectral response, pigment identification may sometimes be possible in a non-destructive manner. It is now one of the most widely used analytical tools for the non-invasive examination of art works. Especially so, because portable equipment with extended sensitivity has become available. Examination in in the near infrared range (up to 2500 nm), may also provide information about vibrational transitions, and thus inform us about oils and proteins used as paint media.

RIS/HI

Reflectance imaging spectroscopy / hyperspectral imaging

This technique transforms the point examinations of reflectance spectroscopy (RS) into an area examination. Reflectance spectroscopy is combined with 'normal' two-dimensional visualisation of the painting by optical imaging (such as photography). This is done in such a way that each pixel of the image taken by the camera is made to represent a complete spectrum of that pixel. Each pixel of the image is associated with its own reflectance spectrum so that the various wavelengths are recorded in a spatial manner. Whereas a regular camera records three different wavelengths of the electromagnetic spectrum (corresponding to the colours red, blue and green), modern reflectance imaging cameras can nowadays easily differentiate between hundreds of wavelengths, resulting in very precise digital images or 'data cubes'.

Reflectance spectroscopy (RS) records the spectral reflectance for a specific circular aperture; a single colour is measured. By analogy a reflectance spectral-imaging system (RIS) records spectral reflectance for a projected scene at a specific spatial resolution; many colours are measured.

Reflectance imaging, often also indicated as hyperspectral imaging, allows the study of the entire image rather than that of individual spots. The characteristics of the spectral signal can be used to get an identification <u>and</u> a localization of the paintings components. These can also be analysed with more quantitative methods such as principle component analysis (PCA). Reflectance imaging spectroscopy allows for selection of specific areas where, based on the spectral response, paint mixtures can be mapped. It also facilitates the reproduction of pigments and pigment mixtures in false colour images.

XRF

X-ray fluorescence spectroscopy

X-ray fluorescence is a non-destructive technique to determine the elemental composition of materials. The composition of the materials in the painting is determined by measuring the fluorescent X-rays emitted from the object when it is excited by a primary x-ray source.

The painting is irradiated on the sampling spot (usually not larger than 1 mm) with a focused X-ray beam. This radiation affects the atoms of the pigments on that spot. The incoming x-ray energy ejects an electron from one of the atom's inner orbital shells. In reaction the atom immediately regains stability by filling the vacancy with an electron from one of the atom's outer, higher energy, orbital shells. The process leads to electronic transitions that result in the reemission of X-rays by the object. The differences in energy between inner and outer orbital shells is emitted from the sample spot in the form of a secondary, fluorescent, X-ray. Each element in a sample spot produces its own set of characteristic fluorescent X-rays that is unique for that specific element. The characteristic X-ray emission from the object (X-ray fluorescence) is used to determine its elemental composition.

Strictly speaking, it is not a method for pigment identification. XRF is a method for elemental analysis. Determination of pigments can be inferred from identification of elements. If a pigment is red and the XRF spectrum shows the presence of mercury (Hg) and sulphur (S), the use of vermilion, a mercuric sulphide (HgS), is very likely.

MA-XRF

Macro-X-ray fluorescence scanning (elemental mapping)

Just as with regular XRF the macro-X-ray fluorescence scanning method is non-invasive, non-destructive and can be performed in situ. This technique transforms the point examinations of x-ray fluorescence spectroscopy (XRF) into an area examination. Both techniques are based on the same principles, but MA-XRF overcomes a significant drawback. The local point information obtained from conventional XRF examinations, is not representative for larger areas of the painting.

It is always difficult to extrapolate the results of those individual point examinations to areas larger than the immediate surroundings of the sampling point. Local analysis gives only local information. MA-XRF provides spatial in formation.

With MA-XRF the x-ray beam is not used for analysing individual points, but is used in scanning larger areas or even the whole painting. This is done by scanning the painting's surface with a focused or collimated x-ray bean of very small dimensions.

The resulting spectra from the secondary X-ray emissions produce thousands and sometimes even millions of data points, that can be rendered as elemental distribution images. The distribution of each chemical element in the entire surface of the painting can be presented in a greyscale map. These greyscale maps can then be combined to form RGB composite images which highlight the presence of different chemical elements in the same area. Maps can simultaneously be made of almost all the elements of the periodic system, ranging from phosphorus (15) to uranium (92). This would facilitate the identification of certain pigments: A map that shows an area of an opaque yellow paint to have much lead (Pb), and another map of that same area that shows the presence of tin (Sn), would indicate that this specific area of the picture must have been painted with lead-tin yellow (Pb_2SnO_4)

The elemental maps can be representative for painted areas of the whole painting, and therefore make it easier to interpret the data. Due to the penetrative nature of the X-rays, not only elements of the uppermost paint layers may be detected, but sometimes also elements below the surface may be identified. Thus, elemental distribution images may reveal hidden sub-surface layers. This could make it possible to detect modifications that the artist made during the painting process, and provide unique insights into the creative process of the artist.

About the Authors

Andrei Anisimov is Assistant Professor in Optical Metrology for Aerospace at Delft University of Technology

Silvia Centeno is Research Scientist at the Metropolitan Museum of Art in New York

Nouchka De Keyser is Junior Scientist at the Rijksmuseum, Amsterdam

Joris Dik is Department Chairman, Materials Science and Engineering at Delft University of Technology

Robert G. Erdmann is Full Professor of Physics and of Conservation and Restoration of Cultural Heritage at the University of Amsterdam and Senior Scientist at the Rijksmuseum, Amsterdam

Marie-Noëlle Grison is PhD student in Art History at the KU Leuven

Roger Groves is Associate Professor of Heritage Diagnostics at Delft University of Technology

Babette Hartwieg is Head of Conservation at the Gemäldegalerie, Staatliche Museen, in Berlin

Erma Hermens is director of the Hamilton Kerr Institute for Easel Painting Conservation and adjunct director Conservation and Heritage Science, Fitzwilliam Museum, University of Cambridge

Katja Kleinert is Curator for Dutch and Flemish Art of the seventeenth century at the Gemäldegalerie, Staatliche Museen, in Berlin

Claudia Laurenze-Landsberg is former paintings conservator at the Gemäldegalerie, Staatliche Museen, in Berlin and a specialist in neutron autoradiography

Annelies van Loon is Paintings Research Scientist at the Rijksmuseum, Amsterdam

Dorothy Mahon is Senior Conservator at the Metropolitan Museum of Art in New York

Vassilis Papadakis is Researcher for Spectral Imaging at the Institute of Molecular Biology and Biotechnology (IMBB) and at the FOundation for Research and Technology-Hellas (FORTH)

A. Tummers, R. G. Erdmann, *Frans Hals or not Frans Hals*, Cultural Heritage Science, https://doi.org/10.1007/978-3-031-59489-2

Anna Tummers is Full Professor in Early Modern Art History at Ghent University
Arie Wallert is Professor emeritus in Technical Art History at the University of Amsterdam and former Senior Scientist at the Rijksmuseum, Amsterdam

Netherlands Organization for Scientific Research (NWO)

GPSR Compliance

The European Union's (EU) General Product Safety Regulation (GPSR) is a set of rules that requires consumer products to be safe and our obligations to ensure this.

If you have any concerns about our products, you can contact us on ProductSafety@springernature.com

In case Publisher is established outside the EU, the EU authorized representative is:

Springer Nature Customer Service Center GmbH
Europaplatz 3
69115 Heidelberg, Germany

Batch number: 08312443

Printed by Printforce, the Netherlands